Lecture Notes in Artificial Intelligence   8825

Subseries of Lecture Notes in Computer Science

T0213713

Vicenç Torra   Yasuo Narukawa
Yasunori Endo (Eds.)

Modeling Decisions for Artificial Intelligence

11th International Conference, MDAI 2014
Tokyo, Japan, October 29-31, 2014
Proceedings

 Springer

Volume Editors

Vicenç Torra
IIIA-CSIC
Bellaterra, Catalonia, Spain
E-mail: vtorra@ieee.org; vtorra@iiia.csic.es

Yasuo Narukawa
Toho Gakuen
Tokyo, Japan,
E-mail: nrkwy@ybb.ne.jp

Yasunori Endo
University of Tsukuba
Tsukuba, Japan
E-mail: endo@risk.tsukuba.ac.jp

ISSN 0302-9743   e-ISSN 1611-3349
ISBN 978-3-319-12053-9   e-ISBN 978-3-319-12054-6
DOI 10.1007/978-3-319-12054-6
Springer Cham Heidelberg New York Dordrecht London

Library of Congress Control Number: 2014949874

LNCS Sublibrary: SL 7 – Artificial Intelligence

© Springer International Publishing Switzerland 2014

Typesetting: Camera-ready by author, data conversion by Scientific Publishing Services, Chennai, India

Printed on acid-free paper

Springer is part of Springer Science+Business Media (www.springer.com)

Preface

This volume contains papers presented at the 11th International Conference on Modeling Decisions for Artificial Intelligence (MDAI 2014), held in Tokyo, Japan, October 29–31. This conference followed MDAI 2004 (Barcelona, Catalonia), MDAI 2005 (Tsukuba, Japan), MDAI 2006 (Tarragona, Catalonia), MDAI 2007 (Kitakyushu, Japan), MDAI 2008 (Sabadell, Catalonia), MDAI 2009 (Awaji Island, Japan), MDAI 2010 (Perpinyà, France), MDAI 2011 (Changsha, China), MDAI 2012 (Girona, Catalonia), and MDAI 2013 (Barcelona, Catalonia) with proceedings also published in the LNAI series (Vols. 3131, 3558, 3885, 4617, 5285, 5861, 6408, 6820, 7647, 8234).

The aim of this conference was to provide a forum for researchers to discuss theory and tools for modeling decisions, as well as applications that encompass decision making processes and information fusion techniques.

The organizers received 38 papers from 16 different countries, from Europe, Asia, and America, 19 of which are published in this volume. Each submission received at least two reviews from the Program Committee and a few external reviewers. We would like to express our gratitude to them for their work. The plenary talks presented at the conference are also included in this volume.

The conference was supported by the Japan Society for Fuzzy Theory and Intelligent Informatics (SOFT), the Catalan Association for Artificial Intelligence (ACIA), the European Society for Fuzzy Logic and Technology (EUSFLAT), the UNESCO Chair in Data Privacy, the Spanish MINECO (TIN2011-15580-E), and the Spanish MEC (ARES - CONSOLIDER INGENIO 2010 CSD2007-00004).

July 2014

Vicenç Torra
Yasuo Narukawa
Yasunori Endo

Organization

Modeling Decisions for Artificial Intelligence – MDAI 2014

General Chairs

Yasunori Endo   University of Tsukuba, Japan

Program Chairs

Vicenç Torra   IIIA-CSIC, Catalonia, Spain
Yasuo Narukawa   Toho Gakuen, Japan

Advisory Board

Bernadette Bouchon-Meunier   Computer Science Laboratory of the University
Paris 6 (LiP6), CNRS, France
Didier Dubois   Institut de Recherche en Informatique de
Toulouse (IRIT), CNRS, France
Lluis Godo   IIIA-CSIC, Spain
Kaoru Hirota   Tokyo Institute of Technology, Japan
Janusz Kacprzyk   Systems Research Institute, Polish Academy of
Sciences, Poland
Sadaaki Miyamoto   University of Tsukuba, Japan
Michio Sugeno   European Centre for Soft Computing, Spain
Ronald R. Yager   Machine Intelligence Institute, Iona College,
USA

Program Committee

Gleb Beliakov   Deakin University, Australia
Gloria Bordogna   Consiglio Nazionale delle Ricerche, Italia
Tomasa Calvo   Universidad Alcala de Henares, Spain
Susana Díaz   Universidad de Oviedo, Spain
Josep Domingo-Ferrer   Universitat Rovira i Virgili, Spain
Jozo Dujmovic   San Francisco State University, USA
Katsushige Fujimoto   Fukushima University, Japan
Michel Grabisch   Université Paris I Panthéon-Sorbonne, France
Enrique Herrera-Viedma   Universidad de Granada, Spain
Aoi Honda   Kyushu Institute of Technology, Japan

Masahiro Inuiguchi	Osaka University, Japan
Xinwang Liu	Southeast University, China
Jun Long	National University of Defense Technology, China
Jean-Luc Marichal	University of Luxembourg, Luxembourg
Radko Mesiar	Slovak University of Technology, Slovakia
Tetsuya Murai	Hokkaido University, Japan
Toshiaki Murofushi	Tokyo Institute of Technology, Japan
Guillermo Navarro-Arribas	Universitat Autònoma de Barcelona, Spain
Michael Ng	Hong Kong Baptist University, Chaina
Gabriella Pasi	Università di Milano Bicocca, Italy
Susanne Saminger-Platz	Jihannes Kepler University, Austria
Sandra Sandri	Instituto Nacional de Pesquisas Espaciais, Brasil
Roman Słowiński	Poznan University of Technology, Poland
László Szilágyi	Sapientia-Hungarian Science University of Transylvania, Hungary
Aida Valls	Universitat Rovira i Virgili, Spain
Vilem Vychodil	Palacky University, Czech Republic
Zeshui Xu	Southeast University, China
Yuji Yoshida	University of Kitakyushu, Japan

Local Organizing Committee Chair

Kenichi Yoshida	University of Tsukuba, Japan
Kazuhiko Tsuda	University of Tsukuba, Japan

Additional Referees

Roger Jardí-Cedó, Malik Imran Daud, Sara Hajian, Montserrat Batet, Yuchi Kanzawa

Supporting Institutions

University of Tsukuba
The Catalan Association for Artificial Intelligence (ACIA)
The European Society for Fuzzy Logic and Technology (EUSFLAT)
The Japan Society for Fuzzy Theory and Intelligent Informatics (SOFT)
The UNESCO Chair in Data Privacy
The Spanish MEC (ARES - CONSOLIDER INGENIO 2010 CSD2007-00004)

Abstracts

Recursive Capacitary Kernel

Hiroyuki Ozaki

Faculty of Economics, Keio University

Abstract. We define a capacitary kernel as a function which maps a current state to a capacity (a non-additive measure) which governs the next period's uncertainty. We then use a capacitary kernel to define the time-consistent recursive objective function for a dynamic optimization problem. We impose and discuss assumptions on the capacitary kernel for this objective function to be well-defined. We also provide some decision-theoretic foundation of this objective function. Furthermore, we develop a dynamic programming technique to solve this optimization problem by exploiting the recursive structure of the objective function defined by the recursive capacitary kernel.

Framework of Entertainment Computing and Its Applications

Junichi Hoshino

University of Tsukuba, Entertainment Computing Lab
jhoshino@esys.tsukuba.ac.jp

Entertainment Computing is one of the promising application domains of MDAI (modeling decisions for artificial intelligence) field. In this talk, I'm going to overview the collaborative creation process of entertainment using rich examples including media art, physical exercise game, learning game, multi-agent Game AI, and social information exchange on hobby.

We also show the examples of the entertainment systems using behavioral and cognitive user modeling:

1) The Fighting Game Character Using Imitation Learning

One of the limitations of computer-based opponents in action games is that the AI character is constructed in advance, and players quickly become bored with their prepared tactics. We built an online coliseum in which a non-player character (NPC) incrementally learns action sequences and combinations of actions, allowing the NPCs to adopt different fighting strategies after fighting with different players. Individual fighting styles can be generated from a unique fighting history. We developed a new action learning engine that automatically analyzes the actions of a human player and extracts the effective fighting sequences. Action control trees are generated automatically and incrementally added to the NPC's action profile.

2) A Wellness Entertainment System using a Trampoline Interface

We describe the wellness entertainment system using a trampoline interface. In this system, we use a mini trampoline as the input device. The system enables the user to move and jump freely in VR space by exaggerated movement corresponding to walking or jumping on the mini trampoline. Improvements in exercise motivation and support for continuous exercise are achieved in our system, since it is possible to enjoy strolling through a virtual space, which is usually difficult to experience, by exercising on the mini trampoline without injury to the user's joints.

3) Disaster Experience Game in a Real World

We describe a disaster experience game system which could instruct about general knowledge and regionally specific disaster risk in a joyful way. The system does not give advice in a unilateral way; instead it helps the user, with an accurate awareness of the real world and then shows the risk information e.g., prevention plans and evacuation maps. Additionally, introducing game elements, the user plays with some level of interaction. Using this system, we created a game application for an earthquake. An assessment experiment of the game was clearly beneficial to not only understand risk perception but support it; it also has the motivation of a muster drill.

4) Communication System for Supporting Information Gathering and Social Interaction in a Niche Market

We describe a communication system by which niche people can obtain cross-cutting information and communicate with other people based on each personality. The system graphically displays the degree and direction of other people's hobbies who are interested in the keyword niche people input, and relation between the knowledge e.g. movies, music, animation, history, geography using nodes. So, we can search friends who have similar interest and direction in hobbies. From a demonstration experiment, we obtained good results that the system could help niche people to gain and exchange useful information.

Uncertain Information Representation for Decision-Making: Emerging Approaches

Ronald R. Yager

Machine Intelligence Institute,
Iona College
New Rochelle, NY 10801
Yager@Panix.Com

Information used in decision making generally comes from multiple sources and is expressed in various modalities. In many cases the information available has some significant associated uncertainty. In order to address this problem there is a need to provide various methods for the representation of different types of uncertain information. Here we shall discuss some recently emerging approaches for attaining this representational capability. One approach we shall discuss is the use of Z-valuations which are based on the use of Z-numbers. These objects allow us to represent information that combines both possibilistic and probabilistic uncertainty. We shall also look at a new approach for modeling fuzzy sets with imprecise membership grades called Pythagorean fuzzy sets. One important issue that arises when using these non-standard representations in decision-making is the comparison of alternative satisfactions, that is we must compare mathematical objects that are not naturally ordered. We consider this important problem.

Table of Contents

Invited Papers

Regular Papers

Aggregation Operators and Decision Making

Inference Systems

Optimization

Clustering and Similarity

Data Mining and Data Privacy

Electronic Road Pricing System for Low Emission Zones to Preserve Driver Privacy

Roger Jardí-Cedó[1], Macià Mut-Puigserver[2], M. Magdalena Payeras-Capellà[2],
Jordi Castellà-Roca[1], and Alexandre Viejo[1]

[1] Dpt. d'Enginyeria Informàtica i Matemàtiques, UNESCO Chair in Data Privacy,
Universitat Rovira i Virgili, Av. Països Catalans 26, E-43007 Tarragona, Spain
{roger.jardi,jordi.castella,alexandre.viejo}@urv.cat
[2] Dpt. de Ciències Matemàtiques i Informàtica, Universitat de les Illes Balears, Ctra.
de Valldemossa, km 7,5. E-07122 Palma de Mallorca, Spain
{macia.mut,mpayeras}@uib.es

Abstract. At present, great cities try to prevent from high levels of pollution and traffic jam by restricting the access of vehicles to centric zones. They are also known as Low-Emission Zones (LEZ). Some of the most important issues of LEZs are the risk of losing privacy of the citizen who drives through the LEZ and a significant error percentage on detection of fraudulent drivers. In this article, an Electronic Road Pricing (ERP) system designed specifically for cities with Low-Emission Zones is proposed. The aim of this system is to detect fraud and to preserve driver privacy. In this case, revocable anonymity makes only fraudulent drivers lose their privacy.

Keywords: electronic road pricing, data privacy, security, low emission zone.

1 Introduction

Cities such as Paris, Barcelona or Rome have circulation problems, with traffic jams and pollution problems due to the huge vehicle concentration in certain areas. The air quality guidelines presented by OMS in 2005, give guidance on "the way to reduce air pollution effects on health". Based on these recommendations, different European directives, such as 2008/50/CE, limit the level of certain environmental pollutants. In order to fulfill this legislation, the different administrations are implanting, among other measures, HOV lanes [3], variable speed or vehicle circulation restrictions in central areas. This last measure, known as **Low-Emission Zone (*LEZ*)** and adopted in many cities like Singapore, Tokyio or Beijing[10,11,5], makes vehicles pay so as to circulate according to certain conditions, such as weight or emissions.

Since some decades ago, Electronic Toll Collection (ETC) has been used in highways, tunnels or bridges in order to expedite toll payments as well as to reduce traffic jams. Likewise, thanks to new technologies such as the GPS and wireless communication, vehicular location-based services (VLBS) have been

V. Torra et al. (Eds.): MDAI 2014, LNAI 8825, pp. 1–13, 2014.
© Springer International Publishing Switzerland 2014

developed with the purpose of providing information to drivers in relation to their geographic location and improving transportation efficiency. Those ETC systems, considered VLBS, are known as **Electronic Road Pricing (ERP)**. They present some improvements such as a more flexible calculus of the fees according to the distance driven, route or time. Moreover, these systems, applied in urban areas, allow managing traffic in central areas through the control of the flow and the density of vehicles, thus reducing traffic jams. This can be achieved by modifying the price of the fees dynamically (the rise on the price of those dense areas suggests drivers to avoid them). However, as it is later shown, these systems have certain privacy problems.

1.1   State of the Art

In recent years, various ERP systems have been proposed in the literature [9,4,1,8,6,7]. All of them require the use of an On-Board Unit (OBU), with GPS and a system of wireless communication, with the aim of getting and sending the Service Provider, SP (see definition in Section 2), information related to the vehicle location and fees to pay. Pricing varies depending on the vehicle path. In [9] and [4], OBU sends information of the path followed to an external server, property of SP, which prices according to its path in every billing period. In [1,8,6,7], fees are calculated locally in each OBU and are sent to SP server in every pricing period. In this case, the information disclosure related to the vehicle location is minimal. In order to achieve it, cryptographic proofs are used to demonstrate that OBU has been honest in the fee calculus and aggregation.

Fraud control is an important aim by ERP systems. Drivers, in order to save money, could act maliciously (i.e. by disconnecting or modifying OBU data). Consequently, mechanisms based on checkpoints, $Chps$ (see definition in Section 2), are implemented with the aim to test their honesty. $Chps$, randomly located in the road and equipped with cameras, register the number plates of all the vehicles that cross them. These pictures are evidences that place a vehicle at a given moment and place, and are used to verify that a vehicle path has not been altered. In order to achieve this, SP and driver interact in the billing period. Fraud detection has a certain probability and depends on the number of $Chps$. Further, the fact of drivers ignoring the number and its location is a basic issue.

The privacy level of drivers and detection of fraud are a compromise. When a high detection level is requested, privacy is affected, that is, SP be capable of rebuilding a vehicle path more precisely as long as the number of $Chps$ is greater. Besides, if $Chps$ are randomly moved every so often, and vehicle paths follow a routine (i.e. going to work), precision could be even greater though privacy would be affected.

1.2   Contributions and Structure of the Document

This paper proposes an ERP system for LEZs with the aim of improving fraud control and honest drivers' privacy through revocable anonymity. Unlike other systems, $Chps$, equipped with cameras, only register fraudulent vehicles, thus keeping honest drivers' privacy. Moreover, the *OBU* of the vehicle does not register its location, plus reconciliation between driver and system in the billing period is not required, and fraud control is non-probabilistic.

The system is presented in Section 2. The protocol is introduced in Section 3. Security is evaluated in Section 4 and conclusions are presented in Section 5.

2   System Model

The proposal considers the following actors.

- *Driver D:* The person who drives a vehicle in a *LEZ*.
- *Vehicle V:* The means of transport registered by a unique *D* (the owner of it) but it may be driven by several *D*s. *V* has an identifier (the vehicle plate) that connects it to the owner.
- *Secure element SE*: A tamper-proof security module installed in each *V* by the traffic competent authority. It performs all sensitive operations to ensure the security requirements.
- *On-board unit OBU:* This device is installed in each *V*. It has more computational power and storage capacity than *SE*. The device connects *SE* with the user and it performs the less sensitive protocol operations. It has a location capabilities (GPS).
- *Service Provider SP:* It offers an ERP service for urban areas thanks to a concession contract with the local public administration (i.e. City Council). This entity, apart from having the right to offer this service, is responsible for managing the system.
- *Checkpoint Chp:* *SP* installs in the restricted zone the checkpoints. The *Chp* aims to control the access of vehicles that enter or leave the zone.
- *Vehicle certification authority VCA:* It provides keys and certificates to *V*s.
- *Punisher authority PA:* The trusted entity that knows the identity of the *V* owner. *PA* will reveal the identity of the *V*'s owner in case of fraud.

2.1   Requirements

The system requirements related with fraud, privacy, authenticity and technology, are described below in order to establish the foundations of the system.

Anti-Fraud Requirements. When a *V* enters or exits a *LEZ* through a *Chp*, both obtain **proof-of-entrance** γ_i or **proof-of-leaving** γ_o. This γ_i contains information to prove that a specific *V* enters the *LEZ* through a specific *Chp* at a specific hour. This *proof* is considered **valid** when it cannot be modified once generated without detection (integrity), when its issuers can prove that it is their generation (authenticity), and also when they cannot deny its authoring (non-repudiation). Each proof is **linked** to a *V* and a *Chp*. The link between a proof and a *V* guarantees that the token cannot be used by another *V*', neither in a voluntary nor in an involuntary way. This avoids the *duplicity* of a proof when a fraudulent *V* tries to use the same proof at the same time. *SP* assures that all the Ds pay correctly. If this is not the case, SP identifies the offending Ds and generates evidences to prove it. The **fraud** is done by a D when she drives in a LEZ without a γ_i, with an invalid γ_i, with a valid γ_i but associated with another V, or if she doesn't pay correctly in the exit. A SP cannot **falsely accuse** an honest D of fraud (an honest D should not be defenseless). A false accusation takes place when a SP unjustly claims that a V does not have a γ_i, has an invalid proof, has a valid proof belonging to another V, or if she doesn't pay correctly in the exit.

Authenticity Requirements. At the entrance and at the exit of a LEZ, Vs and Chps exchange information. When the communication is established, both parts, V and Chp must prove its identity to the other part. This way, each one can be sure that the protocol is executed with the right entity. If this is not the case, this action must be reported.

Privacy Requirements. The fraud control executed by SP can endanger the privacy of the Ds. In this case, the curiosity of SP can cause an excessive monitoring of the system or even can trace the itinerary of a specific V. With the aim of avoiding this excessive control over the Vs by SP, the system must (i) assure the privacy (the identity of D or V cannot be linked to any itinerary); (ii) avoid the traceability (SP mustn't know the itinerary of a V); and (iii) provide revocable anonymity to D (if a D commits fraud, SP needs her identity in order to punish her and for this reason, the identity is revealed).

Functional Requirements. The *communication technology* used to communicate Vs and Chps needs to let Chp communicate with the nearest V. This communication is possible by combining low and medium distance communication technologies, such as Wimax, ZigBee IEEE 802.15.4 or Bluetooth IEEE 802.15.1, using directional antennae or triangulation. The *communication* and the *computation* required by the protocol must be quick enough to allow communication when the Vs are moving. Whatever *interaction* with the D should be easy and agile. The *electronic payment system* required in the system should be anonymous and untraceable. Moreover, it should be quick enough to allow the transaction when the V exits the LEZ.

3   Protocol Description

Before starting the system, the system entities are initialized (3.1: Setup and 3.2: Certification). Moreover, SP prices the LEZ (3.3: Price generation), per time unit and emission category, sending each Chp a list of prices signed by the competent entity. SP repeats these operations every time prices are updated.

SE generates different credentials for V every time it enters a LEZ (3.4: Certificate generation) in order to be able to correctly authenticate with Chps that manage the ins and outs of users from the LEZ.

When a vehicle V enters a LEZ (3.5: Check-in), it communicates with a Chp and they authenticate mutually. When the authentication with V fails, and only in this situation, Chp takes a picture of the V number plate as evidence of the infringement with which a proof-of-entrance incidence ζ_i is generated. ζ_i is sent to PA in order to verify the existence of fraud and so proceed with the corresponding sanction. When the authentication is correct, V gets an proof-of-entrance γ_i, which specifies the entry time.

When a V exits the LEZ (3.6: Check-out), it communicates with a Chp and they authenticate mutually. When the authentication with V is correct, Chp informs V about the exit time and the destination account for payment. With this information, V computes the amount payable for length of stay and its emission category, and makes a transaction through an electronic payment system. The transaction reference is sent to Chp as a proof of payment. Finally, V receives an proof-of-leaving γ_o as a receipt. When the authentication fails, Chp takes a picture of V as an proof-of-leaving evidences ζ_o, which is sent to PA.

The payment verification is done through SP (3.7: Payment verification) a posteriori. For each pair of γ_i and γ_o associated to a same V, SP checks whether the value of the transaction coincides with the pricing, depending on the length of stay and its emission category. If the value is not correct, a proof-of-payment incidence ζ_p with these registers is created and sent to PA.

When PA receives a ζ (3.8: Sanction), it is corroborated. In case of fraud, PA reveals the identity of the V owner (anonymity is revoked), and evidences to refute the accusation by SP are requested. According to these, PA decides whether to sanction the owner.

3.1 Setup

The setup process works as follows:

1. PA obtains from competent authorities (i.e. Police):
   – An asymmetric key pair (Pk_{PA}, Sk_{PA}), its public key certificate $cert_{PA}$, and a certificate repository of the authorities
2. SP and VCA obtain from competent authorities (i.e. city council and a transit authority, respectively):
   – An asymmetric key pair (Pk_{SP}, Sk_{SP}) and (Pk_{VCA}, Sk_{VCA}), its public key certificate $cert_{SP}$ and $cert_{VCA}$, and a certificate repository of the authorities

   The certificate chain length of VCA is 1, and 0 in the case of SP. The $cert_{SP}$ validity period can correspond to the concession lifetime of the service, without exceeding it.
3. VCA:
   i. Defines a set of vehicles $V = \{v_1, v_2, ..., v_{n_V}\}$, where $n_V = |V|$ is the number of vehicles
   ii. Defines a collection of sets $K = \{C_1, C_2, ..., C_{n_K}\}$ partition of V, where $n_K = |K|$, with $|C_i| = n_C$, $\forall i$
   iii. Generates and associates a certification entity VCA_{C_i} to each element of the subset K $(C_1, ..., C_{n_K})$:
      iii.i. An asymmetric key pair $(Pk_{VCA_{C_i}}, Sk_{VCA_{C_i}})$, $\forall i \in \{1, ..., n_K\}$
      iii.ii. A CA certificate $cert_{VCA_{C_i}}$, $\forall i \in \{1, ..., n_K\}$, which has an expiration time c_{exp} and a certificate chain length of 0

4. Each *Chp* applies the following steps:
   i. Obtain a certificate repository of the authorities and entities
   ii. Generate an asymmetric key pair (Pk_{Chp}, Sk_{Chp})
   iii. Securely obtain a public key certificate $cert_{Chp}$ from *SP* containing an extension $cert_{Chp}.loc$ with its location coordinates

3.2   Certification

It is assumed that the *SE* of each *V* has been previously initialized with a certificate repository of the certification authorities, identifying information of the vehicle V_{id} and its technical specifications (number plate, chassis number, owner, brand, model, engine power, or level of emissions of CO_2 and other pollutant gases). The certification process of a *V* is done by *VCA*, before purchasing a vehicle and/or passing the regular technical tests of vehicles:

1. Register *V* in an element of the subset *K* (in a C_i)
2. Download the certification entity VCA_{C_i} associated to C_i (consisting of $Pk_{VCA_{C_i}}$, $Sk_{VCA_{C_i}}$ and $cert_{VCA_{C_i}}$), through a secure channel in the *SE*

3.3   Price Generation

Whenever *SP* wants to modify the fees of a *LEZ*, it performs the next operations:

1. Set the *prices* per unit of time and emission category (i.e. European Emission Standards) and generate a timestamp *ts*
2. Compose *information-of-prices* $\theta = (prices, ts)$
3. Sign θ $(Sign_{SP}(\theta) = \bar{\theta})$ and send $\theta^* = (\theta, \bar{\theta})$ to each *Chp*

3.4   Certificate Generation

This phase is performed every time a *V* is about to enter a *LEZ*. Thanks to the certification entity VCA_{C_i} installed in the *SE* of the *V* in the previous phase, this *SE* is able to perform the following operations in order to generate new public key certificates:

1. Compute an asymmetric key pair (Pk_{V_q}, Sk_{V_q})
2. Generate a public key certificate $cert_{V_q}$ with the following features:
   - An extension $cert_{V_q}.idS$ containing the probabilistic encryption (i.e. by using OAEP padding [2], standardized later in PKCS #1v2 and RFC 2437) of the vehicle identifier V_{id} with the public key of *PA*: $Enc_{Pk_{VCA}}(V_{id})$
   - An extension $cert_{V_q}.em$ containing the pollutant emission category (i.e. European Emission Standards) of the *V*

3.5   Check-in

When a *Chp* placed in the entrance of a *LEZ* detects a *V*, the protocol is applied:

1. Chp has to:
   i. Generate a nonce N_A and compose *information-of-entrance* $\psi = (N_A, \theta^*)$
   ii. Sign ψ: $Sign_{Chp}(\psi) = \bar{\psi}$, and send ψ, $\bar{\psi}$ and its $cert_{Chp}$ to V
2. SE of the V, with the help of the OBU, has to:
   i. Verify the certificate $cert_{Chp}$ and the signature $\bar{\psi}$: $Verif_{Chp}(N_A, \theta^*, \bar{\psi})$
   ii. Verify the signature $\bar{\theta}$: $Verif_{SP}(prices, ts, \bar{\theta})$ and the location coordinates $cert_{Chp}.loc$ of Chp
   iii. Generate a nonce N_B and compute the fingerprint $fing_{Chp}$ of $cert_{Chp}$ (it is computed as the hash function of the certificate and it is used as identifier)
   iv. Compose message $\omega_{V_q} = (\theta^*, N_A, N_B, fing_{Chp})$
   v. Sign ω_{V_q}: $Sign_{V_q}(\omega_{V_q}) = \bar{\omega}_{V_q}$, and send N_B, $\bar{\omega}_{V_q}$ and its $cert_{V_q}$ to Chp
3. Chp has to:
   i. Generate a timestamp ts', and verify the certificate $cert_{V_q}$ and the signature $\bar{\omega}_{V_q}$: $Verif_{V_q}(\theta^*, N_A, N_B, fing_{Chp}, \bar{\omega}_{V_q})$
   ii. If one of the verifications fails, Chp performs the following operations:
      ii.i Generate an incidence number of entrance in_i
      ii.ii Take a photograph ph of V and extract the plate number plt
      ii.iii Compose *proof-of-entrance incidence* $\zeta_i = (in_o, plt, ph, ts', \theta^*, N_A, N_B, fing_{Chp}, \bar{\omega}_{V_q}, cert_{V_q})$
      ii.iv Sign ζ_i: $Sign_{Chp}(\zeta_i) = \bar{\zeta}_i$, and send $\zeta_i^* = (\zeta_i, \bar{\zeta}_i)$ to SP
   iii. If the verifications performed in 3i) are correct, Chp has to:
      iii.i Compute the $fing_{V_q}$ of $cert_{V_q}$ and compose *proof-of-entrance* $\gamma_i = (\theta^*, N_A, N_B, fing_{Chp}, \bar{\omega}_{V_q}, fing_{V_q}, ts')$
      iii.ii Sign γ_i: $Sign_{Chp}(\gamma_i) = \bar{\gamma}_i$ and send ts' and $\bar{\gamma}_i$ to the V
4. If the verifications performed in 3i) are correct, SE of the V, with the help of the OBU, has to:
   i. Verify the signature $\bar{\gamma}_i$: $Verif_{Chp}(\theta^*, N_A, N_B, fing_{Chp}, \bar{\omega}_{V_q}, fing_{V_q}, ts', \bar{\gamma}_i)$
   ii. Verify the freshness of ts': $|ts' - current\ time| < \delta$, where δ is a fixed time
   iii. Store $\gamma_i^* = (\gamma_i, \bar{\gamma}_i)$

3.6 Check-Out

When a Chp placed in the exit of a LEZ detects a V, the next protocol is applied:
1. Chp has to:
   i. Generate a timestamp ts'' and a nonce N_C, and compose *information-of-payment* $\rho = (ts'', N_C, acc)$, where acc identifies the target account, of the electronic payment system assumed, of SP
   ii. Sign ρ: $Sign_{Chp}(\rho) = \bar{\rho}$, and send ρ, $\bar{\rho}$ and its $cert_{Chp}$ to V
2. SE of the V, with the help of the OBU, has to:
   i. Verify the certificate $cert_{Chp}$, the signature $\bar{\rho}$: $Verif_{Chp}(ts'', N_C, acc, \bar{\rho})$, the location coordinates $cert_{Chp}.loc$ of Chp and the freshness of ts'': $|ts'' - current\ time| < \delta$, where δ is a fixed time
   ii. Recover ts' of the last γ_i and compute the length of stay τ a LEZ: $(ts'' - ts') = \tau$

iii. Recover the *prices* included in θ^* of γ_i, and compute the *amount* of money required to pay according to τ, its pollutant emissions and the *prices*

iv. Make a digital transfer of the *amount* of money to the target account *acc* and obtain a transfer reference *trans*

v. Generate a nonce N_D and compute the fingerprint $fing_{Chp}$ of $cert_{Chp}$

vi. Compose message $\omega_{V_q} = (ts'', N_C, N_D, fing_{Chp}, trans)$

vii. Sign ω_{V_q}: $Sign_{V_q}(\omega_{V_q}) = \bar{\omega}_{V_q}$, and send N_D, $trans$, $\bar{\omega}_{V_q}$ and its $cert_{V_q}$ to *Chp*

3. *Chp* has to:

i. Verify the certificate $cert_{V_q}$ and the signature $\bar{\omega}_{V_q}$: $Verif_{V_q}(ts''$, N_C, N_D, $fing_{Chp}$, $trans$, $\bar{\omega}_{V_q})$

ii. If one of the verifications fails, *Chp* performs the following operations:

ii.i Generate an incidence number of leaving in_o

ii.ii Take a photograph ph of the V and extract the number plate plt

ii.iii Compose *proof-of-leaving incidence* $\zeta_o = (in_o, plt, ph, ts'', N_C, N_D, fing_{Chp}, trans, \bar{\omega}_{V_q}, cert_{V_q})$

ii.iv Sign ζ_o: $Sign_{Chp}(\zeta_o) = \bar{\zeta}_o$, and send $\zeta_o^* = (\zeta_o, \bar{\zeta}_o)$ to SP

iii. If the verifications performed in 3i) are correct *Chp* has to:

iii.i Compute the $fing_{V_q}$ of $cert_{V_q}$, and compose *proof-of-leaving* $\gamma_o = (ts'', N_C, N_D, fing_{Chp}, trans, \bar{\omega}_{V_q}, fing_{V_q})$

iii.ii Sign γ_o, $Sign_{Chp}(\gamma_o) = \bar{\gamma}_o$, and send it to the V

4. If the verifications performed in 3i) are correct, SE of the V, with the help of the OBU, has to:

i. Verify the signature $\bar{\gamma}_o$: $Verif_{Chp}(ts'', N_C, N_D, fing_{Chp}, trans, \bar{\omega}_{V_q}, fing_{V_q}, \bar{\gamma}_o)$, and store $\gamma_o^* = (\gamma_o, \bar{\gamma}_o)$

3.7 Payment Verification

Each *Chp* sends SP the different proofs γ_i, γ_o and ζs (ζ_i and ζ_o), generated in the phases 3.5 and 3.6 of the protocol, periodically. SP then forwards the incidences ζ_i and ζ_o to PA. Moreover, SP performs the next operations every so often (in batch) for each set of proofs γ_i and γ_o associated with the same $fing_{V_q}$:

1. Extract ts', ts'', and $cert_{V_q}.em$ from γ_i and γ_o
2. Extract *prices* from θ^*, included in γ_i
3. Extract the reference $trans$ from γ_o
4. Compute the length of stay τ a LEZ: $(ts''\text{-}ts')=\tau$
5. Compute the *amount'* of money required to pay according to τ, $cert_{V_q}.em$ and the *prices*
6. Verify whether the transfer $trans$ was successful and recover the *amount* of money paid
7. Verify whether $amount = amount'$
8. Verify that $trans$ has not been repeated in another γ_o. (i.e. with the help of a time filter according to the γ_o's ts and finding duplicates)
9. If one of the verifications fails then,

i. Generate an incidence number of verification in_v

    ii. Compose *proof-of-verification incidence* ζ_p including γ_i and γ_o of V_q: $\zeta_p=(in_v, \gamma_i, \gamma_o)$. In the case of a reused *trans*, add γ'_o which proves the double spending $\zeta_p = (in_v, \gamma_i, \gamma_o, \gamma'_o)$

    iii. Sign ζ_p $(Sign_{SP}(\zeta_p) = \bar{\zeta}_p)$ and send $\zeta_p{}^*=(\zeta_p, \bar{\zeta}_p)$ to PA

3.8 Sanction

PA performs the following operations according to the origin of the received ζs:

1. In the case of ζ_i or ζ_o:
   i. Verify the signatures and extract the number plate *plt* from the photograph *ph*, included in ζ
2. In the case of ζ_p:
   i. Verify all the signatures included in ζ_p and verify that the signatory of γ_i and γ_o is the same
   ii. Verify the right payment by repeating the steps 1-7 from 3.7
   iii. In case of a double spending incidence, verify that the *trans* of γ_o and γ'_o are equals
   iv. If the incidence is confirmed, recover the identifier V_{id} of V_q by opening the extension $certV_q.idS$ of the certificate $certV_q$, included in the proof γ_i or γ_o (in case of double spending, this is done for the last V to leave the *LEZ*): $Dec_{PA}(certV_q.idS) = V_{id}$
3. Contact the owner of the V from *plt* or V_{id}, inform her about the sanctioning procedure and require her evidence to the contrary to refute the accusation
4. If the owner of the V presents evidence to the contrary, these are verified. If this evidence is not valid, it fines the owner according to the type of infraction

4 Security and Requirements Analysis

The security and system requirements of the protocol are studied in this section, i.e. the entrance proofs are valid, the system preserves users' privacy, and dishonest users are identified. The discussion is organised in three propositions, and each proposition can have several claims to support its fulfilment.

Proposition 1. *The proposed system preserves authenticity, non-repudiation and integrity for the entrance and exit proofs.*

Claim. 1 The creation of fraudulent entrance and exit proofs is computationally unfeasible nowadays.

Entrance proofs have the following form $\gamma_i = (\theta^*, N_A, N_B, fing_{Chp}, \omega\bar{v}_{V_q}, fing_{V_q}, ts')$. The checkpoint signs the entrance proofs γ_i: $Sign_{Chp}(\gamma_i) = \bar{\gamma}_i$ and sends the pair ts' and $\bar{\gamma}_i$ to the vehicle. In the same way, the exit proof $\gamma_o = (ts'', N_C, N_D, fing_{Chp}, trans, \omega\bar{v}_{V_q}, fing_{V_q})$ is signed by the checkpoint, $Sign_{Chp}(\gamma_o)=\bar{\gamma}_o$, and sent to the vehicle. For these reasons, the generation of entrance and exit proofs is nowadays computationally unfeasible, without the knowledge of the secret key used by *Chp* in the signature.

Claim. 2 The *Chp*s, issuers of the entrance and exit proofs, can not deny the emission of these proofs.

Entrance proofs are generated and signed by their issuer (the *Chp*s) and, considering the signature scheme secure, this operation can be only performed by these issuers. Thus, the issuer's identity is linked to the proofs and, for the properties of the electronic signature scheme, it can not deny its authorship.

Claim. 3 The content of the entrance and exit proofs can not be modified by the vehicles.

If we suppose that the signature scheme is secure and that the hash summary function is collision-resistant, if the content of the entrance or exit proofs was modified, the verification of the signature would be incorrect because $Sign_e(m) = E_{Sk_e}(h(m)) = \bar{m}$. In order to pass the verification, the signature would be regenerated from the new entrance or exit ticket. This operation is computationally unfeasible without the knowledge of the checkpoint secret key.

Result 1 *According to the presented proofs in Claims 1, 2 and 3, it can be assured that the protocol satisfies the needed security requirements (authenticity, integrity and non-repudiation) for the proofs to be considered valid.*

Proposition 2. *The toll system presented here preserves the privacy of its users and protects their anonymity, avoiding the traceability of their actions.*

Claim. 4 The system guarantees the anonymity of honest users.

The information that the user transmits to enter the system is $\omega_{V_q} = (N_A, N_B, fing_{cert_{Chp}})$ and its signature. The *Chp* will verify the signature using the certificate $cert_{V_q}$ accompanying the user message. This certificate (generated by the *SE* of the *V* before entering the *LEZ*) identifies the vehicle, but the identification information is protected with an asymmetric encryption, using the public key of *PA*. Thus, *Chp* can verify the signature but it is not able to identify the vehicle. Then, *Chp* generates and transmits to the user $\bar{\gamma}_i$. With this evidence, the vehicle may enter the *LEZ*. The information related to the user identification inside $\bar{\gamma}_i$ is the same included in ω_{V_q}. This means that the *V* enters the *LEZ* without being identified.

When *V* leaves the *LEZ* (We assume that the payment system allows the anonymity of the user. The payment system is beyond the scope of this paper.), the user must send the following information: $\omega_{V_q} = (ts, N_C, N_D, fing_{cert_{Chp}}, trans)$ to the exit *Chp*. Assuming that the payment is anonymous, no one can identify the user through *trans*. Moreover, it is not possible to identify the user through the signature on ω_{V_q} by the reasons explained in the previous paragraph. Consequently, the entrance inside a *LEZ* and the exit from a *LEZ* of honest users are anonymous.

Claim. 5 The toll collection protocol does not allow to trace or to link the actions of the vehicles.

It is not possible to associate the various entries and exits from a *LEZ* of the same vehicle if we use the information generated by the execution of the protocol. This happens because the protocol described in 3.4 is executed each time the V approaches a *LEZ*. This means that the *SE* generates a new $cert_{V_q}$ for the vehicle in each new entrance process. This certificate is the only element that could identify the V. However, considering that the use of the certificate is unique for each entrance/leaving process, nobody can relate the identity of V of this entrance/leaving process with any other entrance/leaving process.

The information that could be repeated in other entrance/leaving process of the same V is the $cert V_q.idS$ field. But, as it is specified in the protocol, $cert V_q.idS$ is computed with a probabilistic encryption, using for example the OAEP padding system, which means that the result of each new encryption credentials is different.

Result 2 *The toll system presented here preserves the privacy in accordance with claims 4 and 5: users can use the system anonymously and each new usage can not be related to any other with respect to the identity of the vehicles.*

Proposition 3. *The toll system has anti-fraud requirements concerning the correctness and verifiability of the evidences generated in the protocol.*

Claim. 6 The system can identify dishonest users thanks to the anonymity revocation property of the protocol.

If users do not properly perform the authentication at the entrance/leaving process of the system, then they can lose the anonymity because the *Chp* takes a picture of the V capturing the number plate. This information is sent to *PA* to act as it is specified in the *Sanction* protocol. In the execution of this protocol, *PA* has the ability to identify the user through the number plate.

If users have not made the proper payment, then *SP* verifies in the *Payment Verification* Protocol that the amount paid corresponds to the tax determined for τ and emissions of V. If the verification fails, this information is sent to *PA* to issue a traffic fine for the user. *PA* verifies the incidence and identifies the user by opening the field $cert V_q.idS$ of the certificate with its private key. Obtaining V_{id} allows the identification and the punishment of the dishonest user.

Claim. 7 The protocol execution generates evidences for an honest user (they are saved in the *OBU*) to prove or disprove the allegations of fraud.

When a user is accused of not performing the authentication correctly, a *pdi* that records the incidence is generated. The user can be accused of using an improper certificate $cert_{V_q}$ or sending an incorrect signature $\omega_{\bar{V}_q}$. In both cases, *PA* contacts her during the *Sanction* process, so she can provide evidences to rebut the charges.

An honest user can retrieve a valid $cert_{V_q}$ from her *OBU*, which matches her vehicle (identified by the number plate) or a signature $\omega_{\bar{V}_q}$ that was successfully computed by the *SE* with the help of the *OBU* during the entrance/leaving process of the *LEZ*. The system requirements state that the *OBU* of a vehicle has to have enough storage capacity to store evidences to rebut the allegations of possible fraud.

In case of a payment incidence, the user has to demonstrate that the payment has been made according to the data stored in $\bar{\rho}$ and $\bar{\gamma}_o$ (both items signed by the *Chp*). Therefore, an honest user will be able to retrieve this information from her *OBU* and send it to *PA* to refute the accusation.

Result 3 *The toll collection scheme keeps fraud under control and it can identify dishonest users. These users receive the appropriate traffic fine. The protocol also allows honest users to get evidences of their correct performance. The evidences are used to rebut any traffic fine due to some kind of malfunction of the system's actors.*

5   Conclusions and Further Works

This paper has presented an *ERP* system for urban areas, which provides a robust fraud control system with a high level of privacy. The entrance/leaving process of the *LEZ* is controlled so that the legitimate tax is computed while the anonymity of the user is preserved. However, if a user commits fraud, she will then be identified by the picture of the number plate taken by the checkpoint in conjunction with the anonymity revocation system of the protocol.

As future work, we consider the extension of the protocol to take more than one *LEZ* into account. We also have the intention of implementing the scheme in order to assess its practical application.

Acknowledgments and Disclaimer. This work was partially supported by the Spanish Government under CO-PRIVACY TIN2011-27076-C03-01, ARES-CONSOLIDER INGENIO 2010 CSD2007-00004 and BallotNext IPT-2012-0603-430000 projects and the FPI grant BES-2012-054780. The authors wish to acknowledge the help provided by Tamar Molina-Asens. Some of the authors are members of the UNESCO Chair in Data Privacy, yet the views expressed in this paper neither necessarily reflect the position of the UNESCO nor commit with that organization.

References

1. Balasch, J., Rial, A., Troncoso, C., Preneel, B., Verbauwhede, I., Geuens, C.: Pretp: Privacy-preserving electronic toll pricing. In: USENIX Security Symposium, pp. 63–78 (2010)
2. Bellare, M., Rogaway, P.: Optimal asymmetric encryption. In: De Santis, A. (ed.) EUROCRYPT 1994. LNCS, vol. 950, pp. 92–111. Springer, Heidelberg (1995)
3. BOE: Resolución int/2836/2013, CVE-DOGC-B-14013017-2014. Núm 6541 - 15.1.2014
4. Chen, X., Lenzini, G., Mauw, S., Pang, J.: A group signature based electronic toll pricing system. In: ARES, pp. 85–93. IEEE Computer Society (2012)

5. Costabile, F., Allegrini, I.: A new approach to link transport emissions and air quality: An intelligent transport system based on the control of traffic air pollution. Environmental Modelling and Software 23(3), 258–267 (2008)
6. Day, J., Huang, Y., Knapp, E., Goldberg, I.: Spectre: spot-checked private ecash tolling at roadside. In: WPES, pp. 61–68. ACM (2011)
7. Garcia, F.D., Verheul, E.R., Jacobs, B.: Cell-based privacy-friendly roadpricing. Computers & Mathematics with Applications 65(5), 774–785 (2013)
8. Meiklejohn, S., Mowery, K., Checkoway, S., Shacham, H.: The phantom tollbooth: Privacy-preserving electronic toll collection in the presence of driver collusion. In: USENIX Security Symposium, pp. 32 (2011)
9. Popa, R.A., Balakrishnan, H., Blumberg, A.J.: Vpriv: Protecting privacy in location-based vehicular services. In: USENIX Security Symposium, pp. 335–350. USENIX Association (2009)
10. Santos, G.: Urban congestion charging: A comparison between London and Singapore. Transport Reviews 25(5), 511–534 (2005)
11. Wolff, H.: Keep your clunker in the suburb: Low-emission zones and adoption of green vehicles. The Economic Journal, n/a–n/a (2014), http://dx.doi.org/10.1111/ecoj.12091

Modeling Complex Nonlinear Utility Spaces Using Utility Hyper-Graphs

Rafik Hadfi and Takayuki Ito

Department of Computer Science and Engineering
Graduate School of Engineering, Nagoya Institute of Technology
Gokiso, Showa-ku, Nagoya 466-8555, Japan
rafik@itolab.nitech.ac.jp, ito.takayuki@nitech.ac.jp

Abstract. There has been an increasing interest in automated negotiation and particularly negotiations that involve interdependent issues, known to yield complex nonlinear utility spaces. However, none of the proposed models was able to tackle the scaling problem as it commonly arises in realistic consensus making situations. In this paper we address this point by proposing a compact representation that minimizes the search complexity in this type of utility spaces. Our representation allows a modular decomposition of the issues and the constraints by mapping the utility space into an issue-constraint hyper-graph with the underlying interdependencies. Exploring the utility space reduces then to a message passing mechanism along the hyper-edges by means of utility propagation. We experimentally evaluate the model using parameterized random nonlinear utility spaces, showing that our mechanism can handle a large family of complex utility spaces by finding the optimal contracts, outperforming previous sampling-based approaches.

Keywords: Utility and decision theory, Optimization methods in AI and decision modeling, Multi-agent systems, Multi-issue Negotiation, Interdependence, Nonlinear Utility, Constraint-based utility spaces, Complexity, Hyper-Graph, Max-Sum, Utility Propagation.

1 Introduction

Automated negotiation is an efficient mechanism to reach agreements among heterogenous and distributed decision makers. In fact, its applications range from coordination and cooperation [1, 2] to task allocation [3, 4], surplus division [5], and decentralized information services [6]. In practical, most of the realistic negotiation problems are characterized by interdependent issues, which yields complex and nonlinear utility spaces [7]. As the search space and the complexity of the problem grow, finding optimal contracts becomes intractable for one single agent. Similarly, reaching an agreement between a group of agents becomes harder.

We propose to tackle the complexity of the utility spaces used in multi-issue negotiation by rethinking the way they are represented. We claim that adopting the adequate representation gives a solid ground to tackle the scaling problem. We address this problem by adopting a representation that allows a modular decomposition of the

V. Torra et al. (Eds.): MDAI 2014, LNAI 8825, pp. 14–25, 2014.

issues-constraints given the idea that constraint-based utility spaces are nonlinear with respect to issues, but linear with respect to the constraints. This allows us to map the utility space into an issue-constraint hyper-graph with the underlying interdependencies. Exploring the utility space reduces then to a message passing mechanism along the hyper-edges by means of utility propagation.

Adopting a graphical representation while reasoning about utilities is not new in both the multi-attribute utility and the multi-issue negotiation literatures. Indeed, the idea of utility graphs could potentially help decomposing highly nonlinear utility functions into sub-utilities of clusters of inter-related items, as in [8, 9]. Similarly, [10] used utility graphs for preferences elicitation and negotiation over binary-valued issues. [11] adopts a weighted undirected graph representation of the constraint-based utility space. However, restricting the graph and the message passing process to constraints' nodes does not allow the representation to be descriptive enough to exploit any potential hierarchical structure of the utility space through a quantitative evaluation of the interdependencies between both issues and constraints. In [12], issues' interdependency are captured by means of similar undirected weighted graphs where a node represents an issue. This representation is restricted to binary interdependencies while real negotiation scenarios involve "bundles" of interdependent issues under one or more specific constraints. In our approach, we do not restrict the interdependency to lower-order constraints but we allow $p-$ary interdependencies to be defined as an hyper-edge connecting p issues. The advantage of our representation is its scalability in the sense that the problem becomes harder for a large number of issues and constraints. But if we can decompose the utility space into independent components, we can exploit it more efficiently using a message passing mechanism.

Another motivation behind the hyper-graph representation is that it allows a layered, hierarchical view of any given negotiation problem. Given such architecture, it is possible to recursively negotiate over the different layers of the problem according to a top-down approach. Even the idea of issue could be abstracted to include an encapsulation of sub-issues, located in sub-utility spaces and represented by cliques in the hyper-graph. Consequently, search processes can help identify optimal contracts for improvement at each level. It is within this perspective that we are proposing our model. The main novelty our work is the efficiency of the new representation when optimizing nonlinear utilities. To the best of our knowledge, our work makes the first attempt to tackle the complexity of such utility spaces using an efficient search heuristic that works and outperforms the previously used sampling-based meta-heuristics. Particularly, the novelty is that we exploit the problem structure (as hyper-graph) as well as randomization. Such performance is required when facing the scaling issues, inherent to complex negotiation. We experimentally evaluated our model using parametrized and random nonlinear utility spaces, showing that it can handle large and complex spaces by finding the optimal contracts while outperforming previous sampling approaches.

The paper is organized as following. In the next section, we propose the basics of our new nonlinear utility space representation. In section 3, we describe the optimal contracts search mechanisms. In section 4, we provide the experimental results. In section 5, we conclude and outline the future work.

2   Nonlinear Utility Spaces

2.1   Formulation

We start from the formulation of nonlinear multi-issue negotiation of [13]. That is, N agents are negotiating over n issues $i_{k \in [1,n]} \in \mathbb{I}$, with $\mathbb{I} = \{i_k\}_{k=1}^n$, forming an $n-$dimensional utility space. The issue k, namely i_k, takes its values from a set $\mathbb{I}_k$ where $\mathbb{I}_k \subset \mathbb{Z}$. A contract c is a vector of issue values $c \in \mathcal{I}$ with $\mathcal{I} = \times_{k=1}^n \mathbb{I}_k$.

An agent's utility function is defined in terms of constraints, making the utility space a constraint-based utility space. That is, a constraint $c_{j \in [1,m]}$ is a region of the total $n-$dimensional utility space. We say that the constraint c_j has value $w(c_j, c)$ for contract c if constraint c_j is satisfied by contract c. That is, when the contract point c falls within the hyper-volume defined by the constraint of c_j, namely $hyp(c_j)$. The utility of an agent for a contract c is thus defined as in (1).

$$u(c) = \sum_{c_{j \in [1,m]},\ c \in hyp(c_j)} w(c_j, c) \tag{1}$$

In the following, we distinguish three types of constraints: Cubic constraints, Bell constraints and Plane constraints, shown in Figure 1. The constraint-based utility formalism is a practical way to reason about preferences subject to restrictions. More details about constraint-based utility spaces and their usage is to be found in [11, 14, 15].

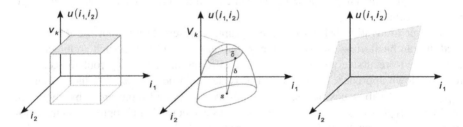

Fig. 1. Cubic, Bell and Plane Constraints

Having a large number of constraints produces a "bumpy" nonlinear utility space with high points whenever many constraints are satisfied and lower points where few or no constraints are satisfied. Figure 2 shows an example of nonlinear utility space for issues i_1 and i_2 taking values in $\mathbb{I}_1 = \mathbb{I}_2 = [0, 100]$, with $m = 500$ constraints and where a constraint involves at least 2 issues.

2.2   Our New Representation

The agent's utility function (1) is nonlinear in the sense that the utility does not have a linear expression against the contract [13]. This is true to the extent that the linearity is evaluated with regard to the contract c. However, from the same expression (1) we can say that the utility is in fact linear, but in terms of the constraints $c_{j \in [1,m]}$. The utility

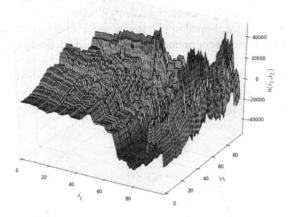

Fig. 2. 2−dimensional nonlinear utility space

space is therefore decomposable according to the c_j constraints. This yields a modular representation of the interactions between the issues and how they locally relate to each other. In fact, $hyp(c_j)$ reflects the idea that the underlying contracts are governed by the bounds defined by c_j once the contracts are projected according to their issues' components.

In this case, the interdependence is not between issues but between constraints. For instance, two constraints c_1 and c_2 can have in common one issue i_k taking values respectively from an interval $\mathbb{I}_{k,c_1}$ if it is in c_1, and values in $\mathbb{I}_{k,c_2}$ if it is in c_2, with $\mathbb{I}_{k,c_1} \neq \mathbb{I}_{k,c_2}$. Finding the value that maximizes the utility of i_k while satisfying both constraints becomes harder due to fact that changing the value of i_k in c_1 changes its instance in c_2 in a cyclic manner. This gets worse with an increasing number of issues, their domains' sizes, and the non-monotonicity of the constraints. Next, we propose to transform (1) into a modular, graphical representation. Since one constraint can involve one or more multiple issues, we adopt a hyper-graph representation.

2.3   From Utility Space to Utility Hypergraph

We assign to each constraint $c_{j\in[1,m]}$, a factor Φ_j, with $\Phi = \{\Phi_j\}_{j=1}^m$. We define the hyper-graph G as $G = (\mathbb{I}, \Phi)$. Nodes in $\mathbb{I}$ define the issues and the hyper-edges in Φ are the factors (constraints). To each factor Φ_j we assign a neighbors' set $\mathcal{N}(\Phi_j) \subset \mathbb{I}$ containing the issues connected to Φ_j (involved in c_j), with $|\mathcal{N}(\Phi_j)| = \varphi_j$. In case $\varphi_j = 2 \ \forall j \in [1,m]$, the problem collapses to a constraints satisfaction problem in a standard graph.

To each factor Φ_j corresponds a φ_j−dimensional matrix, $\mathcal{M}_{\Phi_j}$, where the jth dimension is the discrete interval $[a_k, b_k] = \mathbb{I}_k$, the domain of issue i_k. This matrix contains all the values that could be taken by the issues in $\mathcal{N}(\Phi_j)$. Each factor Φ_j has a function Φ_j defined as a sub-utility function of the issues in $\mathcal{N}(\Phi_j)$, as in (2).

$$\Phi_j : \mathcal{N}(\Phi_j)^{\varphi_j} \to \mathbb{R} \qquad (2)$$
$$\Phi_j(i_1, \ldots, i_k, \ldots, i_{\varphi_j}) \mapsto w(c_j, \boldsymbol{c})$$

As we are dealing with discrete issues, Φ_j is defined by the matrix $\mathcal{M}_{\Phi_j}$. That is, $\Phi_j(i_1, \ldots i_k, \ldots, i_{\varphi_j})$ is simply the $(1, \ldots, k, \ldots, \varphi_j)^{th}$ entry in $\mathcal{M}_{\Phi_j}$ corresponding as well to the value $w(c_j, c)$ mentioned in (1). It is possible to extend the discrete case to the continuous one by allowing continuous issue-values and defining Φ_j as a continuous function.

To provide an example of our representation, let us consider a $10-$dimensional utility space and propose to represent it in a graphical way. That is, as a hyper-graph defined as $G_{10} = (\mathbb{I}, \Phi)$ with $\mathbb{I} = \{i_k\}_{k=1}^{9}$, $\Phi = \{\Phi_j\}_{j=1}^{7}$ and shown in Figure 3.

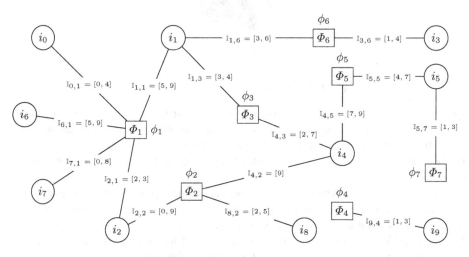

Fig. 3. Issues-Constraints Hypergraph

Each issue i_k has a set $\mathbb{I}_k = \bigcup_{\nu \in \mathcal{N}(k)} \mathbb{I}_{k,\nu}$ where $\mathbb{I}_{k,\nu}$ is an edge connecting i_k to its neighbor $\nu \subset \mathcal{N}(k) \in \Phi$. For example, $\mathbb{I}_1 = \bigcup_{\nu \in \{\Phi_1, \Phi_3, \Phi_6\}} \mathbb{I}_{1,\nu} = \{[5,9], [3,4], [3,6]\}$.

Constraints can have different types in the sense that each type reflects a particular geometric shape. For example, constraints $\Phi_{1,2,3,4}$ could be cubic, $\Phi_{5,6}$ could be defined as planes and Φ_7 defined as a bell. Any combination is in fact possible, and depends only on the problem in hand and how it is being specified. Each constraint is assigned a sub-utility representation used to compute the utility of a contract if it satisfies the corresponding constraint by being located in the underlying hyper-volume. For example, the general utility function Φ_j, defined in (2), could correspond to the functional definition of each constraints, as shown in (3).

$$\Phi_j = \begin{cases} Plane: & \beta_j + \sum_{k=1}^{\varphi_j} \alpha_{j,k} \times v_k(i_k) \qquad (\beta_j, \alpha_{j,k}) \in \mathbb{Z}^2 \\ Cube: & v_j \\ Bell: & V_j \end{cases} \qquad (3)$$

A plane constraint will be defined using its φ_j-dimensional equation, while a cubic constraint will be assigned the value v_j in case the contract is in the cube. The computation of the utility V_j of a bell shaped constraint is performed as in (4) (see Figure 1).

Herein, δ is the Euclidean distance from the center s of the bell constraint to a contract point c. Distances are normalized in $[-1, 1]$.

$$V_j = \begin{cases} \beta_j \left(1 - 2\delta^2\right) & \text{if } \delta < 0.5 & \beta_j \in \mathbb{Z} \\ 2\,\beta_j \left(1 - \delta\right)^2 & \text{if } \delta < 1 & \beta_j \in \mathbb{Z} \\ 0 & \text{else} \end{cases} \tag{4}$$

It is possible to extend the current constraints' to involve Cone constraints [16] or any other type of geometrical shapes.

3   Optimal Contracts

The exploration of the utility hyper-graph is inspired from the sum-product message passing algorithm for belief propagation [17]. However, the multiplicative algebra is changed into an additive algebra to support the utility accumulation necessary for the assessment of the contracts. The messages circulating in the hyper-graph are nothing other than the contracts we are attempting to optimize through utility maximization. Next, we develop the message passing (MP) mechanism operating on the issues and the constraints.

3.1   Message Passing

We consider the issues set $\mathbb{I}$ and a contract point $c = (i_1, \ldots, i_k, \ldots, i_n) \in \mathcal{I}$. We want to find a contract c^* that maximizes the utility function defined in (1). Assuming that Φ_j is the local sub-utility of constraint Φ_j, we distinguish two types of messages: messages sent from issues to constraints, and messages sent from constraints to issues.

From Issue i_k to Constraint Φ_j : In (5), each message $\mu_{i_k \to \Phi_j}$ coming from i_k to Φ_j is the sum of the constraints' messages to i_k coming from constraints other than Φ_j.

$$\mu_{i_k \to \Phi_j}(i_k) = \sum_{\Phi_{j'} \in \mathcal{N}(i_k) \backslash \Phi_j} \mu_{\Phi_{j'} \to i_k}(i_k) \tag{5}$$

From Constraint Φ_j to Issue i_k : Each constraint message (6) is the sum of the messages coming from issues other than i_k, plus the constraint value $\Phi_j(i_1, \ldots, i_k, \ldots, i_n)$, summed over all the possible values of the issues (connected to the constraint Φ) other than the issue i_k.

$$\mu_{\Phi_j \to i_k}(i_k) = \max_{i_1} \ldots \max_{i_{k'} \neq k} \ldots \max_{i_n} \left[\Phi_j(i_1, \ldots, i_k, \ldots, i_n) + \sum_{i_{k'} \in \mathcal{N}(\Phi_j) \backslash i_k} \mu_{i_{k'} \to \Phi_j}(i_k) \right] \tag{6}$$

The MP mechanism starts from the leaves of the hyper-graph, $i.e.$, the issues. At $t = 0$, the content of the initial messages is defined according to (7), with $\phi'_j(i_k)$ being the partial evaluation of i_k in the factor Φ_j.

$$\mu_{i_k \to \Phi_j}(i_k) = 0 \tag{7}$$

$$\mu_{\Phi_j \to i_k}(i_k) = \phi'_j(i_k)$$

The partial evaluation $\phi'_j(i_k)$ of issue i_k in the factor Φ_j is the utility of i_k using Φ_j regardless of any other issue involved in Φ_j. For instance, for cubic and bell constraints, the evaluation is simply $v_j(i_k)$ and $V_j(i_k)$ $\forall k$ as described in (3). If Φ_j is a plane constraint, the partial evaluation of i_k will be $\alpha_{j,k} \times v_k(i_k)$. In this manner, the factor Φ_j will get all the evaluations $(\alpha_{j,k} \times v_k(i_k))$ from its surrounding issues in order to yield the total utility (3) as a sum of the partial evaluations plus the plane constant β_j.

Finally, the optimal contract c^* is found by collecting the optimal issues as in (8).

$$c^* = \left(\arg\max_{i_1} \sum_{\Phi_j \in \mathcal{N}(i_1)} \mu_{\Phi_j \to i_1}(i_1), \ldots, \arg\max_{i_n} \sum_{\Phi_j \in \mathcal{N}(i_n)} \mu_{\Phi_j \to i_n}(i_n) \right) \quad (8)$$

In a negotiation setting, it is more common that the agent requires a collection, or *bundle*, of the optimal contracts rather than one single optimum. In order to find such collection, we should endow (8) with a caching mechanism allowing each node in the hyper-graph to store the messages that have been sent to it from the other nodes. That is, the cached messages will contain the summed-up utility values of the underlying node's instance. This is performed every time the operation *max* is called in (6) so that we can store the settings of the adjacent utility (and contract) that led to the maximum. Once ordered, such data structure allows us to generate an ordered bundle for the bidding process. In the next section, we algorithmically provide the MP mechanism.

3.2   Utility Propagation Algorithm

Main Algorithm. Algorithm 1, operates on the hyper-graph nodes by triggering the MP process. Despite the fact that we have 2 types of nodes (issues and constraints), it is possible to treat them abstractly using $MsgPass$. The resulting bundle is a collection of optimal contracts with utility greater or equal to the agent's reservation value rv.

Algorithm: Utility Propagation

Input: $G = (\mathbb{I}, \Phi), rv, mode, \rho$
Output: Optimal contracts (bundle)

```
1  begin
2      for i = 1 → (ρ × |𝕀 ∪ Φ|) do
3          if mode is Synchronous then
4              foreach ν_src ∈ 𝕀 ∪ Φ do
5                  foreach ν_dest ∈ ν_src.Neighbors() do
6                      ν_src.MsgPass(ν_dest)

7      bundle ← ∅
8      foreach i ∈ 𝕀 do
9          bundle[i] ← ∅
10         ι ← ∪_{j∈i.instances()} [j.min, j.max]
11         μ* ← k* ← −∞
12         μ ← i.getmax()
13         foreach k = 1 → |μ| do
14             if μ* < μ[k] then
15                 μ* ← μ[k], k* ← k
16                 if μ* ≥ rv then
17                     bundle[i] ← bundle[i] ∪ ι[k*]

18     return bundle
```

Algorithm 1. Main Algorithm

Issue to Constraint. The issue's message to a factor (or constraint) is the element-wise sum of all the incoming messages from other factors.

Constraint to Issue. The factor's message to a targeted issue is done by recursively enumerating over all variables that the factor references (6), except the targeted issue. This needs to be performed for each value of the target variable in order to compute the message. If all issues are assigned, the values of the factor and of all other incoming messages are determined, so that their sum term is compared to the prior maximum. Resulting messages, stored in *bundle*, contain the values that maximize the factors' local sub-utility functions. We note that it is possible to avoid the systematic enumeration by adding a local randomization to the issue that the factor is referencing.

Optimal Contracts Collection. The optimal contracts will be the concatenation of the single optimal issue-values. That is, for each issue we will collect the coming messages from the surrounding factors and sum the overlapping utility values in order to obtain the values that maximize the sum of the utilities (8).

4 Experiments

4.1 Settings

Before evaluating the utility propagation algorithm, we identify the criteria that could affect the complexity of the utility space and thus the probability of finding optimal

---

Algorithm: ParamRandHGen

Input: n, m, p
Output: $G(\mathbb{I}, \Phi)$

```
1  begin
2  │   [βmin, βmax] ← [1, 100]      // constants
3  │   [αmin, αmax] ← [0, 1]        // slopes
4  │   [bmin, bmax] ← [0, 9]        // bounds
5  │   Φ ← [∅] × m                  // init constraints set
6  │   for k = 1 → m do
7  │   │   Φ[k].θ ← rand({cube, plane, bell})
8  │   │   if Φ[k].θ = plane then
9  │   │   │   α ← [0] × n
10 │   │   │   α[j] ← rand([αmin, αmax]) ∀i ∈ [1, n]
11 │   │   │   Φ[k].α ← α
12 │   │   if Φ[k].θ ∈ {bell, cube} then
13 │   │   │   // refer to (3) or (4)
14 │   │   Φ[k].β ← rand([βmin, βmax])
15 │   │   μ ← rand([1, n]) , I ← ∅
16 │   │   while |I| ≠ μ do
17 │   │   │   ι ← rand([1, p])
18 │   │   │   if ι ∉ I then
19 │   │   │   │   I ← I ∪ ι
20 │   │   for j = 1 → μ do
21 │   │   │   I[j].a ← rand([bmin, bmax]) I[j].b ← rand([I[j].a + ϵ, bmax])
22 │   │   Φ[k].I ← I
23 │   return Φ
```

Algorithm 2. Utility Hypergraph Generation

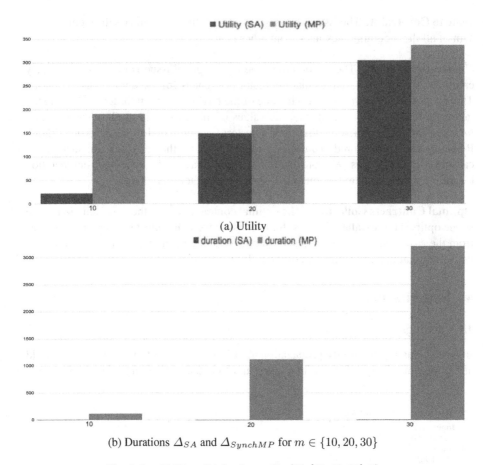

(a) Utility

(b) Durations Δ_{SA} and $\Delta_{SynchMP}$ for $m \in \{10, 20, 30\}$

Fig. 4. SynchMP vs. SA for the profile $(10, [10, 20, 30], 5)$

contract(s). Other than n and m, we distinguish p, defined as the maximal number of issues involved in a constraint. p can be unary ($p = 1$), binary ($p = 2$), ternary ($p = 3$), or p-ary in the general case.

The parametrized generation of a utility space (or utility hyper-graph) should meet the consistency condition $p \le n \le m \times p$, with $n, m, p \in \mathbb{N}^+$, to avoid problems like attempting to have an 8–ary constraints in a 5–dimensional utility space.

4.2   Discussion

After the generation of the hyper-graph using Algorithm 2, the message-passing routines will be evaluated and analyzed microscopically from the agent perspective.

We will compare the MP mechanism in terms of utility and duration to the Simulated Annealing (SA) approach in [13] for optimal contract finding. The SA optimizer will be randomly sampling from the regions that correspond to an overlap of constraints. For instance, generating a random contract satisfying c_j is performed backwardly through

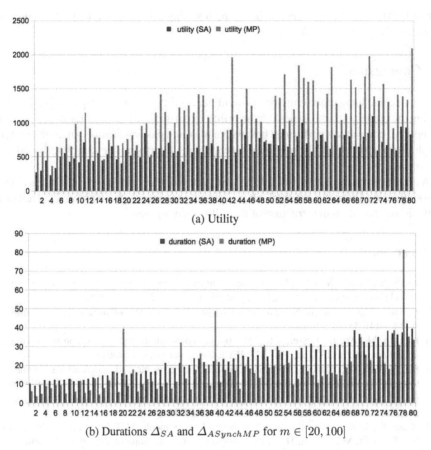

(a) Utility

(b) Durations Δ_{SA} and $\Delta_{ASynchMP}$ for $m \in [20, 100]$

Fig. 5. AsynchMP vs. SA for the profile $(40, [20, \ldots 100], 5)$

random generation of values from $\mathbb{I}_{j,k} \ \forall i_k \in \mathcal{N}(\Phi_j)$. Our comparison criteria is based on the utility/duration performed on a set of profiles of the form (n, m, p), with 100 trials for each profile.

Figure 4 illustrates the performance of SynchMP for $(10, [10, 20, 30], 5)$.

The deterministic aspect of the synchronous message passing algorithm (SynchMP) makes it very slow ($\Delta_{SA} << \Delta_{SynchMP}$) compared to its SA counterpart which exploits the randomization, allowing it to perform "jumps" in the search space. To avoid the enumeration over the local nodes of G, it is possible to add randomization to the way nodes are selected. To introduce an asynchronous mode, AsynchMP, we add another condition after the synchronous mode condition in Algorithm 1, as follows:

if *mode* is *Asynchronous* **then**

$\quad \nu_{src}, \ \nu_{dest} \leftarrow rand_2([1, |V|]), \ \nu_{dest} \neq \nu_{src}$

$\quad \nu_{src}.MsgPass(\nu_{dest})$

For $(40, [20, \ldots 100], 5)$, Figure 5 shows the resulting difference in the performance of AsynchMP compared to SA.

5   Conclusion

A new and modular representation for nonlinear utility spaces is proposed by decomposing the constraints and issues into an utility hyper-graph. The exploration and search for optimal contracts is performed based on a message passing mechanism in the hyper-graph. Results show that the proposed mechanism outperforms the sampling-based optimizers.

As future work, we intend to exploit the structure of the hyper-graphs for hierarchical negotiation. Additionally, we think about studying the interdependence and correlation of the issues based on the structure of the utility hyper-graph.

References

[1] Kraus, S., Sycara, K., Evenchik, A.: Reaching agreements through argumentation: a logical model and implementation. Artificial Intelligence 104(1-2), 1–69 (1998)

[2] Jennings, N.R.: An agent-based approach for building complex software systems. Commun. ACM 44(4), 35–41 (2001)

[3] Krainin, M., An, B., Lesser, V.R.: An application of automated negotiation to distributed task allocation. In: IAT, pp. 138–145. IEEE Computer Society (2007)

[4] Ke, W., Peng, Z., Yuan, Q., Hong, B., Chen, K., Cai, Z.: A method of task allocation and automated negotiation for multi robots. Journal of Electronics (China) 29(6), 541–549 (2012)

[5] Fatima, S.S., Wooldridge, M., Jennings, N.R.: An agenda-based framework for multi-issue negotiation. Artificial Intelligence 152(1), 1–45 (2004)

[6] Kraus, S.: Strategic Negotiation in Multiagent Environments. MIT Press, Cambridge (2001)

[7] Ito, T., Klein, M., Hattori, H.: A multi-issue negotiation protocol among agents with nonlinear utility functions. Multiagent and Grid Systems 4(1), 67–83 (2008)

[8] Chajewska, U., Koller, D.: Utilities as random variables: Density estimation and structure discovery. In: Proceedings of the Sixteenth Annual Conference on Uncertainty in Artificial Intelligence (UAI 2000), pp. 63–71 (2000)

[9] Bacchus, F., Grove, A.: Graphical models for preference and utility. In: Proceedings of the Eleventh Conference on Uncertainty in Artificial Intelligence, UAI 1995, pp. 3–10. Morgan Kaufmann Publishers Inc., San Francisco (1995)

[10] Robu, V., Somefun, D.J.A., Poutre, J.L.: Modeling complex multi-issue negotiations using utility graphs. In: Proceedings of the 4th International Joint Conference on Autonomous Agents and Multi-Agent Systems (AAMAS 2005), pp. 280–287 (2005)

[11] Marsa-Maestre, I., Lopez-Carmona, M.A., Velasco, J.R., de la Hoz, E.: Effective bidding and deal identification for negotiations in highly nonlinear scenarios. In: Proceedings of the 8th International Conference on Autonomous Agents and Multiagent Systems, AAMAS 2009, vol. 2, pp. 1057–1064. International Foundation for Autonomous Agents and Multiagent Systems, Richland (2009)

[12] Fujita, K., Ito, T., Klein, M.: An approach to scalable multi-issue negotiation: Decomposing the contract space based on issue interdependencies. In: Proceedings of the 2010 IEEE/WIC/ACM International Conference on Web Intelligence and Intelligent Agent Technology, WI-IAT 2010, vol. 02, pp. 399–406. IEEE Computer Society, Washington, DC (2010)

[13] Ito, T., Hattori, H., Klein, M.: Multi-issue negotiation protocol for agents: Exploring nonlinear utility spaces. In: Proceedings of the 20th International Joint Conference on Artificial Intelligence (IJCAI-2007), pp. 1347–1352 (2007)

[14] Lopez-Carmona, M.A., Marsa-Maestre, I., De La Hoz, E., Velasco, J.R.: A region-based multi-issue negotiation protocol for nonmonotonic utility spaces. Computational Intelligence 27(2), 166–217 (2011)

[15] Marsa-Maestre, I., Lopez-Carmona, M., Carral, J., Ibanez, G.: A recursive protocol for negotiating contracts under non-monotonic preference structures. Group Decision and Negotiation 22(1), 1–43 (2013)

[16] Fujita, K., Ito, T., Klein, M.: A secure and fair negotiation protocol in highly complex utility space based on cone-constraints. In: Proceedings of the 2009 IEEE/WIC/ACM International Joint Conference on Web Intelligence and Intelligent Agent Technology, WI-IAT 2009, vol. 02, pp. 427–430. IEEE Computer Society, Washington, DC (2009)

[17] Pearl, J.: Probabilistic Reasoning in Intelligent Systems: Networks of Plausible Inference. Morgan Kaufmann Publishers Inc., San Francisco (1988)

On the Distributivity Equation $\mathcal{I}(x,\mathcal{U}_1(y,z)) = \mathcal{U}_2(\mathcal{I}(x,y),\mathcal{I}(x,z))$ for Decomposable Uninorms (in Interval-Valued Fuzzy Sets Theory) Generated from Conjunctive Representable Uninorms

Michał Baczyński and Wanda Niemyska

Institute of Mathematics, University of Silesia,
40-007 Katowice, ul. Bankowa 14, Poland
{michal.baczynski,wniemyska}@us.edu.pl

Abstract. In this paper we continue investigations connected with distributivity of implication operations over decomposable (t-representable) operations. Our main goal is to show the general method of solving the following distributivity equation $\mathcal{I}(x,\mathcal{U}_1(y,z)) = \mathcal{U}_2(\mathcal{I}(x,y),\mathcal{I}(x,z))$, when $\mathcal{U}_1$, $\mathcal{U}_2$ are decomposable uninorms (in interval-valued fuzzy sets theory) generated from two conjunctive representable uninorms. As a byproduct result we show all solutions of some functional equation related to this case.

Keywords: Aggregation operators, Uninorms, Interval-valued fuzzy sets, Distributivity equations, Functional equations.

1 Introduction

The distributivity of (classical) fuzzy implications over different fuzzy logical connectives, like t-norms, t-conorms or uninorms has been studied in the recent past by many authors (see chronologically [1], [31], [14], [28], [29], [13], [2], [3], and [9]). Distributivity equations have a very important role to play in efficient inferencing in approximate reasoning, especially in fuzzy control systems. Given an input "$\widetilde{x}$ is A'", the role of an inference mechanism is to obtain a fuzzy output B' that satisfies some desirable properties. The most important inference schemas are fuzzy relational inference and similarity based reasoning. In the first case, the inferred output B' is obtained either as

(i) $\sup -T$ composition, where T is a t-norm, as in the compositional rule of inference (CRI) of Zadeh (see [33]), or
(ii) $\inf -I$ composition, where I is a fuzzy implication, as in the Bandler-Kohout subproduct (BKS) (see [15]),

of A' and given rules. Since all the rules of an inference engine are exercised during every inference cycle, the number of rules directly affects the computational duration of the overall application.

V. Torra et al. (Eds.): MDAI 2014, LNAI 8825, pp. 26–37, 2014.

To reduce the complexity of fuzzy "IF-THEN" rules, Combs and Andrews [17] proposed an equivalent transformation of the CRI to mitigate the computational cost. In fact, they demanded the following classical tautology

$$(p \wedge q) \to r = (p \to r) \vee (q \to r),$$

written in fuzzy logic language, i.e., using t-norms, t-conorms and fuzzy implications. Subsequently, there were many discussions (see [16], [20], [27]), most of them pointed out the need for a theoretical investigation required for employing such equations. Later, the similar method but for similarity based reasoning was presented by Jayaram [26]. For an overview of the most important of these methods see [12, Chapter 8].

Recently, in [4], [5], [6] (for the full article see [10]), [8] and [11] we have discussed the following distributivity equations

$$\mathcal{I}(x,\mathcal{T}_1(y,z)) = \mathcal{T}_2(\mathcal{I}(x,y),\mathcal{I}(x,z)),$$
$$\mathcal{I}(\mathcal{S}(x,y),z) = \mathcal{T}(\mathcal{I}(x,z),\mathcal{I}(y,z)),$$

for t-representable (decomposable) t-norms and t-conorms (in interval-valued fuzzy sets theory) generated from continuous Archimedean operations. In fact, in these articles, we have obtained the solutions for each of the following functional equations, respectively:

$$f(u_1 + v_1, u_2 + v_2) = f(u_1, u_2) + f(v_1, v_2), \tag{A}$$
$$g(\min(u_1 + v_1, a), \min(u_2 + v_2, a)) = g(u_1, u_2) + g(v_1, v_2), \tag{B}$$
$$h(\min(u_1 + v_1, a), \min(u_2 + v_2, a)) = \min(h(u_1, u_2) + h(v_1, v_2), b), \tag{C}$$
$$k(u_1 + v_1, u_2 + v_2) = \min(k(u_1, u_2) + k(v_1, v_2), b), \tag{D}$$

where $a, b > 0$ are fixed real numbers, $f \colon L^\infty \to [0, \infty]$, $g \colon L^a \to [0, \infty]$, $h \colon L^a \to [0, b]$, $k \colon L^\infty \to [0, b]$ are unknown functions and

$$L^\infty = \{(u_1, u_2) \in [0, \infty]^2 \mid u_1 \geq u_2\},$$
$$L^a = \{(u_1, u_2) \in [0, a]^2 \mid u_1 \geq u_2\}.$$

More precisely, the solutions of Eq. (A) are presented in [4, Proposition 3.2], the solutions of Eq. (B) are presented in [5, Proposition 4.2], the solutions of Eq. (C) are presented in [10, Proposition 5.2] and the solutions of Eq. (D) are presented in [8, Proposition 3.2].

In this paper we continue these investigations, but for the following functional equation

$$\mathcal{I}(x,\mathcal{U}_1(y,z)) = \mathcal{U}_2(\mathcal{I}(x,y),\mathcal{I}(x,z)), \tag{D-UU}$$

when $\mathcal{U}_1, \mathcal{U}_2$ are decomposable uninorms on $\mathcal{L}^I$ generated from two conjunctive representable uninorms and $\mathcal{I}$ is an unknown function. In fact, from mathematical point of view, we discuss the solutions of the following functional equation

$f(u_1 + v_1, u_2 + v_2) = f(u_1, u_2) + f(v_1, v_2)$, for all $(u_1, u_2), (v_1, v_2) \in L^{\overline{\infty}}$, where $L^{\overline{\infty}} = \{(x_1, x_2) \in [-\infty, \infty]^2 \mid x_1 \leq x_2\}$, with the assumption $(-\infty) + \infty = \infty + (-\infty) = -\infty$ in both sets of domain and codomain of a function f.

Please note that solutions for this Eq. (D-UU) for classical representable uninorms, have been presented in [3] (see also [7]). Moreover, solutions of distributivity equations for different classes of classical implications have been obtained by Ruiz-Aguilera and Torrens in [28] and [29].

2 Interval-Valued Fuzzy Sets

One possible extension of fuzzy sets theory is interval-valued fuzzy sets theory introduced, independently, by Sambuc [30] and Gorzałczany [25], in which to each element of the universe a closed subinterval of the unit interval is assigned – it can be used as an approximation of the unknown membership degree. Let us define

$$L^I = \{(x_1, x_2) \in [0, 1]^2 \mid x_1 \leq x_2\},$$
$$(x_1, x_2) \leq_{L^I} (y_1, y_2) \Longleftrightarrow x_1 \leq y_1 \wedge x_2 \leq y_2.$$

In the sequel, if $x \in L^I$, then we denote it by $x = [x_1, x_2]$. In fact, $\mathcal{L}^I = (L^I, \leq_{L^I})$ is a complete lattice with units $0_{\mathcal{L}^I} = [0, 0]$ and $1_{\mathcal{L}^I} = [1, 1]$.

Definition 2.1. *An interval-valued fuzzy set on X is a mapping $A \colon X \to L^I$.*

3 Implications and Uninorms

We assume that the reader is familiar with the classical results concerning basic fuzzy logic connectives, but we briefly mention some of the results employed in the rest of the work.

One possible definition of an implication on $\mathcal{L}^I$ is based on the well-accepted notation introduced by Fodor and Roubens [23] (see also [12] and [19]).

Definition 3.1. *Let $\mathcal{L} = (L, \leq_L)$ be a complete lattice. A function $\mathcal{I} \colon L^2 \to L$ is called a fuzzy implication on $\mathcal{L}$ if it is decreasing with respect to the first variable, increasing with respect to the second variable and fulfills the following conditions:*

$$\mathcal{I}(0_{\mathcal{L}}, 0_{\mathcal{L}}) = \mathcal{I}(1_{\mathcal{L}}, 1_{\mathcal{L}}) = \mathcal{I}(0_{\mathcal{L}}, 1_{\mathcal{L}}) = 1_{\mathcal{L}}, \qquad \mathcal{I}(1_{\mathcal{L}}, 0_{\mathcal{L}}) = 0_{\mathcal{L}}. \qquad (1)$$

Uninorms (on the unit interval) were introduced by Yager and Rybalov in 1996 (see [32]) as a generalization of triangular norms and conorms. For the recent overview of this family of operations see [22].

Definition 3.2. *Let $\mathcal{L} = (L, \leq_L)$ be a complete lattice. An associative, commutative and increasing operation $\mathcal{U} \colon L^2 \to L$ is called a uninorm on $\mathcal{L}$, if there exists $e \in L$ such that $\mathcal{U}(e, x) = \mathcal{U}(x, e) = x$, for all $x \in L$.*

Remark 3.3. (i) The neutral element e corresponding to a uninorm $\mathcal{U}$ is unique. Moreover, if $e = 0_{\mathcal{L}}$, then $\mathcal{U}$ is a t-conorm and if $e = 1_{\mathcal{L}}$, then $\mathcal{U}$ is a t-norm.

(ii) For a uninorm $\mathcal{U}$ on any $\mathcal{L}$ we get $\mathcal{U}(0_{\mathcal{L}},0_{\mathcal{L}}) = 0_{\mathcal{L}}$ and $\mathcal{U}(1_{\mathcal{L}},1_{\mathcal{L}}) = 1_{\mathcal{L}}$.

(iii) For a uninorm U on $([0,1], \leq)$ we get $U(0,1) \in \{0,1\}$.

(iv) For a uninorm $\mathcal{U}$ on $\mathcal{L}^I$ with the neutral element $e \in L^I \setminus \{0_{\mathcal{L}^I}, 1_{\mathcal{L}^I}\}$ we get $\mathcal{U}(0_{\mathcal{L}^I}, 1_{\mathcal{L}^I}) \in \{0_{\mathcal{L}^I}, 1_{\mathcal{L}^I}\}$ or $\mathcal{U}(0_{\mathcal{L}^I}, 1_{\mathcal{L}^I}) \| e$, i.e., $\mathcal{U}(0_{\mathcal{L}^I}, 1_{\mathcal{L}^I})$ is not comparable with e (cf. [18] and [21]).

(v) In general, for any lattice $\mathcal{L}$, if $\mathcal{U}(0_{\mathcal{L}}, 1_{\mathcal{L}}) = 0_{\mathcal{L}}$, then it is called conjunctive and if $\mathcal{U}(0_{\mathcal{L}}, 1_{\mathcal{L}}) = 1_{\mathcal{L}}$, then it is called disjunctive.

In the literature one can find several classes of uninorms (see [24]). Uninorms that can be represented as in Theorem 3.4 are called representable uninorms.

Theorem 3.4 ([24, Theorem 3]). *For a function $U\colon [0,1]^2 \to [0,1]$ the following statements are equivalent:*

(i) *U is a strictly increasing and continuous on $]0,1[^2$ uninorm with the neutral element $e \in]0,1[$ such that U is self-dual, except in points $(0,1)$ and $(1,0)$, with respect to a strong negation N with the fixed point e, i.e.,*

$$U(x,y) = N(U(N(x), N(y))), \qquad x,y \in [0,1]^2 \setminus \{(0,1),(1,0)\}.$$

(ii) *U has a continuous additive generator, i.e., there exists a continuous and strictly increasing function $h\colon [0,1] \to [-\infty, \infty]$, such that $h(0) = -\infty$, $h(e) = 0$ for $e \in]0,1[$ and $h(1) = \infty$, which is uniquely determined up to a positive multiplicative constant, such that for all $x,y \in [0,1]$ either*

$$U(x,y) = \begin{cases} 0 & \text{if } (x,y) \in \{(0,1),(1,0)\}, \\ h^{-1}(h(x) + h(y)), & \text{otherwise,} \end{cases}$$

when U is conjunctive, or

$$U(x,y) = \begin{cases} 1 & \text{if } (x,y) \in \{(0,1),(1,0)\}, \\ h^{-1}(h(x) + h(y)), & \text{otherwise,} \end{cases}$$

when U is disjunctive.

Remark 3.5 (cf. [3]). If a representable uninorm U is conjunctive, then $U(x,y) = h^{-1}(h(x) + h(y))$ holds for all $x,y \in [0,1]$ with the assumption

$$(-\infty) + \infty = \infty + (-\infty) = -\infty. \tag{A-}$$

If a representable uninorm U is disjunctive, then $U(x,y) = h^{-1}(h(x) + h(y))$ holds for all $x,y \in [0,1]$ with the assumption

$$(-\infty) + \infty = \infty + (-\infty) = \infty. \tag{A+}$$

Now we shall consider the following special class of uninorms on $\mathcal{L}^I$.

Definition 3.6 (see [18] and [21]). *A uninorm $\mathcal{U}$ on $\mathcal{L}^I$ is called decomposable (or t-representable) if there exist uninorms U_1, U_2 on $([0,1], \leq)$ such that*

$$\mathcal{U}([x_1, x_2], [y_1, y_2]) = [U_1(x_1, y_1), U_2(x_2, y_2)], \qquad [x_1, x_2], [y_1, y_2] \in L^I,$$

and $U_1 \leq U_2$. In this case we will write $\mathcal{U} = (U_1, U_2)$.

It should be noted that not all uninorms on $\mathcal{L}^I$ are decomposable (see [21]).

Lemma 3.7 ([21, Lemma 8]). *If $\mathcal{U}$ on $\mathcal{L}^I$ is a decomposable uninorm, then $\mathcal{U}(0_{\mathcal{L}^I}, 1_{\mathcal{L}^I}) = 0_{\mathcal{L}^I}$ or $\mathcal{U}(0_{\mathcal{L}^I}, 1_{\mathcal{L}^I}) = 1_{\mathcal{L}^I}$ or $\mathcal{U}(0_{\mathcal{L}^I}, 1_{\mathcal{L}^I}) = [0, 1]$.*

Therefore it is not possible that for decomposable uninorm $\mathcal{U} = (U_1, U_2)$ on $\mathcal{L}^I$ we have that U_1 is disjunctive and U_2 is conjunctive.

Lemma 3.8 (cf. [21, Theorems 5 and 6]). *If $\mathcal{U} = (U_1, U_2)$ on $\mathcal{L}^I$ is a decomposable uninorm with the neutral element $e = [e_1, e_2]$, then $e_1 = e_2$ is the neutral element of U_1 and U_2.*

Lemma 3.9. *Let a function $\mathcal{I} \colon (\mathcal{L}^I)^2 \to \mathcal{L}^I$ satisfy (1) and Eq. (D-UU) with some uninorms $\mathcal{U}_1$, $\mathcal{U}_2$. Then $\mathcal{U}_1$ is conjunctive if and only if $\mathcal{U}_2$ is conjunctive.*

Proof. Firstly, substituting $x = y = 1_{\mathcal{L}^I}$ and $z = 0_{\mathcal{L}^I}$ into (D-UU), we get

$$\mathcal{I}(1_{\mathcal{L}^I}, \mathcal{U}_1(1_{\mathcal{L}^I}, 0_{\mathcal{L}^I})) = \mathcal{U}_2(\mathcal{I}(1_{\mathcal{L}^I}, 1_{\mathcal{L}^I}), \mathcal{I}(1_{\mathcal{L}^I}, 0_{\mathcal{L}^I})). \tag{2}$$

If $\mathcal{U}_1$ is conjunctive, then $\mathcal{U}_1(1_{\mathcal{L}^I}, 0_{\mathcal{L}^I}) = 0_{\mathcal{L}^I}$ and by (1) we get $\mathcal{I}(1_{\mathcal{L}^I}, 0_{\mathcal{L}^I}) = \mathcal{U}_2(1_{\mathcal{L}^I}, 0_{\mathcal{L}^I})$, thus $0_{\mathcal{L}^I} = \mathcal{U}_2(1_{\mathcal{L}^I}, 0_{\mathcal{L}^I})$, i.e., $\mathcal{U}_2$ is also a conjunctive uninorm.

Instead, if $\mathcal{U}_1$ is disjunctive, then $\mathcal{U}_1(1_{\mathcal{L}^I}, 0_{\mathcal{L}^I}) = 1_{\mathcal{L}^I}$ and we get from (2) and (1) that $1 = \mathcal{U}_2(1_{\mathcal{L}^I}, 0_{\mathcal{L}^I})$, i.e., $\mathcal{U}_2$ is also a disjunctive uninorm. □

The above results allow us to investigate Eq. (D-UU) for decomposable uninorms which are generated from the same conjunctive representable uninorms. We deal with this idea in the next section for distributivity equation.

4 General Method for Solving Distributivity Eq. (D-UU) for Decomposable Uninorms

In this section we show how we can obtain all solutions, in particular fuzzy implications, of our main distributive equation Eq. (D-UU)

$$\mathcal{I}(x, \mathcal{U}_1(y, z)) = \mathcal{U}_2(\mathcal{I}(x, y), \mathcal{I}(x, z)), \qquad x, y, z \in L^I,$$

where $\mathcal{I}$ is an unknown function and uninorms $\mathcal{U}_1$ and $\mathcal{U}_2$ on $\mathcal{L}^I$ are decomposable and generated from (classical) uninorms U_1, U_2 and U_3, U_4, respectively. Assume that the projection mappings on $\mathcal{L}^I$ are defined as the following:

$$pr_1([x_1, x_2]) = x_1, \qquad pr_2([x_1, x_2]) = x_2, \qquad \text{for } [x_1, x_2] \in L^I.$$

At this situation our distributive equation (D-UU) has the following form

$$\mathcal{I}([x_1,x_2],[U_1(y_1,z_1),U_2(y_2,z_2)])$$
$$=[U_3(pr_1(\mathcal{I}([x_1,x_2],[y_1,y_2])),pr_1(\mathcal{I}([x_1,x_2],[z_1,z_2]))),$$
$$U_4(pr_2(\mathcal{I}([x_1,x_2],[y_1,y_2])),pr_2(\mathcal{I}([x_1,x_2],[z_1,z_2])))],$$

for all $[x_1,x_2],[y_1,y_2],[z_1,z_2] \in L^I$. As a consequence we obtain the following two equations

$$pr_1(\mathcal{I}([x_1,x_2],[U_1(y_1,z_1),U_2(y_2,z_2)]))$$
$$= U_3(pr_1(\mathcal{I}([x_1,x_2],[y_1,y_2])),pr_1(\mathcal{I}([x_1,x_2],[z_1,z_2]))),$$

$$pr_2(\mathcal{I}([x_1,x_2],[U_1(y_1,z_1),U_2(y_2,z_2)]))$$
$$= U_4(pr_2(\mathcal{I}([x_1,x_2],[y_1,y_2])),pr_2(\mathcal{I}([x_1,x_2],[z_1,z_2]))),$$

which are satisfied for all $[x_1,x_2],[y_1,y_2],[z_1,z_2] \in L^I$. Now, let us fix arbitrarily $[x_1,x_2] \in L^I$ and define two functions $g^1_{[x_1,x_2]}, g^2_{[x_1,x_2]} : L^I \to L^I$ by

$$g^1_{[x_1,x_2]}(\cdot) := pr_1 \circ \mathcal{I}([x_1,x_2],\cdot), \qquad g^2_{[x_1,x_2]}(\cdot) := pr_2 \circ \mathcal{I}([x_1,x_2],\cdot),$$

where $\circ$ denotes the standard composition of functions. Thus we have shown that if $\mathcal{U}_1$ and $\mathcal{U}_2$ on $\mathcal{L}^I$ are decomposable, then

$$g^1_{[x_1,x_2]}([U_1(y_1,z_1),U_2(y_2,z_2)]) = U_3(g^1_{[x_1,x_2]}([y_1,y_2]),g^1_{[x_1,x_2]}([z_1,z_2])),$$
$$g^2_{[x_1,x_2]}([U_1(y_1,z_1),U_2(y_2,z_2)]) = U_4(g^2_{[x_1,x_2]}([y_1,y_2]),g^2_{[x_1,x_2]}([z_1,z_2])).$$

Let us assume that $U_1 = U_2$ and $U_3 = U_4$ are conjunctive representable uninorms generated from generators h_1 and h_3, respectively. Furthermore, let us assume (A-) in the spaces of h_1 and h_3 codomains. Using the representation theorem of representable, conjunctive uninorms, i.e., Theorem 3.4 and Remark 3.5, we can transform our problem to the following equation (for a simplicity we deal only with g^1 now):

$$g^1_{[x_1,x_2]}([h_1^{-1}(h_1(y_1) + h_1(z_1)),h_1^{-1}(h_1(y_2) + h_1(z_2))])$$
$$= h_3^{-1}(h_3(g^1_{[x_1,x_2]}([y_1,y_2])) + h_3(g^1_{[x_1,x_2]}([z_1,z_2]))).$$

Let us put $h_1(y_1) = u_1$, $h_1(y_2) = u_2$, $h_1(z_1) = v_1$ and $h_1(z_2) = v_2$. Of course $u_1,u_2,v_1,v_2 \in [-\infty,\infty]$ (with the assumption (A-) on that space). Moreover $[y_1,y_2],[z_1,z_2] \in L^I$, thus $y_1 \leq y_2$ and $z_1 \leq z_2$. The generator h_1 is strictly increasing, so $u_1 \leq u_2$ and $v_1 \leq v_2$. If we put

$$f_{[x_1,x_2]}(u_1,u_2) := h_3 \circ g^1_{[x_1,x_2]}([h_1^{-1}(u_1),h_1^{-1}(u_2)]), \quad u_1,u_2 \in [-\infty,\infty], \; u_1 \leq u_2,$$

then we get the following functional equation

$$f_{[x_1,x_2]}(u_1 + v_1, u_2 + v_2) = f_{[x_1,x_2]}(u_1,u_2) + f_{[x_1,x_2]}(v_1,v_2), \tag{3}$$

where $(u_1, u_2), (v_1, v_2) \in L^{\overline{\infty}}$ and $f_{[x_1, x_2]} \colon L^{\overline{\infty}} \to [-\infty, \infty]$ is an unknown function. In the same way, we can repeat all the above calculations, but for the function g^2, to obtain the following functional equation

$$f^{[x_1, x_2]}(u_1 + v_1, u_2 + v_2) = f^{[x_1, x_2]}(u_1, u_2) + f^{[x_1, x_2]}(v_1, v_2), \qquad (4)$$

where $f^{[x_1, x_2]} \colon L^{\overline{\infty}} \to [-\infty, \infty]$ is an unknown function given by

$$f^{[x_1, x_2]}(u_1, u_2) := h_3 \circ g^2_{[x_1, x_2]}([h_1^{-1}(u_1), h_1^{-1}(u_2)]), \quad u_1, u_2 \in [-\infty, \infty], \; u_1 \leq u_2.$$

Observe that (3) and (4) are exactly the same functional equation. In the next section we present main mathematical result which shows what are the solutions of the above equation.

5   Some New Results Pertaining to Functional Equations

Recently, in [3] we have solved the additive Cauchy functional equation:

$$f(x + y) = f(x) + f(y), \qquad x, y \in [-\infty, \infty],$$

for an unknown function $f \colon [-\infty, \infty] \to [-\infty, \infty]$. It should be noted that the main problem in this context was with the adequate definition of the additions $\infty + (-\infty)$ and $(-\infty) + \infty$. Using [3, Proposition 4.1] we are able to prove the following main mathematical result. Since we are limited in number of pages we omit the proof.

Proposition 5.1. *Let* $L^{\overline{\infty}} = \{(u_1, u_2) \in [-\infty, \infty]^2 \mid u_1 \leq u_2\}$. *For a function* $f \colon L^{\overline{\infty}} \to [-\infty, \infty]$ *the following statements are equivalent:*

(i) f satisfies the functional equation

$$f(u_1 + v_1, u_2 + v_2) = f(u_1, u_2) + f(v_1, v_2) \qquad (F)$$

for all $(u_1, u_2), (v_1, v_2) \in L^{\overline{\infty}}$, *with the assumption* (A-), *i.e.,* $(-\infty) + \infty = \infty + (-\infty) = -\infty$ *in both sets of domain and codomain.*

(ii) Either $f = -\infty$, or $f = 0$, or $f = \infty$ or

$$f(u, v) = \begin{cases} -\infty, & u = -\infty, \\ 0, & u > -\infty, \end{cases} \quad or \quad f(u, v) = \begin{cases} -\infty, & v = -\infty, \\ 0, & v > -\infty, \end{cases}$$

or

$$f(u, v) = \begin{cases} -\infty, & u \in \{-\infty, \infty\}, \\ 0, & u \in \mathbb{R}, \end{cases} \quad or \quad f(u, v) = \begin{cases} \infty, & u = -\infty, \\ 0, & u > -\infty, \end{cases}$$

or

$$f(u, v) = \begin{cases} \infty, & v = -\infty, \\ 0, & v > -\infty, \end{cases} \quad or \quad f(u, v) = \begin{cases} \infty, & u \in \{-\infty, \infty\}, \\ 0, & u \in \mathbb{R}, \end{cases}$$

or

$$f(u,v) = \begin{cases} -\infty, & u = -\infty, \\ \infty, & u > -\infty, \end{cases} \quad or \quad f(u,v) = \begin{cases} -\infty, & v = -\infty, \\ \infty, & v > -\infty, \end{cases}$$

or

$$f(u,v) = \begin{cases} -\infty, & u \in \{-\infty, \infty\}, \\ \infty, & u \in \mathbb{R}, \end{cases} \quad or \quad f(u,v) = \begin{cases} -\infty, & v \in \{-\infty, \infty\}, \\ \infty, & v \in \mathbb{R}, \end{cases}$$

or

$$f(u,v) = \begin{cases} -\infty, & u = -\infty \text{ or } v = \infty, \\ \infty, & u,v \in \mathbb{R}, \end{cases} \quad or \quad f(u,v) = \begin{cases} -\infty, & u = -\infty, \\ 0, & u \in \mathbb{R}, \\ \infty, & u = \infty, \end{cases}$$

or

$$f(u,v) = \begin{cases} -\infty, & v = -\infty, \\ 0, & u \in \mathbb{R}, \\ \infty, & (u = -\infty \text{ and } v > -\infty) \text{ or } u = \infty, \end{cases}$$

or

$$f(u,v) = \begin{cases} -\infty, & v = -\infty, \\ 0, & u > -\infty, \\ \infty, & u = -\infty \text{ and } v > -\infty, \end{cases}$$

or there exists a unique additive function $c \colon \mathbb{R} \to \mathbb{R}$ such that

$$f(u,v) = \begin{cases} -\infty, & u \in \{-\infty, \infty\}, \\ c(u), & u \in \mathbb{R}, \end{cases} \quad or \quad f(u,v) = \begin{cases} -\infty, & v \in \{-\infty, \infty\}, \\ c(v), & v \in \mathbb{R}, \end{cases}$$

or

$$f(u,v) = \begin{cases} \infty, & u \in \{-\infty, \infty\}, \\ c(u), & u \in \mathbb{R}, \end{cases} \quad or \quad f(u,v) = \begin{cases} \infty, & v \in \{-\infty, \infty\}, \\ c(v), & v \in \mathbb{R}, \end{cases}$$

or

$$f(u,v) = \begin{cases} -\infty, & u = -\infty \text{ or } v = \infty, \\ c(v), & u,v \in \mathbb{R}, \end{cases}$$

or

$$f(u,v) = \begin{cases} \infty, & u = -\infty \text{ or } v = \infty, \\ c(v), & u,v \in \mathbb{R}, \end{cases}$$

or

$$f(u,v) = \begin{cases} -\infty, & u = -\infty, \\ c(u), & u \in \mathbb{R}, \\ \infty, & u = \infty, \end{cases} \quad or \quad f(u,v) = \begin{cases} -\infty, & v = -\infty, \\ c(v), & v \in \mathbb{R}, \\ \infty, & v = \infty, \end{cases}$$

or

$$f(u,v) = \begin{cases} -\infty, & u = -\infty, \\ c(v), & u,v \in \mathbb{R}, \\ \infty, & u > -\infty \text{ and } v = \infty, \end{cases}$$

or

$$f(u,v) = \begin{cases} -\infty, & v = -\infty, \\ c(u), & u \in \mathbb{R}, \\ \infty, & (u = -\infty \text{ and } v > -\infty) \text{ or } u = \infty, \end{cases}$$

or

$$f(u,v) = \begin{cases} -\infty, & v = -\infty, \\ c(v), & u,v \in \mathbb{R}, \\ \infty, & (u = -\infty \text{ and } v > -\infty) \text{ or } (u > -\infty \text{ and } v = \infty), \end{cases}$$

or

$$f(u,v) = \begin{cases} -\infty, & u \in \{-\infty, \infty\}, \\ c(v), & u,v \in \mathbb{R}, \\ \infty, & u \in \mathbb{R} \text{ and } v = \infty, \end{cases}$$

or

$$f(u,v) = \begin{cases} -\infty, & v \in \{-\infty, \infty\}, \\ c(v), & u,v \in \mathbb{R}, \\ \infty, & u = -\infty \text{ and } v \in \mathbb{R}, \end{cases}$$

or there exist unique additive functions $c_1, c_2 \colon \mathbb{R} \to \mathbb{R}$ *such that*

$$f(u,v) = \begin{cases} -\infty, & u = -\infty \text{ or } v = \infty, \\ c_1(u-v) + c_2(v), & u,v \in \mathbb{R}, \end{cases}$$

or

$$f(u,v) = \begin{cases} \infty, & u = -\infty \text{ or } v = \infty, \\ c_1(u-v) + c_2(v), & u,v \in \mathbb{R}, \end{cases}$$

or

$$f(u,v) = \begin{cases} -\infty, & u = -\infty, \\ c_1(u-v) + c_2(v), & u,v \in \mathbb{R}, \\ \infty, & u > -\infty \text{ and } v = \infty, \end{cases}$$

or

$$f(u,v) = \begin{cases} -\infty, & v = -\infty, \\ c_1(u-v) + c_2(v), & u,v \in \mathbb{R}, \\ \infty, & (u = -\infty \text{ and } v > -\infty) \\ & \text{or } (u > -\infty \text{ and } v = \infty), \end{cases}$$

or

$$f(u, v) = \begin{cases} -\infty, & u \in \{-\infty, \infty\}, \\ c_1(u - v) + c_2(v), & u, v \in \mathbb{R}, \\ \infty, & u \in \mathbb{R} \ and \ v = \infty, \end{cases}$$

or

$$f(u, v) = \begin{cases} -\infty, & v \in \{-\infty, \infty\}, \\ c_1(u - v) + c_2(v), & u, v \in \mathbb{R}, \\ \infty, & u = -\infty \ and \ v \in \mathbb{R}, \end{cases}$$

for all $(u, v) \in L^{\overline{\infty}}$.

6   Some Remarks on Solutions of Eq. (D-UU)

Now, using main result from previous section i.e., Proposition 5.1, we are able to solve equations (3) and (4), i.e., we can obtain the description of the two projections of the vertical section $\mathcal{I}([x_1, x_2], \cdot)$, for fixed $[x_1, x_2] \in L^{\overline{\infty}}$, of the solutions of our main distributive equation (D-UU). As we noted earlier, Eq. (F) is the other version of equations (3) and (4). Since we are limited in number of pages we show one such solution for the first projection. Let us fix arbitrarily $[x_1, x_2] \in L^{\overline{\infty}}$. If $f_{[x_1, x_2]} = -\infty$, then

$$h_3 \circ g^1_{[x_1, x_2]}([h_1^{-1}(u), h_1^{-1}(v)]) = -\infty,$$

for all $u, v \in [-\infty, \infty]$, $u \leq v$, thus $g^1_{[x_1, x_2]}([h_1^{-1}(u), h_1^{-1}(v)]) = 0$, which means that for all $[y_1, y_2] \in L^I$ we have $g^1_{[x_1, x_2]}([y_1, y_2]) = 0.$, i.e.,

$$pr_1 \circ \mathcal{I}([x_1, x_2], [y_1, y_2]) = 0.$$

Since in Proposition 5.1 we have 36 possible solutions (separately for $g^1_{[x_1, x_2]}$ and $g^2_{[x_1, x_2]}$), we should have 1296 different solutions of Eq. (D-UU) (for a fixed $[x_1, x_2] \in L^{\overline{\infty}}$). Of course not every combination of these solutions give a correct value in the space L^I. For instance if $pr_1 \circ \mathcal{I}([x_1, x_2], [y_1, y_2]) = 0$ and $pr_2 \circ \mathcal{I}([x_1, x_2], [y_1, y_2]) = 1$, for every $[x_1, x_2]$, then our (constant) solution is correct: $\mathcal{I}([x_1, x_2], [y_1, y_2]) = [0, 1]$. But if $pr_1 \circ \mathcal{I}([x_1, x_2], [y_1, y_2]) = e_2$ and

$$pr_2 \circ \mathcal{I}([x_1, x_2], [y_1, y_2]) = \begin{cases} 0, & y_1 \in \{0, 1\}, \\ e_2, & y_1 \in (0, 1), \end{cases} \text{ for every } [x_1, x_2], \text{ then our solution}$$

is incorrect, since $\mathcal{I}([x_1, x_2], [y_1, y_2]) = [e_2, 0]$ for $y_1 \in \{0, 1\}$ and any y_2.

Remark 6.1. We should also notice that not all correct solutions may form vertical section of $\mathcal{I}$, when $\mathcal{I}$ is a fuzzy implication. For instance if we had

$$pr_1 \circ \mathcal{I}([x_1, x_2], [y_1, y_2]) = \begin{cases} 1, & y_2 = 0, \\ e_2, & y_2 > 0, \end{cases}$$

for some fixed $[x_1, x_2]$, then the function $\mathcal{I}$ is not increasing with respect to the second variable, thus it is not a fuzzy implication in the sense of Definition 3.1. Actually, if we had $g(\cdot) = pr_1 \circ \mathcal{I}([x_1, x_2], \cdot)$ or $g(\cdot) = pr_2 \circ \mathcal{I}([x_1, x_2], \cdot)$ for some $[x_1, x_2]$, then we may consider this function as a projection of vertical section of a fuzzy implication if and only if $g([1, 1]) = 1$ and g is increasing. In fact, only 8 functions among 36 solutions of the equation (3) may satisfy those conditions. We have checked relations between those 8 solutions and as a conclusion we get that we have only 35 combinations of them, which may serve as a vertical section of a fuzzy implication I. In our future work we will consider these problems in details.

Moreover, in our next investigations we will concentrate on the situation when $\mathcal{U}_1, \mathcal{U}_2$ are decomposable uninorms generated from two disjunctive representable uninorms. The case when classical $U_1 \neq U_2$ also lies in the area of our interest.

References

1. Baczyński, M.: On a class of distributive fuzzy implications. Internat. J. Uncertain. Fuzziness Knowledge-Based Systems 9, 229–238 (2001)
2. Baczyński, M.: On the distributivity of fuzzy implications over continuous and Archimedean triangular conorms. Fuzzy Sets and Systems 161, 1406–1419 (2010)
3. Baczyński, M.: On the distributivity of fuzzy implications over representable uninorms. Fuzzy Sets and Systems 161, 2256–2275 (2010)
4. Baczyński, M.: On the Distributivity of Implication Operations over t-Representable t-Norms Generated from Strict t-Norms in Interval-Valued Fuzzy Sets Theory. In: Hüllermeier, E., Kruse, R., Hoffmann, F. (eds.) IPMU 2010, Part I. CCIS, vol. 80, pp. 637–646. Springer, Heidelberg (2010)
5. Baczyński, M.: On the distributive equation for t-representable t-norms generated from nilpotent and strict t-norms. In: Galichet, S., et al. (eds.) Proc. EUSFLAT-LFA 2011, Aix-les-Bains, France, July 18-22, pp. 540–546 (2011)
6. Baczyński, M.: Distributivity of Implication Operations over t-Representable T-Norms Generated from Nilpotent T-Norms. In: Fanelli, A.M., Pedrycz, W., Petrosino, A. (eds.) WILF 2011. LNCS, vol. 6857, pp. 25–32. Springer, Heidelberg (2011)
7. Baczyński, M.: A Note on the Distributivity of Fuzzy Implications over Representable Uninorms. In: Greco, S., Bouchon-Meunier, B., Coletti, G., Fedrizzi, M., Matarazzo, B., Yager, R.R. (eds.) IPMU 2012, Part II. CCIS, vol. 298, pp. 375–384. Springer, Heidelberg (2012)
8. Baczyński, M.: Distributivity of Implication Operations over T-Representable T-Norms Generated from Continuous and Archimedean T-Norms. In: Greco, S., Bouchon-Meunier, B., Coletti, G., Fedrizzi, M., Matarazzo, B., Yager, R.R. (eds.) IPMU 2012, Part II. CCIS, vol. 298, pp. 501–510. Springer, Heidelberg (2012)
9. Baczyński, M.: On two distributivity equations for fuzzy implications and continuous, Archimedean t-norms and t-conorms. Fuzzy Sets and Systems 211, 34–54 (2013)
10. Baczyński, M.: Distributivity of implication operations over t-representable t-norms in interval-valued fuzzy set theory: the case of nilpotent t-norms. Inform. Sci. 257, 388–399 (2014)

11. Baczyński, M.: The Equation $\mathcal{I}(\mathcal{S}(x,y),z) = \mathcal{T}(\mathcal{I}(x,z),\mathcal{I}(y,z))$ for t-representable t-conorms and t-norms Generated from Continuous, Archimedean Operations. In: Masulli, F., et al. (eds.) WILF 2013. LNCS, vol. 8256, pp. 131–138. Springer, Heidelberg (2013)
12. Baczyński, M., Jayaram, B.: Fuzzy Implications. STUDFUZZ, vol. 231. Springer, Heidelberg (2008)
13. Baczyński, M., Jayaram, B.: On the distributivity of fuzzy implications over nilpotent or strict triangular conorms. IEEE Trans. Fuzzy Syst. 17(3), 590–603 (2009)
14. Balasubramaniam, J., Rao, C.J.M.: On the distributivity of implication operators over T and S norms. IEEE Trans. Fuzzy Syst. 12, 194–198 (2004)
15. Bandler, W., Kohout, L.J.: Semantics of implication operators and fuzzy relational products. Internat. J. Man-Mach. Stud. 12, 89–116 (1980)
16. Combs, W.E.: Author's reply. IEEE Trans. Fuzzy Syst. 7, 371–373, 477–478 (1999)
17. Combs, W.E., Andrews, J.E.: Combinatorial rule explosion eliminated by a fuzzy rule configuration. IEEE Trans. Fuzzy Syst. 6, 1–11 (1998)
18. Deschrijver, G., Kerre, E.E.: Uninorms in L^*-fuzzy set theory. Fuzzy Sets and Systems 148, 243–262 (2004)
19. Deschrijver, G., Cornelis, C., Kerre, E.E.: Implication in intuitionistic and interval-valued fuzzy set theory: construction, classification and application. Internat. J. Approx. Reason. 35, 55–95 (2004)
20. Dick, S., Kandel, A.: Comments on "Combinatorial rule explosion eliminated by a fuzzy rule configuration". IEEE Trans. Fuzzy Syst. 7, 475–477 (1999)
21. Drygaś, P.: On a class of operations on interval-valued fuzzy sets. In: Atanassov, K.T., et al. (eds.) New Trends in Fuzzy Sets, Intuitionistic Fuzzy Sets, Generalized Nets and Related Topics, vol. I, Foundations, pp. 67–83. SRI PAS/IBS PAN, Warsaw (2013)
22. Fodor, J., De Baets, B.: Uninorm Basics. In: Wang, P.P. (ed.) Fuzzy Logic: A Spectrum of Theoretical and Practical Issues. STUDFUZZ, vol. 215, pp. 49–64. Springer, Heidelberg (2007)
23. Fodor, J., Roubens, M.: Fuzzy Preference Modelling and Multicriteria Decision Support. Kluwer Academic Publishers, Dordrecht (1994)
24. Fodor, J.C., Yager, R.R., Rybalov, A.: Structure of uninorms. Int. J. Uncertainty Fuzziness Knowledge-Based Syst. 5, 411–427 (1997)
25. Gorzałczany, M.B.: A method of inference in approximate reasoning based on interval-valued fuzzy sets. Fuzzy Sets and Systems 21, 1–17 (1987)
26. Jayaram, B.: Rule reduction for efficient inferencing in similarity based reasoning. Internat. J. Approx. Reason. 48, 156–173 (2008)
27. Mendel, J.M., Liang, Q.: Comments on "Combinatorial rule explosion eliminated by a fuzzy rule configuration". IEEE Trans. Fuzzy Syst. 7, 369–371 (1999)
28. Ruiz-Aguilera, D., Torrens, J.: Distributivity of strong implications over conjunctive and disjunctive uninorms. Kybernetika 42, 319–336 (2006)
29. Ruiz-Aguilera, D., Torrens, J.: Distributivity of residual implications over conjunctive and disjunctive uninorms. Fuzzy Sets and Systems 158, 23–37 (2007)
30. Sambuc, R.: Fonctions Φ-floues. Application á l'aide au diagnostic en pathololologie thyroidienne. PhD thesis, Univ. Marseille, France (1975)
31. Trillas, E., Alsina, C.: On the law $[(p \wedge q) \rightarrow r] = [(p \rightarrow r) \vee (q \rightarrow r)]$ in fuzzy logic. IEEE Trans. Fuzzy Syst. 10, 84–88 (2002)
32. Yager, R.R., Rybalov, A.: Uninorm aggregation operators. Fuzzy Sets and Systems 80, 111–120 (1996)
33. Zadeh, L.A.: Outline of a new approach to the analysis of complex systems and decision processes. IEEE Trans. on Syst. Man and Cyber. 3, 28–44 (1973)

Aggregation of Dynamic Risk Measures
in Financial Management

Yuji Yoshida

Faculty of Economics and Business Administration, University of Kitakyushu
4-2-1 Kitagata, Kokuraminami, Kitakyushu 802-8577, Japan
yoshida@kitakyu-u.ac.jp

Abstract. This paper discusses aggregation of dynamic risks in financial management. The total risks in dynamic systems are usually estimated from risks at each time. This paper discusses what kind of aggregation methods are possible for dynamic risks. Coherent risk measures and their possible aggregation methods are investigated. This paper presents aggregation of dynamic coherent risks by use of generalized deviations. A few examples are also given.

1 Introduction

In the classical economic theory, the variance and the standard deviation have been used as risk indexes. Recently quantile-based risk criteria are employed widely in financial management. The concept of risk is different in its application fields. In engineering risks are considered in the both upper and lower areas from a true value since the risk is usually represented as the errors of the data to the true value. On the other hand in economics the concept of risk is given in a different way from the risk in engineering. The risk in economics is discussed only in an area of low rewards since the risk is connected deeply to losses and bankruptcy in financial management.

In this paper, we focus on the estimation of dynamic risks in financial management. The total estimation of dynamic risks are important for the stability of financial systems. The total risks in dynamic systems are usually estimated from risks at each time. The most popular methods for the total risks are the weighted arithmetic mean and the maximum of the risks over all periods. The method with the weighted arithmetic mean is sometimes insensitive to find the serious risks in dangerous situations ([10]). On the other hand regarding the method with the maximum it may happen to lose the chance to find out the other potential risks regarding the dynamic system since we observe only the largest risk through all periods. We can give ad hoc methods to construct a total risk from risks at each time. However is the total risk consistent as a risk measure? We need to investigate whether the total risk inherits propertires as a risk mesure from risks at each time. From the view point of aggregation operators ([1] and [9, Section 4.1]), this paper discusses what kind of aggregation methods are possible for dynamic risks.

V. Torra et al. (Eds.): MDAI 2014, LNAI 8825, pp. 38–49, 2014.

In Section 2 we investigate coherent risks and their possible direct aggregation methods. In Section 3 we discuss generalized deviations and their aggregation methods. In Section 4 we present aggregated dynamic coherent risks by use of generalized deviations. A few examples are also given.

2 Coherent Risk Measures

In recent financial management, the risk indexes derived from percentiles are used widely to estimate risks regarding losses and bankruptcy. Let (Ω, P) be a probability space, where P is a non-atomic probability measure. Let $\mathcal{X}$ be a set of integrable real random variables on Ω. The expectation of a random variable $X (\in \mathcal{X})$ is written by $E(X) := \int_\Omega X \, dP$.

Example 2.1 (Risk indexes defined by percentiles, Jorion [4], Tasche [7]).

(i) Value-at-risk (VaR): Let $X (\in \mathcal{X})$ be a real random variable on Ω for which there exist a non-empty open interval I and a strictly increasing and onto continuous distribution function $x (\in I) \mapsto F_X(x) := P(X < x)$. Then, the *value-at-risk (VaR)* at a risk-level probability p is given by the p-percentile of the distribution function F_X as follows:

$$\mathrm{VaR}_p(X) := \begin{cases} \inf I & \text{if } p = 0 \\ \sup\{x \in I \mid F_X(x) \le p\} & \text{if } 0 < p < 1 \\ \sup I & \text{if } p = 1. \end{cases} \quad (2.1)$$

(ii) Average value-at-risk (AVaR): Take $\mathcal{X}$ in the same way as (i). The *average value-at-risk (AVaR)* at a risk-level probability p is given by

$$\mathrm{AVaR}_p(X) := \begin{cases} \inf I & \text{if } p = 0 \\ \dfrac{1}{p} \displaystyle\int_0^p \mathrm{VaR}_q(X) \, dq & \text{if } 0 < p \le 1. \end{cases} \quad (2.2)$$

Let $\mathbb{R}$ be the set of all real numbers. Rockafellar and Uryasev [5] and Artzner et al. [2,3] introduce the following concept regarding risk measures.

Definition 2.1. A map $R : \mathcal{X} \mapsto \mathbb{R}$ is called a *(coherent) risk measure* on $\mathcal{X}$ if it satisfies the following conditions (R.a) – (R.e):

(R.a) $R(X) \le R(Y)$ for $X, Y \in \mathcal{X}$ satisfying $X \ge Y$. (*monotonicity*)
(R.b) $R(X + \theta) = R(X) - \theta$ for $X \in \mathcal{X}$ and real numbers θ.
(R.c) $R(\lambda X) = \lambda R(X)$ for $X \in \mathcal{X}$ and nonnegative real numbers λ. (*positive homogeneity*)
(R.d) $R(X + Y) \le R(X) + R(Y)$ for $X, Y \in \mathcal{X}$. (*sub-additivity*)
(R.e) $\lim_{k \to \infty} R(X_k) = R(X)$ for $\{X_k\} \subset \mathcal{X}$ and $X \in \mathcal{X}$ such that $\lim_{k \to \infty} X_k = X$ almost surely. (*continuity*)

The property (R.b) in Definition 2.1 is called *translation invariance* in financial management. We can easily check the following lemma for Example 2.1.

Lemma 2.1. *An index $R = -\text{VaR}$ given by the value-at-risk satisfies the conditions of Definition 2.1 except for the sub-additivity (R.d). However an index $R = -\text{AVaR}_p$ given by the average value-at-risk is a risk measure in the sense of Definition 2.1.*

Let T be a positive integer. Now we introduce risk measures for a *stochastic sequence*, where a random event at time $t(= 1, 2, \cdots, T)$ is denoted by a real random variable $X_t(\in \mathcal{X})$. In this paper, we represent the stochastic sequence simply as a *random vector* $\boldsymbol{X} = (X_1, X_2, \cdots, X_T)$. We discuss aggregation of risk measures $R_1(X_1), R_2(X_2), \cdots, R_T(X_T)$ for a stochastic sequence of random variables $X_1, X_2, \cdots, X_T$. Denote a vector space of random variables in $\mathcal{X}$ by the product space $\mathcal{X}^T$. For random variables $\boldsymbol{X} = (X_1, X_2, \cdots, X_T) \in \mathcal{X}^T$ and $\boldsymbol{Y} = (Y_1, Y_2, \cdots, Y_T) \in \mathcal{X}^T$, a partial order $\boldsymbol{X} \geq \boldsymbol{Y}$ implies $X_t \geq Y_t$ for all $t = 1, 2, \cdots, T$. We introduce the following definition from Definition 2.1.

Definition 2.2. A map $\boldsymbol{R} : \mathcal{X}^T \mapsto \mathbb{R}$ is called a *(coherent) risk measure* on $\mathcal{X}^T$ if it satisfies the following conditions (R.a) – (R.e):

(R.a) $\boldsymbol{R}(\boldsymbol{X}) \leq \boldsymbol{R}(\boldsymbol{Y})$ for $\boldsymbol{X}, \boldsymbol{Y} \in \mathcal{X}^T$ satisfying $\boldsymbol{X} \geq \boldsymbol{Y}$. (*monotonicity*)
(R.b) $\boldsymbol{R}(\boldsymbol{X}+\boldsymbol{\theta}) = \boldsymbol{R}(\boldsymbol{X})-\theta$ for $\boldsymbol{X} \in \mathcal{X}^T$ and real vectors $\boldsymbol{\theta} = (\theta, \theta, \cdots, \theta) \in \mathbb{R}^T$. (*translation invariance*)
(R.c) $\boldsymbol{R}(\lambda \boldsymbol{X}) = \lambda \boldsymbol{R}(\boldsymbol{X})$ for $\boldsymbol{X} \in \mathcal{X}^T$ and nonnegative real numbers λ. (*positive homogeneity*)
(R.d) $\boldsymbol{R}(\boldsymbol{X} + \boldsymbol{Y}) \leq \boldsymbol{R}(\boldsymbol{X}) + \boldsymbol{R}(\boldsymbol{Y})$ for $\boldsymbol{X}, \boldsymbol{Y} \in \mathcal{X}^T$. (*sub-additivity*)
(R.e) $\lim_{k\to\infty} \boldsymbol{R}(\boldsymbol{X}_k) = \boldsymbol{R}(\boldsymbol{X})$ for $\{\boldsymbol{X}_k\} \subset \mathcal{X}^T$ and $\boldsymbol{X} \in \mathcal{X}^T$ such that $\lim_{k\to\infty} \boldsymbol{X}_k = \boldsymbol{X}$ almost surely. (*continuity*)

We note that $\boldsymbol{R}(\boldsymbol{0}) = 0$ and $\boldsymbol{R}(\boldsymbol{\theta}) = -\theta$ for real vectors $\boldsymbol{0} = (0, 0, \cdots, 0) \in \mathbb{R}^T$ and $\boldsymbol{\theta} = (\theta, \theta, \cdots, \theta) \in \mathbb{R}^T$. The risk criterion $\boldsymbol{R}$ of a random variable $\boldsymbol{X} = (X_1, X_2, \cdots, X_T) \in \mathcal{X}^T$ is given by aggregation of risk indexes $R_1(X_1), R_2(X_2), \cdots, R_T(X_T)$. Let a set of weighting vectors $\mathcal{W}^T := \{(w_1, w_2, \cdots, w_T) \mid w_t \geq 0 \, (t = 1, 2, \cdots, T) \text{ and } \sum_{t=1}^{T} w_t = 1\}$. The following proposition can be checked easily.

Proposition 2.1. *Let R_t be a risk measure on $\mathcal{X}$ at time $t = 1, 2, \cdots, T$. The following (i) – (iii) hold.*

(i) The weighted average: Let a weighting vector $(w_1, w_2, \cdots, w_T) \in \mathcal{W}^T$. Define a map $\boldsymbol{R} : \mathcal{X}^T \mapsto \mathbb{R}$ by

$$\boldsymbol{R}(\boldsymbol{X}) := \sum_{t=1}^{T} w_t R_t(X_t) \tag{2.3}$$

for $\boldsymbol{X} = (X_1, X_2, \cdots, X_T) \in \mathcal{X}^T$. Then $\boldsymbol{R}$ is a risk measure on $\mathcal{X}^T$.

(ii) The order weighted average (Torra [8]): *Let* $(w_1, w_2, \cdots, w_T) \in \mathcal{W}^T$ *be a weighting vector satisfying* $w_1 \geq w_2 \geq \cdots \geq w_T \geq 0$. *Define a map* $\boldsymbol{R}$: $\mathcal{X}^T \mapsto \mathbb{R}$ *by*

$$\boldsymbol{R}(\boldsymbol{X}) := \sum_{t=1}^{T} w_t R_{(t)}(X_{(t)}) \tag{2.4}$$

for $\boldsymbol{X} = (X_1, X_2, \cdots, X_T) \in \mathcal{X}^T$, *where* $R_{(t)}(X_{(t)})$ *is the t-th largest risk values in* $\{R_1(X_1), R_2(X_2), \cdots, R_T(X_T)\}$. *Then* $\boldsymbol{R}$ *is a risk measure on* $\mathcal{X}^T$.

(iii) The maximum: *Define a map* $\boldsymbol{R} : \mathcal{X}^T \mapsto \mathbb{R}$ *by*

$$\boldsymbol{R}(\boldsymbol{X}) := \max\{R_1(X_1), R_2(X_2), \cdots, R_T(X_T)\} \tag{2.5}$$

for $\boldsymbol{X} = (X_1, X_2, \cdots, X_T) \in \mathcal{X}^T$. *Then* $\boldsymbol{R}$ *is a risk measure on* $\mathcal{X}^T$.

When we construct aggregation $\boldsymbol{R}$ directly from of risk indexes $R_1(X_1)$, $R_2(X_2), \cdots, R_T(X_T)$, it is difficult to find other methods except for the methods (i) − (iii) in Proposition 2.1 from the view point of aggregation operators ([1] and [9, Section 4.1]).

Example 2.2 (Average value-at-risks). By Proposition 2.1, the following (2.6) − (2.8) are risk measures induced from Example 2.1:

$$\boldsymbol{R}(\boldsymbol{X}) = \sum_{t=1}^{T} w_t(-\text{AVaR}_{p_t}(X_t)) = -\sum_{t=1}^{T} w_t \text{AVaR}_{p_t}(X_t), \tag{2.6}$$

$$\boldsymbol{R}(\boldsymbol{X}) = \sum_{t=1}^{T} w_t(-\text{AVaR}_{p_{(t)}}(X_{(t)})) = -\sum_{t=1}^{T} w_t \text{AVaR}_{p_{(t)}}(X_{(t)}), \tag{2.7}$$

$$\boldsymbol{R}(\boldsymbol{X}) = \max\{-\text{AVaR}_{p_1}(X_1), -\text{AVaR}_{p_2}(X_2), \cdots, -\text{AVaR}_{p_T}(X_T)\} \tag{2.8}$$

for random variables $\boldsymbol{X} = (X_1, X_2, \cdots, X_T) \in \mathcal{X}^T$, where p_t $(0 < p_t < 1)$ is a given risk-level probability at time $t = 1, 2, \cdots, T$ and $-\text{AVaR}_{p_{(t)}}(X_{(t)})$ is the t-th largest risk values in $\{-\text{AVaR}_{p_1}(X_1), -\text{AVaR}_{p_2}(X_2), \cdots, -\text{AVaR}_{p_T}(X_T)\}$.

Let n be a positive integer. When we aggregate n risk indexes for a random variable X, we can use the following corollary derived from Proposition 2.1.

Corollary 2.1. *Let* R_i *be a risk measure on* $\mathcal{X}$ *for item* $i = 1, 2, \cdots, n$. *The following (i) – (iii) hold.*

(i) The weighted average: *Let a weighting vector* $(w_1, w_2, \cdots, w_n) \in \mathcal{W}^n$. *Define a map* $R : \mathcal{X} \mapsto \mathbb{R}$ *by*

$$R(X) := \sum_{i=1}^{n} w_i R_i(X) \tag{2.9}$$

for $X \in \mathcal{X}$. *Then* R *is a risk measure on* $\mathcal{X}$.

(ii) The order weighted average: *Let $(w_1, w_2, \cdots, w_n) \in \mathcal{W}^n$ be a weighting vector satisfying $w_1 \geq w_2 \geq \cdots \geq w_n \geq 0$. Define a map $R : \mathcal{X} \mapsto \mathbb{R}$ by*

$$R(X) := \sum_{i=1}^{n} w_i R_{(i)}(X) \tag{2.10}$$

for $X \in \mathcal{X}$, where $R_{(i)}(X)$ is the i-th largest risk values in $\{R_1(X), R_2(X), \cdots, R_n(X)\}$. Then R is a risk measure on $\mathcal{X}$.

(iii) The maximum: *Define a map $R : \mathcal{X} \mapsto \mathbb{R}$ by*

$$R(X) := \max\{R_1(X), R_2(X), \cdots, R_n(X)\} \tag{2.11}$$

for $X \in \mathcal{X}$. Then R is a risk measure on $\mathcal{X}$.

In the next section we discuss relations between risk measures and deviations to introduce other kinds of aggregation of risk measures.

3   Deviation Measures

Risk measure is related to deviation measures ([6]). In this section we introduce deviation measures to investigate indirect approaches which are different from direct methods in the previous section. Denote $L^2(\Omega)$ and $L^1(\Omega)$ the space of square integrable real random variables on Ω and the space of integrable real random variables on Ω respectively. We use a notation $a_- := \max\{-a, 0\}$ for real numbers a.

Example 3.1 (Classical deviations). The following criteria are classical deviations in financial management, engineering and so on.

(i) Let the space $\mathcal{X} = L^2(\Omega)$. The *standard deviation* of a random variable $X(\in \mathcal{X})$ is defined by $\sigma(X) := E((X - E(X))^2)^{1/2}$.

(ii) Let the space $\mathcal{X} = L^1(\Omega)$. The *absolute deviation* of a random variable $X(\in \mathcal{X})$ is defined by $W(X) := E(|X - E(X)|)$.

(iii) Let the space $\mathcal{X} = L^2(\Omega)$. The *lower standard semi-deviation* of a random variable $X(\in \mathcal{X})$ is defined by $\sigma_-(X) := E(((X - E(X))_-)^2)^{1/2}$.

(iv) Let the space $\mathcal{X} = L^1(\Omega)$. The *lower absolute semi-deviation* of a random variable $X(\in \mathcal{X})$ is defined by $W_-(X) := E((X - E(X))_-)$.

Recently Rockafellar et al. [6] has studied the following concept regarding deviations.

Definition 3.1. Let $\mathcal{X}$ be a set of real random variables on Ω. A map $D : \mathcal{X} \mapsto [0, \infty)$ is called a *deviation measure* on $\mathcal{X}$ if it satisfies the following conditions (D.a) – (D.e):

(D.a) $D(X) \geq 0$ and $D(\theta) = 0$ for $X \in \mathcal{X}$ and real numbers θ. (*positivity*)

(D.b) $D(X+\theta) = D(X)$ for $X \in \mathcal{X}$ and real number θ. (*translation invariance*)

(D.c) $D(\lambda X) = \lambda D(X)$ for $X \in \mathcal{X}$ and nonnegative real numbers λ. (*positive homogeneity*)

(D.d) $D(X + Y) \leq D(X) + D(Y)$ for $X, Y \in \mathcal{X}$. (*sub-additivity*)

(D.e) $\lim_{k \to \infty} D(X_k) = D(X)$ for $\{X_k\} \subset \mathcal{X}$ and $X \in \mathcal{X}$ such that $\lim_{k \to \infty} X_k = X$ almost surely. (*continuity*)

Hence, we have the following lemma for Example 3.1.

Lemma 3.1. *The standard deviation σ, the absolute deviation W, the lower standard semi-deviation σ_- and the lower absolute semi-deviation W_- are deviation measures in the sense of Definition 3.1.*

Proof. We have $|a + b| \leq |a| + |b|$ and $(a+b)_- \leq a_- + b_-$ for $a, b \in \mathbb{R}$. We can easily check this lemma with these inequalities and Schwartz's inequality. □

For a deviation measure D, we put

$$N(X) := \frac{D(X) + D(-X)}{2} \tag{3.1}$$

for $X \in \mathcal{X}$. Then N is a semi-norm on $\mathcal{X}$, i.e, it satisfies the following conditions (N.a) – (N.c):

(N.a) $N(X) \geq 0$ and $N(0) = 0$ for $X \in \mathcal{X}$. (*positivity*)

(N.b) $N(\lambda X) = |\lambda| N(X)$ for $X \in \mathcal{X}$ and real numbers λ. (*homogeneity*)

(N.c) $N(X + Y) \leq N(X) + N(Y)$ for $X, Y \in \mathcal{X}$. (*sub-additivity*)

We find from (3.1) that we can aggregate deviation measures in a similar way to norms on the space $\mathcal{X}$. Let $\mathcal{D}(\mathcal{X})$ denote the family of all deviation measures on $\mathcal{X}$. Then the following proposition shows $\mathcal{D}(\mathcal{X})$ becomes a *convex cone*, and it indicates a hint to construct a deviation criterion D of a random variable X from deviations $D_1(X)$ and $D_2(X)$ estimated by two viewpoints $D_1(\cdot)$ and $D_2(\cdot)$.

Proposition 3.1

(i) Let $D \in \mathcal{D}(\mathcal{X})$ and a nonnegative real number λ. Then $\lambda D \in \mathcal{D}(\mathcal{X})$.

(ii) Let $D_1, D_2 \in \mathcal{D}(\mathcal{X})$. Then $D_1 + D_2 \in \mathcal{D}(\mathcal{X})$.

The sum, the scalar multiplication and the shift on the vector space $\mathcal{X}^T$ are defined as follows: We put $\boldsymbol{X} + \boldsymbol{Y} = (X_1 + Y_1, X_2 + Y_2, \cdots, X_T + Y_T), \lambda \boldsymbol{X} = (\lambda X_1, \lambda X_2, \cdots, \lambda X_T)$ and $\boldsymbol{X} + \boldsymbol{\theta} = (X_1 + \theta, X_2 + \theta, \cdots, X_T + \theta)$ for $\boldsymbol{X} = (X_1, X_2, \cdots, X_T) \in \mathcal{X}^T, \boldsymbol{Y} = (Y_1, Y_2, \cdots, Y_T) \in \mathcal{X}^T$ and real numbers λ and real vectors $\boldsymbol{\theta} = (\theta, \theta, \cdots, \theta) \in \mathbb{R}^T$. We introduce the following definition for random vectors from Definition 3.1.

Definition 3.2. A map $\boldsymbol{D} : \mathcal{X}^T \mapsto \mathbb{R}$ is called a *deviation measure* on $\mathcal{X}^T$ if it satisfies the following conditions (D.a) – (D.e):

(D.a) $D(X) \geq 0$ and $D(\theta) = 0$ for $X \in \mathcal{X}^T$ and real vectors $\theta = (\theta, \theta, \cdots, \theta) \in \mathbb{R}^T$. (*positivity*)

(D.b) $D(X + \theta) = D(X)$ for $X \in \mathcal{X}^T$ and real vectors $\theta = (\theta, \theta, \cdots, \theta) \in \mathbb{R}^T$. (*translation invariance*)

(D.c) $D(\lambda X) = \lambda D(X)$ for $X \in \mathcal{X}^T$ and nonnegative real numbers λ. (*positive homogeneity*)

(D.d) $D(X + Y) \leq D(X) + D(Y)$ for $X, Y \in \mathcal{X}^T$. (*sub-additivity*)

(D.e) $\lim_{k \to \infty} D(X_k) = D(X)$ for $\{X_k\} \subset \mathcal{X}^T$ and $X \in \mathcal{X}^T$ such that $\lim_{k \to \infty} X_k = X$ almost surely. (*continuity*)

The following proposition shows methods to construct a deviation D on $\mathcal{X}^T$ from deviations $D_1(X_1), D_2(X_2), \cdots, D_T(X_T)$ for a random vector $X = (X_1, X_2, \cdots, X_T) \in \mathcal{X}^T$.

Theorem 3.1. *Let D_t be a deviation measure on $\mathcal{X}$ at time $t = 1, 2, \cdots, T$. Let d be a real number satisfying $1 \leq d < \infty$. The following (i) – (iii) hold.*

(i) The generalized weighted average: *Let a weighting vector $(w_1, w_2, \cdots, w_T) \in \mathcal{W}^T$. Define a map $D : \mathcal{X}^T \mapsto [0, \infty)$ by*

$$D(X) := \left(\sum_{t=1}^{T} w_t D_t(X_t)^d \right)^{1/d} \tag{3.2}$$

for $X = (X_1, X_2, \cdots, X_T) \in \mathcal{X}^T$. Then D is a deviation measure on $\mathcal{X}^T$.

(ii) The generalized order weighted average: *Let $(w_1, w_2, \cdots, w_T) \in \mathcal{W}^T$ be a weighting vector satisfying $w_1 \geq w_2 \geq \cdots \geq w_T \geq 0$. Define a map $D : \mathcal{X}^T \mapsto [0, \infty)$ by*

$$D(X) := \left(\sum_{t=1}^{T} w_t D_{(t)}(X_{(t)})^d \right)^{1/d} \tag{3.3}$$

for $X = (X_1, X_2, \cdots, X_T) \in \mathcal{X}^T$, where $D_{(t)}(X_{(t)})$ is the t-th largest deviation values in $\{D_1(X_1), D_2(X_2), \cdots, D_T(X_T)\}$. Then D is a deviation measure on $\mathcal{X}^T$.

(iii) The maximum: *Define a map $D : \mathcal{X}^T \mapsto [0, \infty)$ by*

$$D(X) := \max\{D_1(X_1), D_2(X_2), \cdots, D_T(X_T)\} \tag{3.4}$$

for $X = (X_1, X_2, \cdots, X_T) \in \mathcal{X}^T$. Then D is a deviation measure on $\mathcal{X}^T$.

Proof. (i) We can check this proposition easily with Minkowski's inequality. (ii) Let (t) denote indexes for the t-th largest deviation values in $\{D_1(X_1 + Y_1), D_2(X_2 + Y_2), \cdots, D_T(X_T + Y_T)\}$. By Minkowski's inequality, we get

$$D(X + Y) \leq \left(\sum_{t=1}^{T} w_t \left(D_{(t)}(X_{(t)}) + D_{(t)}(Y_{(t)}) \right)^d \right)^{1/d}$$

$$\leq \left(\sum_{t=1}^{T} w_t D_{(t)}(X_{(t)})^d \right)^{1/d} + \left(\sum_{t=1}^{T} w_t D_{(t)}(Y_{(t)})^d \right)^{1/d}$$

$$\leq D(X) + D(Y).$$

We can easily check the other conditions. $\qquad\qquad\qquad\qquad\qquad\square$

Let n be a positive integer. When we aggregate n deviation indexes for a random variable X, we can use the following corollary derived from Proposition 3.1.

Corollary 3.1. *Let D_i be a deviation measure on $\mathcal{X}$ for item $i = 1, 2, \cdots, n$. Let d be a real number satisfying $1 \leq d < \infty$. The following (i) – (iii) hold.*

(i) The generalized weighted average: *Let a weighting vector $(w_1, w_2, \cdots, w_n) \in \mathcal{W}^n$. Define a map $D : \mathcal{X} \mapsto [0, \infty)$ by*

$$D(X) := \left(\sum_{i=1}^{n} w_i D_i(X)^d \right)^{1/d} \tag{3.5}$$

for $X \in \mathcal{X}$. Then D is a deviation measure on $\mathcal{X}$.

(ii) The generalized order weighted average: *Let $(w_1, w_2, \cdots, w_n) \in \mathcal{W}^n$ be a weighting vector satisfying $w_1 \geq w_2 \geq \cdots \geq w_n \geq 0$. Define a map $D : \mathcal{X} \mapsto [0, \infty)$ by*

$$D(X) := \left(\sum_{i=1}^{n} w_i D_{(i)}(X)^d \right)^{1/d} \tag{3.6}$$

for $X \in \mathcal{X}$, where $D_{(i)}(X)$ is the i-th largest deviation values in $\{D_1(X), D_2(X), \cdots, D_n(X)\}$. Then D is a deviation measure on $\mathcal{X}$.

(iii) The maximum: *Define a map $D : \mathcal{X} \mapsto [0, \infty)$ by*

$$D(X) := \max\{D_1(X), D_2(X), \cdots, D_n(X)\} \tag{3.7}$$

for $X \in \mathcal{X}$. Then D is a deviation measure on $\mathcal{X}$.

4   Construction of Risk Measures by Use of Deviation Measures

In this section we construct coherent risk measures for random vectors by use of deviation measures. Now we introduce the following definition for random vectors.

Definition 4.1. A map $E : \mathcal{X}^T \mapsto \mathbb{R}$ is called an *expectation measure* on $\mathcal{X}^T$ if it satisfies the following conditions (E.a) – (E.d):

(E.a) $\boldsymbol{E}(\boldsymbol{\theta}) = \theta$ for real vectors $\boldsymbol{\theta} = (\theta, \theta, \cdots, \theta) \in \mathbb{R}^T$.

(E.b) $\boldsymbol{E}(\lambda \boldsymbol{X}) = \lambda \boldsymbol{E}(\boldsymbol{X})$ for $\boldsymbol{X} \in \mathcal{X}^T$ and real numbers λ. (*homogeneity*)

(E.c) $\boldsymbol{E}(\boldsymbol{X} + \boldsymbol{Y}) = \boldsymbol{E}(\boldsymbol{X}) + \boldsymbol{E}(\boldsymbol{Y})$ for $\boldsymbol{X}, \boldsymbol{Y} \in \mathcal{X}^T$. (*additivity*)

(E.d) $\lim_{k \to \infty} \boldsymbol{E}(\boldsymbol{X}_k) = \boldsymbol{E}(\boldsymbol{X})$ for $\{\boldsymbol{X}_k\} \subset \mathcal{X}^T$ and $\boldsymbol{X} \in \mathcal{X}^T$ such that $\lim_{k \to \infty}$ $\boldsymbol{X}_k = \boldsymbol{X}$ almost surely. (*continuity*)

The following lemma shows the relation between deviation measures D on $\mathcal{X}$ and risk measures R on $\mathcal{X}$.

Lemma 4.1

(i) *Let D be a deviation measure on $\mathcal{X}$. Suppose*

$$D(X) \le E(X) - \operatorname*{ess\,inf}_{\omega} X(\omega) \quad for \ X \in \mathcal{X}. \tag{4.1}$$

Define

$$R(X) := D(X) - E(X)$$

for $X \in \mathcal{X}$. Then R is a risk measure on $\mathcal{X}$.

(ii) *Let R be a risk measure on $\mathcal{X}$. Suppose*

$$R(X) + E(X) \ge 0 \quad for \ X \in \mathcal{X}. \tag{4.2}$$

Define

$$D(X) := R(X) + E(X)$$

for $X \in \mathcal{X}$. Then D is a deviation measure on $\mathcal{X}$.

Proof. (i) From (D.b) $-$ (D.d), we can easily check (R.b) $-$ (R.d). (R.a) Let $X, Y \in \mathcal{X}$ satisfying $X \ge Y$. Let $Z := X - Y \ge 0$. Then from the assumption, we have $D(Z) \le D(Z) + \operatorname{ess\,inf}_\omega Z(\omega) \le E(Z)$. Then $R(Z) \le 0$. Then from (R.d) we obtain $R(X) = R(Y + Z) \le R(Y) + R(Z) \le R(Y)$. Thus we also get (R.a).

(ii) From (R.b) $-$ (R.d), we can easily check (D.b) $-$ (D.d). (D.a) Let $X \in \mathcal{X}$. From the assumption we have $D(X) = R(X) + E(X) \ge 0$. Let θ be a real number. From (R.c) we have $R(0) = 0$ and from (R.b) we also have $R(\theta) = R(0 + \theta) = R(0) - \theta = -\theta$. Therefore we obtain $D(\theta) = R(\theta) + E(\theta) = -\theta + \theta = 0$. Thus this lemma holds. □

Remark. The lower standard semi-deviation σ_- and the lower absolute semi-deviation W_- satisfy the condition (4.1) in Lemma 4.1(i). On the other hand, $-\mathrm{AVaR}_p$ is a risk measure which satisfies the condition (4.2) in Lemma 4.1(i) if $\lim_{x \downarrow \inf I} x F_X(x) = 0$.

Extending Lemma 4.1, the following lemma shows the relation between deviation measures $\boldsymbol{D}$ on $\mathcal{X}^T$ and risk measures $\boldsymbol{R}$ on $\mathcal{X}^T$.

Lemma 4.2

(i) *Let D be a deviation measure on $\mathcal{X}^T$. Suppose*

$$D(X) \leq E(X) - \operatorname*{ess\,inf}_{\omega} \min_{1 \leq t \leq T} X_t(\omega) \quad \text{for } X \in \mathcal{X}^T. \tag{4.3}$$

Define

$$R(X) := D(X) - E(X)$$

for $X \in \mathcal{X}^T$. Then R is a risk measure on $\mathcal{X}^T$.

(ii) *Let R be a risk measure on $\mathcal{X}^T$. Suppose*

$$R(X) + E(X) \geq 0 \quad \text{for } X \in \mathcal{X}^T. \tag{4.4}$$

Define

$$D(X) := R(X) + E(X)$$

for $X \in \mathcal{X}^T$. Then D be a deviation measure on $\mathcal{X}^T$.

Proof. The proof is in the same way as Lemma 4.1. □

From this lemma, we can derive indirect construction methods for risk measures for stochastic sequences.

Theorem 4.1. *Let a weighting vector $(v_1, v_2, \cdots, v_T) \in \mathcal{W}^T$ and let an expectation measure*

$$E(X) = \sum_{t=1}^{T} v_t E(X_t)$$

for $X = (X_1, X_2, \cdots, X_T) \in \mathcal{X}^T$, Assume $R_t(X_t) + E(X) \geq 0$ for $X = (X_1, X_2, \cdots, X_T) \in \mathcal{X}^T$ and $t = 1, 2, \cdots, T$. Let d be a real number satisfying $1 \leq d < \infty$. The following (i) and (ii) hold.

(i) *The weighted average: Let a weighting vector $(w_1, w_2, \cdots, w_T) \in \mathcal{W}^T$. Define a map $R : \mathcal{X}^T \mapsto \mathbb{R}$ by*

$$R(X) := \left(\sum_{t=1}^{T} w_t (R_t(X_t) + E(X))^d \right)^{1/d} - E(X) \tag{4.5}$$

for $X = (X_1, X_2, \cdots, X_T) \in \mathcal{X}^T$. Then R is a risk measure on $\mathcal{X}^T$.

(ii) *The order weighted average: Let $(w_1, w_2, \cdots, w_T) \in \mathcal{W}^T$ be a weighting vector satisfying $w_1 \geq w_2 \geq \cdots \geq w_T \geq 0$. Define a map $R : \mathcal{X}^T \mapsto \mathbb{R}$ by*

$$R(X) := \left(\sum_{t=1}^{T} w_t (R_{(t)}(X_{(t)}) + E(X))^d \right)^{1/d} - E(X) \tag{4.6}$$

for $X = (X_1, X_2, \cdots, X_T) \in \mathcal{X}^T$, where $R_{(t)}(X_{(t)}) + E(X_{(t)})$ is the t-th largest risk values in $\{R_1(X_1) + E(X_1), R_2(X_2) + E(X_2), \cdots, R_T(X_T) + E(X_T)\}$. Then R is a risk measure on $\mathcal{X}^T$.

Proof. (i) First we have $R_t(X_t) + \boldsymbol{E}(\boldsymbol{X}) \geq 0$ for $\boldsymbol{X} = (X_1, X_2, \cdots, X_T) \in \mathcal{X}^T$. Let

$$\boldsymbol{D}(\boldsymbol{X}) = \left(\sum_{t=1}^{T} w_t (R_t(X_t) + \boldsymbol{E}(\boldsymbol{X}))^d\right)^{1/d} \tag{4.7}$$

for $\boldsymbol{X} = (X_1, X_2, \cdots, X_T) \in \mathcal{X}^T$. We can easily check $\boldsymbol{D}$ satisfies (D.a) – (D.c) in Definition 3.2 since $R_t(\theta) = -\theta$, $\boldsymbol{E}(\boldsymbol{\theta}) = \theta$ and $R_t(X_t + \theta) = R_t(X_t) - \theta$ for $\boldsymbol{X} \in \mathcal{X}^T$ and real vectors $\boldsymbol{\theta} = (\theta, \theta, \cdots, \theta) \in \mathbb{R}^T$. Then by Minkowski's inequality we obtain that $\boldsymbol{D}$ is a deviation measure on $\mathcal{X}^T$.

Next we fix any random vector $\boldsymbol{X} = (X_1, X_2, \cdots, X_T) \in \mathcal{X}^T$. Put a constant $c = \text{ess inf}_\omega \min_t X_t(\omega)$. Then we have $X_t - c \geq 0$ for $t = 1, 2, \cdots, T$. Since R_t is a risk measure, from (R.a) – (R.c) in Definition 2.1 we get $R_t(X_t) + c = R_t(X_t - c) \leq R_t(0) = 0$ for $t = 1, 2, \cdots, T$. Thus it holds that $R_t(X_t) \leq -c$ for $t = 1, 2, \cdots, T$ and $X_t \in \mathcal{X}$. Hence we have

$$\boldsymbol{D}(\boldsymbol{X}) - \boldsymbol{E}(\boldsymbol{X}) = \left(\sum_{t=1}^{T} w_t (R_t(X_t) + \boldsymbol{E}(\boldsymbol{X}))^d\right)^{1/d} - \boldsymbol{E}(\boldsymbol{X})$$
$$\leq \left(\sum_{t=1}^{T} w_t (-c + \boldsymbol{E}(\boldsymbol{X}))^d\right)^{1/d} - \boldsymbol{E}(\boldsymbol{X})$$
$$= -c = -\text{ess} \inf_\omega \min_t X_t(\omega).$$

Thus by Lemma 4.1(i) we obtain that $\boldsymbol{R} = \boldsymbol{D} - \boldsymbol{E}$ is a risk measure on $\mathcal{X}^T$. We can check (ii) in the same way. □

Let n be a positive integer. When we have n risk indexes for a random variable X, we can apply Theorem 4.1 to aggregation of these risk indexes.

Corollary 4.1. *Let R_i be a risk measure on $\mathcal{X}$ satisfying $R_i(\cdot) + E(\cdot) \geq 0$ on $\mathcal{X}$ for item $i = 1, 2, \cdots, n$. Let d be a real number satisfying $1 \leq d < \infty$. The following (i) and (ii) hold.*

(i) *The weighted average: Let a weighting vector $(w_1, w_2, \cdots, w_n) \in \mathcal{W}^n$. Define a map $R : \mathcal{X} \mapsto \mathbb{R}$ by*

$$R(X) := \left(\sum_{i=1}^{n} w_i (R_i(X) + E(X))^d\right)^{1/d} - E(X) \tag{4.8}$$

for $X \in \mathcal{X}$. Then R is a risk measure on $\mathcal{X}$.

(ii) *The order weighted average: Let $(w_1, w_2, \cdots, w_n) \in \mathcal{W}^n$ be a weighting vector satisfying $w_1 \geq w_2 \geq \cdots \geq w_n \geq 0$. Define a map $R : \mathcal{X} \mapsto \mathbb{R}$ by*

$$R(X) := \left(\sum_{i=1}^{n} w_i (R_{(i)}(X) + E(X))^d\right)^{1/d} - E(X) \tag{4.9}$$

for $X \in \mathcal{X}$, where $R_{(i)}(X)$ is the i-th largest risk values in $\{R_1(X), R_2(X), \cdots, R_n(X)\}$. Then R is a risk measure on $\mathcal{X}$.

We obtain the following example since $\mathrm{AVaR}_p(\cdot) \le E(\cdot)$ holds on $\mathcal{X}$ for probabilities p $(0 < p < 1)$.

Example 4.1 (Dynamic average value-at-risks). Let d be a real number satisfying $1 \le d < \infty$. Let p_t $(0 < p_t < 1)$ is a risk-level probability at time $t = 1, 2, \cdots, T$. The following (i) and (ii) hold.

(i) The weighted average: Let a weighting vector $(w_1, w_2, \cdots, w_T) \in \mathcal{X}$. Define a map $\boldsymbol{R} : \mathcal{X}^T \mapsto \mathbb{R}$ by

$$\boldsymbol{R}(\boldsymbol{X}) := \left(\sum_{t=1}^{T} w_t (-\mathrm{AVaR}_{p_t}(X_t) + \boldsymbol{E}(\boldsymbol{X}))^d \right)^{1/d} - \boldsymbol{E}(\boldsymbol{X}) \qquad (4.10)$$

for $\boldsymbol{X} = (X_1, X_2, \cdots, X_T) \in \mathcal{X}^T$. Then $\boldsymbol{R}$ is a risk measure on $\mathcal{X}^T$.

(ii) The order weighted average: Let $(w_1, w_2, \cdots, w_T) \in \mathcal{X}$ be a weighting vector satisfying $w_1 \ge w_2 \ge \cdots \ge w_T \ge 0$. Define a map $\boldsymbol{R} : \mathcal{X}^T \mapsto \mathbb{R}$ by

$$\boldsymbol{R}(\boldsymbol{X}) := \left(\sum_{t=1}^{T} w_t (-\mathrm{AVaR}_{p_{(t)}}(X_{(t)}) + \boldsymbol{E}(\boldsymbol{X}))^d \right)^{1/d} - \boldsymbol{E}(\boldsymbol{X}) \qquad (4.11)$$

for $\boldsymbol{X} = (X_1, X_2, \cdots, X_T) \in \mathcal{X}^T$, where $-\mathrm{AVaR}_{p_t}(X_t) + E(X_t)$ is the t-th largest risk values in $\{ -\mathrm{AVaR}_{p_1}(X_1) + E(X_1), -\mathrm{AVaR}_{p_2}(X_2) + E(X_2), \cdots, -\mathrm{AVaR}_{p_T}(X_T) + E(X_T) \}$. Then $\boldsymbol{R}$ is a risk measure on $\mathcal{X}^T$.

References

1. Aczél, J.: On weighted mean values. Bulletin of the American Math. Society 54, 392–400 (1948)
2. Artzner, P., Delbaen, F., Eber, J.-M., Heath, D.: Thinking coherently. Risk 10, 68–71 (1997)
3. Artzner, P., Delbaen, F., Eber, J.-M., Heath, D.: Coherent measures of risk. Mathematical Finance 9, 203–228 (1999)
4. Jorion, P.: Value-a- Risk: The New Benchmark for Managing Financial Risk, 3rd edn. McGraw-Hill, New York (2006)
5. Rockafellar, R.T., Uryasev, S.P.: Optimization of conditional value-at-risk. Journal of Risk 2, 21–42 (2000)
6. Rockafellar, R.T., Uryasev, S.P., Zabarankin, M.: Generalized deviations in risk analysis. J. Finance and Stochastics 10, 51–74 (2006)
7. Tasche, D.: Expected shortfall and beyond. Journal of Banking Finance 26, 1519–1533 (2002)
8. Torra, V.: The weighted OWA operator. Int. J. of Intel. Syst. 12, 153–166 (1997)
9. Torra, V., Narukawa, Y.: Modeling Decisions. Springer, Heidelberg (2007)
10. Yoshida, Y.: A dynamic risk allocation of value-at-risks with portfolioss. Journal of Advanced Computational Intelligence and Intelligent Informatics 16, 800–806 (2012)

Single-Preference Consensus Measures Based on Models of Ecological Evenness

Gleb Beliakov, Simon James*, and Laura Smith

School of Information Technology
Deakin University
221 Burwood Hwy, Burwood 3125, Australia
sjames@deakin.edu.au

Abstract. We investigate the relationship between consensus measures used in different settings depending on how voters or experts express their preferences. We propose some new models for single-preference voting, which we derive from the evenness concept in ecology, and show that some of these can be placed within the framework of existing consensus measures using the discrete distance. Finally, we suggest some generalizations of the single-preference consensus measures allowing the incorporation of more general notions of distance.

Keywords: Consensus measures, decision making, preferences, evenness measures, specifity, distances.

1 Introduction

The idea of consensus is relevant to various decision domains, including voting, group decision making and multi-criteria evaluation. It is closely tied to notions such as 'majority' and 'agreement' and can be used to provide an authoritative stamp on decisions or persuade individuals to alter their preferences during a mediation process.

In popular usage, the term *consensus* is sometimes implied to mean only the unanimous agreement amongst individuals, however the notion of *soft consensus* [7,9], i.e. with the degree of consensus ranging from 0 to 1, has been applied in decision processes where the level of agreement between individuals needs to be quantified.

Each of the scenarios in Fig. 1 involve multiple decision makers expressing their preferences over 4 alternatives or candidates, $\{A, B, C, D\}$. They can be seen to comprise a spectrum in terms of the precision and detail required from the decision makers (from most to least detail).

For the evaluation of alternatives on a numerical or completely ordered scale, consensus measures such as those developed in [2,3,8,28] can be used to give an idea of how much the voters agree with respect to each candidate after providing their scores, either by looking at pairwise differences or the difference between

---

* Corresponding author.

V. Torra et al. (Eds.): MDAI 2014, LNAI 8825, pp. 50–59, 2014.
© Springer International Publishing Switzerland 2014

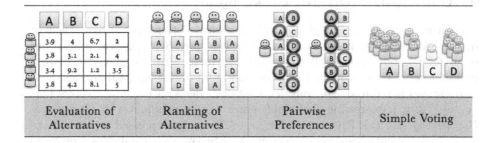

| Evaluation of Alternatives | Ranking of Alternatives | Pairwise Preferences | Simple Voting |

Fig. 1. Various situations with multiple alternatives/candidates and experts/voters where consensus evaluations may be useful

each input and an average score. We can see in this example that the consensus regarding A's scores would be higher than that around C.

Consensus for ranked alternatives was investigated in [5] and some of these methods have been generalized to the case of weak orders in [12–15]. Such consensus measures here focus on differences between the ranks allocated to each of the candidates/alternatives or the agreement in terms of pairwise comparisons. The consensus between the first two voters is clearly higher than between the first and fourth.

Since complete rankings are sometimes difficult to elicit from experts or decision makers, a popular approach in group decision making has been to obtain pairwise preferences from decision makers as part of the consensual process. Consensus measures over these preference relations have been developed by a number of authors (e.g. in [1, 7, 18, 31]), which, as with the evaluation of alternatives, also look at pairwise differences between experts and the differences between each expert and the overall opinion.

Finally, we can consider the very simple case of single-preference voting. We have noted recently in [4] that the indices used to evaluate the concept of evenness in ecology, which operate on proportional input vectors, share some properties with consensus measures. These evenness indices capture the differences in species populations and usually reach a minimum when a single species dominates. Here we propose that the negation or reciprocal of these models (a single species dominating being analogous to a dominating opinion or unilateral preference) could be appropriate for evaluating consensus where the inputs represent voting proportions.

The article will be set out as follows. In Section 2, we give an overview of consensus measures and note that models used in different settings can be considered as special cases of a general framework, differing in terms of the distance defined between decision makers' preferences. In Section 3, we propose a number of consensus measures for the case of single-preference voting, then in Section 4 we show how they fit into the general consensus framework and how they could be extended. We provide some summarizing comments in the Conclusion.

2 Consensus Measures Background

Consensus measures are used to give some idea of the level of agreement or overall similarity for a set of inputs, whether they be numerical evaluations or preference relations. If all inputs are the same, then we should achieve a perfect level of consensus, and as more and more of the inputs differ, the level of consensus is reduced. We consider a set of m decision-makers or voters $V = \{v_1, v_2, \ldots, v_m\}$ expressing their preferences over n candidates or alternatives $U = \{u_1, u_2, \ldots, u_n\}$. Preferences $f(v_i, u_j) \in \mathcal{P}$ can be expressed as: evaluations $x_{ij} \in [0, 1]$, indicating an overall score awarded by voter i to candidate j; rankings $r_{ij} \in \{1, 2, \ldots, n\}$, showing that voter i ranked candidate j as the (r_{ij})-th best candidate; pairwise preferences $a_{jk}^i \in \{0, 1\}$ indicating whether or not voter i prefers candidate j to candidate k; and single votes $s_i \in \{1, 2, \ldots, n\}$ showing which candidate is preferred by voter i.

For each situation, there are various existing generalizations and extensions. For example, evaluations can be expressed as intervals [32], rankings can be expressed as *weak* (rather than complete) orders [15], and pairwise preferences can be expressed as either additive or multiplicative fuzzy degrees of preferences [18]. For the moment we will contain our investigations to the simplified cases, although we note that many of the consensus measures have been developed with these generalized frameworks in mind.

Table 1 lists some of the properties relating to consensus measures. In [5, 13], $C1 - C3$ were considered as the minimum requirements for consensus[1] while further properties have been studied in [2, 14, 28].

Existing consensus measures can often be considered as special cases of the following general models (see [3, 8, 26, 28]).

$$C_{\langle M, f \rangle}(\mathcal{P}_{V \times U}) = 1 - \frac{\displaystyle \underset{i,j=1, i \neq j}{\overset{m}{M}} d(f(v_i, U), f(v_j, U))}{\max_{V \times U} \triangle} \tag{1}$$

$$C_{\langle M_1, M_2, d \rangle}(\mathcal{P}_{V \times U}) = 1 - \frac{\displaystyle \underset{i=1}{\overset{m}{M_1}} d\left(f(v_i, U), \underset{j=1}{\overset{m}{M_2}} f(v_j, U) \right)}{\max_{V \times U} \triangle} \tag{2}$$

where M, M_1, M_2 are averaging aggregation functions, d is a dissimilarity or distance function[3][4], and $\max_{V \times U} \triangle$ is a scaling factor, either used to normalize d to $[0, 1]$ or to ensure that a level of zero consensus is possible for all m.

---

[1] These were considered to be essential for sets of ordered preferences. For sets of real inputs given over the unit interval, we have proposed the basic requirements of unanimity $(C(t, t, \ldots, t) = 1)$ and that minimum consensus be reached for $C(0, 1) = C(1, 0) = 0$ when there are two inputs [2].

[3] We note that consensus can also be framed in terms of the inverse notion of similarity, however similarities, in turn, are often based on distances (see [10, 14]).

[4] In [3] we also considered consensus measures in the framework of multi-distances [20] with $C = 1 - D$ where D is a scaled multiple-argument distance function.

Table 1. Properties for consensus measures $\mathcal{C} : \mathcal{P} \to [0,1]$ where $\mathcal{P}$ is the set of voters and their preferences pertaining to a set of alternatives $V \times U$

C1 Unanimity	It holds that $\mathcal{C}(\mathcal{P}_{V \times U}) = 1$, if and only if $f(v_i, u_j) = f(v_k, u_j)$ for all j;
C2 Anonymity	It holds that $\mathcal{C}(\mathcal{P}_{V \times U}) = \mathcal{C}(\mathcal{P}_{V_\sigma \times U})$ where V_σ represents any permutation of the voters $\{v_{\sigma(1)}, v_{\sigma(2)}, \ldots, v_{\sigma(m)}\}$;
C3 Neutrality	It holds that $\mathcal{C}(\mathcal{P}_{V \times U}) = \mathcal{C}(\mathcal{P}_{V \times U_\sigma})$ where U_σ represents any permutation of the alternatives $\{u_{\sigma(1)}, u_{\sigma(2)}, \ldots, u_{\sigma(n)}\}$;
C4 Maximum dissension	The consensus value reaches a minimum when the voters can be partitioned into two equally sized groups $|V_1| = |V_2|$ with preferences in V_1 being as far as possible from V_2;
C5 Reciprocity	The consensus value is the same if we reverse the preference ordering (or take the negation[2] of each evaluation) for each voter;
C6 Replication invariance	Duplicating the set of voters or inputs does not alter the level of consensus;
C7 Monotonicity with respect to the majority	When there exists a subset of unanimous voters V_{maj} comprising half of the population, any movement of the remaining voters 'closer' to the preferences expressed in V_{maj} should not decrease the level of consensus.

In the first case, we aggregate the differences between each pair of voters v_i, v_j in terms of their preferences over the set of alternatives in U, while in the second case we look at the differences between each voter and an *overall* opinion (given by M_2).

For numerical inputs, standard choices for d include absolute difference $d(x,y) = |x - y|$ or squared difference $d(x,y) = (x - y)^2$. The overall distance between the evaluations of two decision makers can then be found either by summing these differences over each alternative or using a distance defined for multiple dimensions such as the Euclidean distance.

A common way to assess differences between voters who provide a complete (or weak) ranking of alternatives is to codify their preferences into a vector of ranks r_{ij} and then calculate distances between the vectors [15]. Common choices include the Euclidean distance, Manhattan distance or cosine difference. Spearman's rank correlation is also one such measure that takes into account the differences in the ranks for each alternative [5].

For pairwise preferences, a natural way to measure differences between voters is to count the number of pairs on which they disagree. This amounts to what is referred to as Kendall's τ [5] (where the orderings are complete) or the Kemeny distance (for weak orders where there are ties). García-Lapresta and

Pérez-Román have also suggested a weighted version of the Kemeny distance that accounts for the position of the ranks on which the voters disagree[5] [13,17].

We now consider the use of models based on evenness indices in ecology for single-preference voting.

3   Consensus for Single-Preference Voting

The problem of measuring evenness and the associated notion of diversity has been studied in ecology for over 60 years [24]. Many of these indices have their origins in the concept of equitability developed in economics [11,27] and the calculation of entropy from information science [23]. A number of desired properties relating to these indices have been investigated (e.g. in [25,29]). While many of these relate to continuity, independence or boundary conditions, of particular interest to us is the requirement that "Evenness should decrease when abundance is transferred from a less abundant species to a more abundant one." [29]. This is a particularly pertinent property to consensus, essentially extending condition $C7$.

In [4] we noted that evenness indices obtained from a calculation of species dominance do not satisfy the maximum dissension property $C4$ because they reach their minimum when a single species dominates - rather than for cases of maximum variability between populations. For example, E_1 in Table 2 is based on Simpson's dominance index, $\sum_{j=1}^{n} p_j^2$. This very notion of dominance however can be seen to correspond with that of consensus in the case of simple voting where each p_j represents the proportion of voters voting for the j-th candidate; reaching a maximum when one of the $p_j = 1$ and a minimum when all p_j are equal.

We therefore propose a number of single-preference consensus measures in Table 2 alongside their associated evenness indices. The equations for C_{E_2} and C_{McI} are also based on Simpson's dominance index, however with alternative scaling operations regulating the behavior between the maximum and minimum consensus levels. The evenness indices used for C_O and C_{Heip} are of a similar form, however they calculate dominance in a different way.

We draw special attention to C_{sp}, derived from an evenness measure presented in [22] based on Yager's *specificity* [33]. Specificity captures the extent to which a possibility distribution contains a single element. The $p_{(j)}$ notation denotes a reordering of the p_j such that $p_{(1)} \geq p_{(2)} \geq \ldots \geq p_{(n)}$ and the w_j are non-increasing weights such that $w_2 > 0$ and $\sum_{j=2}^{n} w_j \leq 1$. For the consensus to increase when smaller populations are redistributed to larger populations, however, the

---

Table 2. Consensus measures derived from evenness indices using $1 - E(p_1, p_2, \ldots, p_n)$ with different choices for E

Consensus Measure	Evenness Calculation	Reference
$C_{E_1} = \dfrac{\left(n \sum_{j=1}^{n} p_j^2\right) - 1}{n-1}$	$E_1 = \dfrac{1 - \sum_{j=1}^{n} p_j^2}{1 - 1/n}$	[19]
$C_{E_2} = \dfrac{\left(n \sum_{j=1}^{n} p_j^2\right) - 1}{\left((n-1) \sum_{j=1}^{n} p_j^2\right)}$	$E_2 = \dfrac{1}{n \sum_{j=1}^{n} p_j^2}$	[30]
$C_O = \dfrac{n\left(1 - \sum_{j=1}^{n} \min(p_j, 1/n)\right)}{n-1}$	$E_O = \dfrac{\sum_{j=1}^{n} \min(p_j, 1/n) - \frac{1}{n}}{1 - \frac{1}{n}}$	[6]
$C_{Heip} = \dfrac{\left(n \prod_{j=1}^{n} p_j^{p_j}\right) - 1}{(n-1) \prod_{j=1}^{n} p_j^{p_j}}$	$E_{Heip} = \dfrac{\frac{1}{\prod_{j=1}^{n} p_j^{p_j}} - 1}{n - 1}$	[16]
$C_{McI} = \dfrac{\sqrt{n \sum_{j=1}^{n} p_j^2} - 1}{\sqrt{n} - 1}$	$E_{McI} = \dfrac{\left(\sum_{j=1}^{n} p_j\right) - \sqrt{\sum_{j=1}^{n} p_j^2}}{\left(\sum_{j=1}^{n} p_j\right) - \frac{\left(\sum_{j=1}^{n} p_j\right)}{\sqrt{n}}}$	[21]
$C_{sp} = p_{(1)} - \sum_{j=2}^{n} w_i p_{(j)}$	$E_{sp} = 1 - \left(p_{(1)} - \sum_{j=2}^{n} w_i p_{(j)}\right)$	[22, 33]

weights should be non-decreasing. We also enforce that $\sum_{j=2}^{n} w_j = 1$ to ensure that minimum consensus is reached when all p_j are equal.

As well as exhibiting the redistribution-based behavior analogous to property $C7$ for the context of simple voting[6], all of the consensus measures in Table 2 satisfy the basic requirements $C1 - C3$ given in Table 1. They also reach the minimum consensus value of 0 when preferences for the candidates are equally distributed, which is consistent with the maximum dissension property $C4$. Although reciprocity ($C5$) does not apply here since we can't take the negation or reverse ordering, the use of proportions ensures replication invariance ($C6$) will automatically hold.

Tables 3-4 show some example input sets with the resulting consensus evaluations. In Table 3, we consider just the case of two candidates and 10 voters, while for Table 4 we show evaluations for a varying number of candidates and the proportion of voters supporting each.

---

[6] Shannon's information entropy $Ent(\mathbf{p}) = \sum_{i=1}^{n} p_j \log(p_j)$ also exhibits this behavior, summing the terms $p_j \log(p_j)$ rather than $p_j p_j$. E_{Heip} effectively calculates dominance using $\exp(Ent(\mathbf{p}))$.

Table 3. Consensus evaluations using measures based on evenness for 10 voters and 2 candidates

	Candidates		Consensus measures					
	A	B	C_{E_1}	C_{E_2}	C_O	C_{Heip}	C_{McI}	C_{sp}
	5	5	0	0	0	0	0	0
Voters	6	4	0.04	0.077	0.2	0.04	0.048	0.2
for each	7	3	0.16	0.276	0.4	0.158	0.186	0.4
candidate	8	2	0.36	0.529	0.6	0.351	0.401	0.6
	9	1	0.64	0.78	0.8	0.616	0.677	0.8
	10	0	1	1	1	1	1	1

The values in Table 3 give some idea of how the measures graduate between minimum consensus and maximum consensus as the majority steadily increases. While C_O and C_{sp} are linear, the remaining measures are affected more by redistributions of voters when the disparity between supporters of A and B is already high than when it is low.

Table 4. Consensus evaluations for varying number of candidates and proportional support

	Candidates					Consensus measures					
n	A	B	C	D	E	C_{E_1}	C_{E_2}	C_O	C_{Heip}	C_{McI}	C_{sp}
3	0.4	0.32	0.28			0.011	0.033	0.1	0.016	0.015	0.107
3	0.5	0.4	0.1			0.13	0.310	0.35	0.216	0.167	0.3
3	0.5	0.25	0.25			0.063	0.167	0.25	0.086	0.083	0.25
4	0.4	0.4	0.1	0.1		0.12	0.353	0.4	0.234	0.166	0.245
4	0.6	0.2	0.1	0.1		0.227	0.540	0.467	0.343	0.296	0.482
5	0.4	0.3	0.1	0.1	0.1	0.1	0.357	0.375	0.217	0.148	0.276
5	0.6	0.3	0.05	0.03	0.02	0.317	0.699	0.625	0.570	0.410	0.539
5	0.8	0.1	0.05	0.03	0.02	0.567	0.868	0.75	0.725	0.654	0.763

For the examples shown in Table 4, we note that while property $C7$ holds for fixed n, sometimes the evaluations for varying n are not necessarily consistent, e.g. from the input set $(0.4, 0.4, 0.1, 0.1)$, if the voters for D change their preference to candidate A, this should increase the level of consensus - however because of the different scaling that happens depending on the number of candidates, this is not the case except for C_{sp} and C_{McI}. This problem can be avoided if the distances are not scaled, however the sacrifice will be that the property of maximum dissension ($C4$) will not hold.

In the following we suggest that these alternative models for measuring consensus could also be extended to other domains.

4 Extending Single-Preference Consensus Models to Other Settings

As well as being conceptually appealing from the viewpoint of redistributing voters toward a majority candidate, Simpson's dominance index can also be interpreted in light of Eq. (1). We model the difference between two voters' preferences using the discrete distance, i.e. $d_0(x,y) = 0$ if $x = y$ and 1 otherwise. Using d_0, the proportion of pairs (including self-identical pairs) that will have coinciding preferences and hence zero distance will be precisely Simpson's dominance index, so subtracting this from 1 (with appropriate scaling) will result in Eq. (1) being equivalent to C_{E_1}.

Indeed, the results obtained for C_{E_1} in Table 1 are equivalent to those which would be obtained for equivalent scenarios in the different decision making settings using Eq. (1), i.e. the distance between the preferences $A > B$ and $B > A$ is 1, and for evaluations on a numerical scale we can consider p_1 and p_2 to reflect the proportion of 0s and 1s respectively.

On the other hand, C_O and C_{sp} coincide with consensus measures of the form in Eq. (2). Assuming that the candidate with the majority is the overall selection, these values represent the relative proportion[7] of voters that agree with this decision.

We therefore can look to extend these models to settings with differing notions of distance. For instance, C_{E_1} can be expressed,

$$C_{E_1} = 1 - \frac{n \sum_{i,j=1, i \neq j}^{n} d_0(u_i, u_j) p_i p_j}{n - 1}.$$

By changing the distance d_0, we recover other special cases of Eq. (1), simply replacing the count of pairs which differ by $d(u_i, u_j)$ with the relative proportions.

It might also be interesting to adapt the equations from Table 1 which are obtained from pairs in a similar fashion. For example, in C_{Heip} we can replace the terms $p_j^{p_j}$ with $\left(\sum_{i=1}^{n} (1 - d(u_i, u_j)) p_i \right)^{p_j}$, i.e., for each of the p_j indices, we include not only the population that supports j but also all those 'close' to supporting j. We can take similar approaches for C_O and C_{E_2}. We recover the original cases whenever d is the discrete distance.

5 Conclusion

We have presented a number of consensus measures based on indices studied in ecology for measuring consensus in a single-preference scenario and shown that some of these can be considered within a more general consensus framework, establishing their relationship to measures used in other decision making settings.

---

[7] I.e. after scaling by the minimum proportion possible.

We suggested some ways for these measures to be extended to more general consensus domains.

As well as being useful for single-preference settings or where there is no structure to the decision makers' alternatives and preferences, these consensus measures are also useful for providing an auxiliary consensus measurement. While Herrera et al. distinguish between consensus over alternatives or consensus with the overall decision (e.g. see [8]), and while order-based consensus measures often look at differences between rankings, the equations in Table 2 can be used to indicate the overall consensus with respect to a particular position. For instance, in the second setting of Fig. 1, it may be useful to know that there is high agreement when ranking A in first place, although the decision makers are varied with respect to the other positions.

References

1. Alonso, S., Pérez, I.J., Cabrerizo, F.J., Herrera-Viedma, E.: A linguistic consensus model for web 2.0 communities. Applied Soft Computing 13(1), 149–157 (2013)
2. Beliakov, G., Calvo, T., James, S.: Consensus measures constructed from aggregation functions and fuzzy implications. Knowledge-Based Systems 55, 1–8 (2014)
3. Beliakov, G., James, S., Calvo, T.: Aggregating fuzzy implications to measure group consensus. In: Proc. Joint IFSA World Congress and NAFIPS Annual Meeting, Edmonton, Canada, pp. 1016–1021 (2013)
4. Beliakov, G., James, S., Nimmo, D.: Can indices of ecological evenness be used to measure consensus? In: Proc. of FUZZIEEE, Beijing, China (2014)
5. Bosch, R.: Characterizations of Voting Rules and Consensus Measures, Ph.D. Thesis. Universiteit van Tilburg (2006)
6. Bulla, L.: An index of evenness and its associated diversity measure. Oikos 70, 167–171 (1994)
7. Cabrerizo, F., Moreno, J., Pérez, I., Herrera-Viedma, E.: Analyzing consensus approaches in fuzzy group decision making: advantages and drawbacks. Soft Computing 14, 451–463 (2010)
8. Cabrerizo, F.J., Alonso, S., Pérez, I.J., Herrera-Viedma, E.: On consensus measures in fuzzy group decision making. In: Torra, V., Narukawa, Y. (eds.) MDAI 2008. LNCS (LNAI), vol. 5285, pp. 86–97. Springer, Heidelberg (2008)
9. Cabrerizo, F.J., Chiclana, F., Ureña, M.R., Herrera-Videma, E.: Challenges and open questions in soft consensus models. In: Proc. Joint IFSA World Congress and NAFIPS Annual Meeting, Edmonton, Canada, pp. 944–949 (2013)
10. Chiclana, F., Tapia García, J.M., del Moral, M.J., Herrera-Viedma, E.: A statistical comparative study of different similarity measures of consensus in group decision making. Information Sciences 221, 110–123 (2013)
11. Dalton, H.: The measurement of the inequality of incomes. The Economic Journal 30, 348–361 (1920)
12. García-Lapresta, J.L., Pérez-Román, D.: Consensus measures generated by weighted Kemeny distances on weak orders. In: Proceedings of ISDA, Cairo, Egypt, pp. 463–468 (2010)
13. García-Lapresta, J.L., Pérez-Román, D.: Measuring consensus in weak orders. In: Herrera-Viedma, E., García-Lapresta, J., Kacprzyk, J., Fedrizzi, M., Nurmi, H., Zadrożny, S. (eds.) Consensual Processes. STUDFUZZ, vol. 267, pp. 213–234. Springer, Heidelberg (2011)

14. García-Lapresta, J.L., Pérez-Román, D.: Consensus-based hierarchical agglomerative clustering in the context of weak orders. In: Proceedings of IFSA World Congress, Edmonton, Canada, pp. 1010–1015 (2013)
15. García-Lapresta, J.L., Román, D.P.: Some measures of consensus generated by distances on weak orders. In: Proceedings of XIV Congreso Español sobre Tecnologías y Lógica Fuzzy, Cuencas Mineras, September 17-19, pp. 477–483 (2008)
16. Heip, C.: A new index measuring evenness. Journal of Marine Biological Association of the United Kingdom 54(3), 555–557 (1974)
17. Heiser, W.J., D'Ambrosio, A.: Clustering and prediction of rankings within a Kemeny distance framework. In: Algorithms from and for Nature and Life, pp. 19–31. Springer International Publishing (2013)
18. Herrera-Viedma, E., Herrera, F., Chiclana, F.: A consensus model for multiperson decision making with different preference structures. IEEE Transactions on Systems, Man and Cybernetics – Part A: Systems and Humans 32, 394–402 (2002)
19. Krebs, C.J.: Ecological Methodology. Harper Collins, New York (1989)
20. Martín, J., Mayor, G.: Multi-argument distances. Fuzzy Sets and Systems 167, 92–100 (2011)
21. Pielou, E.C.: An Introduction to Mathematical Ecology. Wiley-Interscience, New York (1969)
22. Ricotta, C.: A recipe for unconventional evenness measures. Acta Biotheoretica 52, 95–104 (2004)
23. Shannon, C.E., Weaver, W.: The mathematical theory of communication. Univ. of Illinois Press, Urbana (1949)
24. Simpson, E.H.: Measurement of diversity. Nature 163, 688 (1949)
25. Smith, B., Wilson, J.B.: A consumer's guide to evenness indices. Oikos 76, 70–82 (1996)
26. Szmidt, E., Kacprzyk, J.: Analysis of consensus under intuitionistic fuzzy preferences. In: Proc. Int. Conf. in Fuzzy Logic and Technology, Leicester, pp. 79–82, De Montfort University (2001)
27. Taillie, C.: Species equitibility: a comparative approach. In: Ecological Diversity in Theory and Practice, pp. 51–62. Int. Coop. Publ. House, Fairland (1979)
28. Tastle, W.J., Wierman, M.J., Dumdum, U.R.: Ranking ordinal scales using the consensus measure. Issues in Information Systems 185, 96–102 (2006)
29. Tuomisto, H.: An updated consumer's guide to evenness and related indices. Oikos 121, 1203–1218 (2012)
30. Williams, C.B.: Patterns in the balance of nature. Academic Press, London (1964)
31. Xu, Z., Cai, X.: Group consensus algorithms based on preference relations. Information Sciences 181, 150–162 (2011)
32. Xu, Z., Cai, X.: On consensus of group decision making with interval utility values and interval preference orderings. Group Decision and Negotiation 22, 997–1019 (2013)
33. Yager, R.R.: On the specificity of a possibility distribution. Fuzzy Sets and Systems 50, 279–292 (1992)

Distributed Decision-Theoretic Active Perception for Multi-robot Active Information Gathering

Jennifer Renoux[1], Abdel-Illah Mouaddib[1], and Simon LeGloannec[2]

[1] University of Caen, Lower-Normandy, France
`firtname.lastname@unicaen.fr`
[2] Airbus Defence and Space, Val de Reuil, France
`firstname.lastname@cassidian.com`

Abstract. Multirobot systems have made tremendous progress in exploration and surveillance. In that kind of problem, agents are not required to perform a given task but should gather as much information as possible. However, information gathering tasks usually remain passive. In this paper, we present a multirobot model for active information gathering. In this model, robots explore, assess the relevance, update their beliefs and communicate the appropriate information to relevant robots. To do so, we propose a distributed decision process where a robot maintains a belief matrix representing its beliefs and beliefs about the beliefs of the other robots. This decision process uses entropy and Kullback-Leibler in a reward function to access the relevance of their beliefs and the divergence with each other. This model allows the derivation of a policy for gathering information to make the entropy low and a communication policy to reduce the divergence. An experimental scenario has been developed for an indoor information gathering mission.

1 Introduction

Robotic systems are increasingly used in surveillance and exploration applications. In the future robots may assist humans and eventually replace them in dangerous areas. In these particular research fields the main goal of the system is not to reach a given goal but to gather information. The system needs to create a complete and accurate view of the situation. This built view may be used afterwards by some agents - human or artificial - to make some decisions and perform some actions. Therefore the information gathering system must be able to identify lacking information and take the necessary steps to collect it. However it is obviously not productive that all the robots in the system try to collect all possible information, just as it is not possible for the robots to communicate all the information they have all the time. They should select pieces of information to collect or to communicate depending on what they already know and what other agents already know. Developing methods to allow robots to decide how to act and what to communicate is a decision problem under uncertainty. Partially Observable Markov Decision Processes (POMDPs) are traditionally used to deal with this kind of problem, more particularly Decentralized POMDPs (DEC-POMDPs) that are an extension of POMDPs for multiagent systems. However the classic POMDP framework is not designed to have information gathering as a target : information gathering

V. Torra et al. (Eds.): MDAI 2014, LNAI 8825, pp. 60–71, 2014.

is usually a means of reaching another goal. Some extensions have been developed for mono-agent systems but the plunge to multi-agent systems has not been taken. We suggest in this paper a formal definition of the relevance of a piece of information as well as a new model dedicated to multiagent information gathering that is able to explore actively its environment and communicate relevant information.

Section 2 presents some background knowledge and other studies that are relevant to our problem. Section 3 presents the proposed model to do active sensing with a multi-robot system. It defines a agent-oriented relevance degree and describes the Partially Observable Markov Decision Process used in the system. Finally, section 4 presents an implementation of the model on a simple indoor sensing problem.

2 Background and Related Work

Relevance

Agents situated in an environment perceive a huge amount of data and they need to process those data to extract higher-level features. However the interest of a feature for an agent depends on several parameters such as the situation, the problem to be dealt etc. Since it is counterproductive to communicate neither to perform an action to collect non interesting information, agents need to quantify the importance of a piece of information according to the current situation. This degree of importance is the relevance of information. Borlund [1] defined two types of relevance : system-oriented relevance and agent-oriented relevance. System-oriented relevance analyzes relevance regarding a request. The better the match between information and the request, the greater the degree of relevance. System-oriented relevance is used and studied in several Information Retrieval Systems [2]. Agent-oriented relevance defines a link between information and an agents needs. Information is relevant if it matches a specific need, if it is informative and useful for an agent which receives it. However the need may not be explicit. Floridi [3] suggested a base of epistemic relevance and Roussel and Cholvy [4] deepened this study in the case of BDI agents and multimodal logic. However those studies are based on propositional logic and are not applicable for reasoning with uncertain knowledge. Therefore we need to define a relevance theory that may be used in uncertain reasoning.

Active Information Gathering

Using relevance, a robot is able to decide if a piece of information is interesting or not. Therefore it is able to perform active information gathering. Active information gathering defines the fact that an agent will act voluntarily to gather information and not just perceive passively its environment. In this context the agent has to make decisions in an environment that it cannot perceive completely. One of the best and commonly used models to deal with that kind of problem is Partially Observable Markov Decision Process. Some studies have already been carried out to perform active perception using POMDPs. Ponzoni et al. [5] suggested a mixed criterion for active perception as a mission to recognize some vehicles. The criterion is based on an entropy function and a classical reward function to mix active perception and more classical goals of POMDPs.

Meanwhile Araya-Lopez et al. [6] suggested a more general approach to use any reward function based on belief state in POMDPs. These two approaches proved the feasibility of such a system where information gathering is the goal. However they are both mono-agent and are not applicable to a multiagent system. To our knowledge there is no model of multiagent system for active information gathering.It is obvious that information gathering would be more efficient if it is done by several agents instead of a single one. However, it is important that agents are able to coordinate themselves to make the gathering efficient.

Multiagent Active Information Gathering

The problem of multiagent active information gathering relates to multiagent decision under uncertainty: a set of agents have to control together a decision process to collect information. However, no agent is able to control the whole process. Different equivalent frameworks extending POMDPs have been developed to deal with multiagent decision problems under uncertainty [7].

Solving a multiagent POMDP is a problem NEXP-complete [8]. Even if algorithms and heuristics have been suggested to overcome this complexity [14], those frameworks are usually not applicable to real problems. To overcome this issue, Spaan et al. [15] suggested a system based on POMDP to enable a Network Robot System to classify particular features by acting to get the best information possible. In this study, the authors decided to model the active information gathering thanks to classifying actions in order to avoid using a reward function based on entropy, which would increase the complexity of planning. However Araya-Lopez et al. [6] proved that it is possible to reuse techniques from the standard POMDP literature with a reward function based on belief states as would be reward function using entropy. On top of that, in the system suggested by Spaan et al., all the agents have to make the classification steps and build a complete view of the environment. However, in usual active information gathering problems, it is not useful that each agent of the system has this complete view as long as the system view is complete. Therefore, agents need to communicate with each others in order to avoid repetitive exploration. Communication in multiagent POMDP framework is usually assumed to be free and instantaneous. However such assumption is not possible in real problems. Communication is an action that has a cost and must be decided. Roth et al. [9] presented an algorithm to take into account the communication cost in multiagent POMDPs. In this paper, the communication is considered only during execution and should improve the performance of the system : if it is useful for the system, an agent communicates all its history of observations. There is no decision concerning the observations to communicate. Information gathering is once again a means to reach a goal and not the goal in itself.

3   The Model

3.1   Definition of an Agent-Based Relevance

Let's assume an agent a_i situated in an environment $\mathcal{E}$. The environment is modeled as a set of features of interest. Each feature is described using a random variable X_k which can take values in its domain $DOM(X_k)$. The agent a_i has some beliefs $\mathcal{B}_i^{\mathcal{E}}$ concerning the features of interest modeled as probability distributions over the $X_k \in \mathcal{E}$.

$$\mathcal{B}_{i,t}^{\mathcal{E}} = \{b_{i,t}^k, \forall X_k \in \mathcal{E}\}$$

with $b_{i,t}^k$ being the probability distribution of agent a_i over the variable X_k at time t. Let's assume an agent receives observations concerning the features of interest. When receiving a new observation, agent a_i updates its beliefs as follows : $\mathcal{B}_{i,t+1}^{\mathcal{E}} = update(\mathcal{B}_{i,t}^{\mathcal{E}}, o_k)$ [10]

First of all we considered that observations received are true. As a matter of fact, an observation cannot be relevant if it is a false observation [3]. We discuss about assumption and the way it is used in the decision process in section 3.2. Considering this assumption, an observation o_k is relevant for an agent a_i if it matches the following criteria:

1. agent a_i is interested in the subject of the observation o_k, that is to say X_k
2. the observation o_k is new for agent a_i
3. if the observation o_k is not new, it should render agent's a_i beliefs more accurate

The first point is dealt with the way we represent agent's beliefs : if agent a_i is interested in X_k then X_k is in agent's a_i beliefs. We assume that an observation o_k is new for agent a_i if beliefs $\mathcal{B}_{i,t+1}^{\mathcal{E}}$ and $\mathcal{B}_{i,t}^{\mathcal{E}}$ are distant from each other. The dissimilarity between two probability distributions is measured by the Kullback-Leibler ratio.

Definition 1. *An observation o_k is new for agent a_i at time t if and only if*

$$D_{KL}(\mathcal{B}_{i,t}^{\mathcal{E}}||\mathcal{B}_{i,t+1}^{\mathcal{E}}) > \epsilon \tag{1}$$

ϵ *is a fixed threshold and $D_{KL}(\mathcal{B}_{i,t}^{\mathcal{E}}||\mathcal{B}_{i,t+1}^{\mathcal{E}})$ is the Kullback-Leibler ratio and defined by*

$$D_{KL}(\mathcal{B}_{i,t}^{\mathcal{E}}||\mathcal{B}_{i,t+1}^{\mathcal{E}}) = \sum_{X_k \in \mathcal{E}} \sum_{k=1}^{n} b_{i,t}^k(x_k) \log \frac{b_{i,t}^k(x_k)}{b_{i,t+1}^k(x_k)} \tag{2}$$

where $b_{i,t}^k(x_k)$ is the belief of agent a_i that the random variable X_k takes value x_k.

To model the accuracy of a belief $\mathcal{B}_{i,t}^{\mathcal{E}}$, we use an entropy measure.

Definition 2. *Belief $\mathcal{B}_{i,t+1}^{\mathcal{E}}$ is more precise than belief $\mathcal{B}_{i,t}^{\mathcal{E}}$ if and only if*

$$H(\mathcal{B}_{i,t+1}^{\mathcal{E}}) < H(\mathcal{B}_{i,t}^{\mathcal{E}}) \tag{3}$$

with $H(\mathcal{B}_{i,t}^{\mathcal{E}}) = -\sum_{X_k \in \mathcal{E}} \sum_{k=1}^{n} b_{i,t}(x_k) log(b_{i,t}(x_k))$.

Given the previous definitions we may define the degree of relevance as shown below :

Definition 3. *The degree of relevance of an observation o_k concerning a random variable X_k for an agent a_i, noted $rel_i(o_k)$, is given by*

$$rel_i(o_k) = \alpha D_{KL}(\mathcal{B}_{i,t}^{\mathcal{E}}||\mathcal{B}_{i,t+1}^{\mathcal{E}}) + \beta(H(\mathcal{B}_{i,t}^{\mathcal{E}}) - H(\mathcal{B}_{i,t+1}^{\mathcal{E}})) + \delta \tag{4}$$

with $\mathcal{B}_{i,t+1}^{\mathcal{E}} = update(\mathcal{B}_{i,t}^{\mathcal{E}}, o_k)$, α and β being weights and δ being a translation factor to ensure the relevance is positive.

3.2   Decision Process for Multiagent Active Information Gathering

Let a multiagent system be defined as a tuple
$< \mathcal{E}, \mathcal{AG}, \mathcal{B}, \mathcal{D} >$ with $\mathcal{E}$ being the environment as defined previously, $\mathcal{AG}$ being the
set of agents, $\mathcal{B}$ being the set of all agent's beliefs on the environment and $\mathcal{D}$ being the
set of all agent's decision functions $\mathcal{D} = \{\mathcal{D}_i, \forall i \in \mathcal{AG}\}$. Each $\mathcal{D}_i$ is represented as a
Factored Partially Observable Markov Decision Process (FPOMDP)[11].

Set of Actions. We consider two type of actions : look for the value of a particular
random variable ($Explore$-type actions) and communicate an observation to an agent
($Communicate$-type actions):
$$\mathcal{A} = \{Exp(X_k), \forall X_k \in \mathcal{E}\} \cup$$
$$\{Comm(o, ag), \forall o \in \mathcal{O}, \forall ag \in \mathcal{AG}\}$$

The size of the action set is :
$$|\mathcal{A}| = |\mathcal{A}_{Explore}| + |\mathcal{A}_{Communicate}| \tag{5}$$
$$|\mathcal{A}| = |\mathcal{E}| + |\mathcal{O}| \times |\mathcal{AG}| \tag{6}$$

Set of Observations. In a Partially Observable Markov Decision Process, an agent
doesn't know exactly the current state of the system. It only receives observations when
performing actions, which are only indications about the current state. So the agent may
estimate the current state from the history of observations it received. When performing an $Explore$-type action, the agent receives an observation concerning the random
variable it is trying to sense. This $\mathcal{O}_{Explore}$ set of observations depends on the problem considered. When performing a $Communicate$-type action the agent receives an
observation stating that the message has been properly sent or not. Therefore two observations are possible for any $Communicate$-type action:
$$\mathcal{O}_{Communicate} = \{okMsg, nokMsg\}$$

Maintaining a Belief State. The agent doesn't know the exact current state of the
system. It only has observations about it. Therefore it should maintain some beliefs
concerning this current state. In the context of multiagent information gathering, an
agent should not only have beliefs about the state of the environment but also about the
other agents. As a matter of fact, to avoid agents exploring the same areas and to enable
them to choose the most relevant observation to communicate, they should model the
knowledge of other agents in their own belief state. Thus we defined an *extended belief
state* as following :

Definition 4. *Let an extended belief state of an agent a_i at time t be defined as following:*
$$\mathcal{B}_{i,t} = \mathcal{B}_{i,t}^{\mathcal{E}} \cup \mathcal{B}_{i,t}^{j,\mathcal{E}} \tag{7}$$
*with $\mathcal{B}_{i,t}^{\mathcal{E}} = \{b_{i,t}^{i,k}, \forall X_k \in \mathcal{E}\}$ being the beliefs of agent a_i concerning the environment
$\mathcal{E}$ and $\mathcal{B}_{i,t}^{j,\mathcal{E}} = \{b_{i,t}^{j,k}, \forall X_k \in \mathcal{E}\}$ being the beliefs of agent a_i concerning the beliefs of
agent a_j concerning the environment.*

Let's note that $B_{i,t}^{j,\mathcal{E}}$ is an approximation of $B_{j,t}^{\mathcal{E}}$. We may use a matrix representation. Rows represent the different random variables describing the environment and columns represent agent's i beliefs on each agent's beliefs, including itself :

$$\mathcal{B}_{i,t} = \begin{pmatrix} b_{i,t}^{1,1} & \cdots & b_{i,t}^{j,1} & \cdots & b_{i,t}^{i,1} & \cdots & b_{i,t}^{|\mathcal{AG}|,1} \\ \vdots & & \vdots & & \vdots & & \vdots \\ b_{i,t}^{1,k} & \cdots & b_{i,t}^{j,k} & \cdots & b_{i,t}^{i,k} & \cdots & b_{i,t}^{|\mathcal{AG}|,k} \\ \vdots & & \vdots & & \vdots & & \vdots \\ b_{i,t}^{1,|\mathcal{E}|} & \cdots & b_{i,t}^{j,|\mathcal{E}|} & \cdots & b_{i,t}^{i,|\mathcal{E}|} & \cdots & b_{i,t}^{|\mathcal{AG}|,|\mathcal{E}|} \end{pmatrix} \tag{8}$$

To keep an accurate representation of the current state of the system an agent has to update its beliefs regularly. An update will occur in three cases :

1. the agent receives a new observation from its sensors after an $Explore$ action. It updates its own beliefs concerning the environment : $\mathcal{B}_{i,t+1}^{\mathcal{E}}$.
2. the agent receives a new observation from agent a_j. It updates its own beliefs $\mathcal{B}_{i,t+1}^{\mathcal{E}}$ as well as its beliefs concerning agent a_j : $\mathcal{B}_{i,t+1}^{j,\mathcal{E}}$.
3. the agent sends an observation to agent a_j. It updates its beliefs concerning agent a_j : $\mathcal{B}_{i,t+1}^{j,\mathcal{E}}$.

In all cases the update $\mathcal{B}_{i,t+1}^{x,\mathcal{E}} = update(\mathcal{B}_{i,t}^{x,\mathcal{E}}, o_k)$, o_k being the observation received, is made as usual in Partially Observable Markov Decision Processes :

$$\mathcal{B}_{i,t+1}(s') = \frac{\omega(o_k, s', a) \sum_{s \in \mathcal{S}} p(s'|s,a)\mathcal{B}_{i,t}(s)}{\sum_{s \in \mathcal{S}} \sum_{s'' \in \mathcal{S}} \omega(o_k, s'', a)p(s''|s,a)b_{i,t}^{i,k}} \tag{9}$$

In the matrix representation, the columns that may be updated are:

$$\mathcal{B}_{i,t+1} = \begin{pmatrix} b_{i,t}^{1,1} & \cdots & \mathbf{b_{i,t+1}^{j,1}} & \cdots & \mathbf{b_{i,t+1}^{i,1}} & \cdots & b_{i,t}^{|\mathcal{AG}|,1} \\ \vdots & & \vdots & & \vdots & & \vdots \\ b_{i,t}^{1,k} & \cdots & \mathbf{b_{i,t+1}^{j,k}} & \cdots & \mathbf{b_{i,t+1}^{i,k}} & \cdots & b_{i,t}^{|\mathcal{AG}|,k} \\ \vdots & & \vdots & & \vdots & & \vdots \\ b_{i,t}^{1,|\mathcal{E}|} & \cdots & \mathbf{b_{i,t+1}^{j,|E|}} & \cdots & \mathbf{b_{i,t+1}^{i,|E|}} & \cdots & b_{i,t}^{|\mathcal{AG}|,|\mathcal{E}|} \end{pmatrix}$$

Reward Function. The best action to perform at a given time is set by a policy, computed considering a relevance based reward function. This reward function defines the reward an agent may receive by performing action a in state s. However in an information gathering context we are not interested in reaching some special state of the system but gathering and communicating relevant observations. Therefore the reward function is defined on the belief states of the agent and not on the real states of the system. An agent is rewarded if it collects observations that are relevant for itself and if it communicates observations that are relevant for other agents. As mentioned in section 3.1, an observation must be true to be relevant. Since agents only have beliefs concerning the world, they cannot ensure that an observation is true. However they should not exchange observations that reinforce their existing beliefs, regardless of their veracity. Therefore we need to find a compromise between the agents belief concerning the observation and the relevance of this observation. To do so we weight the relevance of a given observation by the agents belief concerning its truth. This belief is given by the probability of receiving the observation in the state s considered multiplied by the agent's belief that the state s is the current state. On top of that, agents should ensure that they have homogeneous beliefs. As a matter of fact, agents having different beliefs means that one agent at least is wrong. Therefore the reward function for communication includes a term to ensure that the different between agent's own beliefs and its beliefs on beliefs of other agents are low. The reward function is thus defined as follows:

$$
\begin{aligned}
&R(\mathcal{B}_{i,t}, Exp(X_k)) = \\
&\sum_{s \in \mathcal{S}} \sum_{o_p \in \mathcal{O}} \mathcal{B}_{i,t}(s)\omega(o_p, s, a)rel_i(o_p) - C_{Exp(X_k)}
\end{aligned}
\tag{10}
$$

$$
\begin{aligned}
&R(\mathcal{B}_{i,t}, Comm(o_p, a_j)) = \\
&\sum_{s \in \mathcal{S}} \mathcal{B}_{i,t}(s)\omega(o_p, s, a)rel_j(o_p) \\
&+\gamma(D_{KL}(\mathcal{B}_{i,t}^{\mathcal{E}}||\mathcal{B}_{i,t}^{j,\mathcal{E}}) - D_{KL}(\mathcal{B}_{i,t+1}^{\mathcal{E}}||\mathcal{B}_{i,t+1}^{j,\mathcal{E}})) \\
&-C_{Comm(o_p,a_j)}
\end{aligned}
\tag{11}
$$

with $C_{Explore(X_k)}$ and $C_{Communicate(o_k,a_j)}$ being the costs of taking the $Explore$ or $Communicate$ action and $\mathcal{B}_{i,t}(s)$ being the belief of agent a_i that state s is the current state. This cost may represent battery loss, bandwith used etc.

Resolution. To solve the POMDP presented above, we can rely on classic MDP algorithms. As a matter of fact, actions are epistemic and so don't modify the real state of the system. Therefore it is possible to transform our POMDP into a Belief MDP defined as a tuple $< \Delta, \mathcal{A}, \tau >$ where :

- Δ is the new state space. It correspond directly to the belief state space in the initial POMDP. $\Delta = \mathcal{B}_i$
- $\mathcal{A}$ is the same state of actions as previously defined
- τ is the new transition function defined as following

Theorem 1. *The transition function τ of the Belief MDP is defined as following :*

$$
\tau(\mathcal{B}_{i,t}, a, \mathcal{B}_{i,t+1}) =
$$

$$
\begin{cases}
\sum_{s \in \mathcal{S}} \sum_{o_k \in U_t} \omega(o_k, s, a)\mathcal{B}_{i,t}(s) & if \ U_t \neq \emptyset \\
0 & otherwise
\end{cases}
$$

where $U_t = \{o_k \in \mathcal{O}, \text{ such as, } \mathcal{B}_{i,t+1} = update(\mathcal{B}_{i,t}, o_k)\}$ is the set of all observations enabling the transition from state $\mathcal{B}_{i,t}$ to state $\mathcal{B}_{i,t+1}$, $\omega(o_k, s_t, a)$ is the observation function of the POMDP and $\mathcal{B}_{i,t}(s_t)$ is the belief of agent a_i that the current state is s_t.

Proof. If there is no observation such as $\mathcal{B}_{i,t+1} = update(\mathcal{B}_{i,t}, o_k)$, it is not possible to transfer from belief state $\mathcal{B}_{i,t}$ to belief state $\mathcal{B}_{i,t+1}$. Therefore, $\tau(\mathcal{B}_{i,t}, a, \mathcal{B}_{i,t+1}) = 0$. If there exists at least one observation o_k such as $\mathcal{B}_{i,t+1} = update(\mathcal{B}_{i,t}, o_k)$ we have the following equations :

$$
\begin{aligned}
\tau(\mathcal{B}_{i,t}, a, \mathcal{B}_{i,t+1}) &= P(\mathcal{B}_{i,t+1}|\mathcal{B}_{i,t}, a) \\
&= \sum_{o_k \in U_t} P(o_k|\mathcal{B}_{i,t}, a) \\
&= \sum_{s \in \mathcal{S}} \sum_{o_k \in U_t} P(o_k|s, a)\mathcal{B}_{i,t}(s) \\
&= \sum_{s \in \mathcal{S}} \sum_{o_k \in U_t} \omega(o_k, s, a)\mathcal{B}_{i,t}(s)
\end{aligned}
$$

The value function corresponding to this Belief MDP is defined as following:

$$
V(\mathcal{B}_{i,t}) = R(\mathcal{B}_{i,t}) + \max_{a \in \mathcal{A}} \int_{\mathcal{B}'_{i,t}} \tau(\mathcal{B}_{i,t}, a, \mathcal{B}'_{i,t})V(\mathcal{B}'_{i,t}) \tag{12}
$$

Using discretization techniques (by discretizing the probability distributions) we may transform equation 12 in :

$$
V(\mathcal{B}_{i,t}) = R(\mathcal{B}_{i,t}) + \max_{a \in \mathcal{A}} \sum_{\mathcal{B}'_{i,t} \in Samples} \tau(\mathcal{B}_{i,t}, a, \mathcal{B}'_{i,t})V(\mathcal{B}'_{i,t}) \tag{13}
$$

Then, any technique from the literature may be used to solve this belief-MDP [12] [13].

4 Experiments

Simulated Robots. The suggested model was implemented on a simple scenario. Two robots have to explore an environment made of 4 different zones connected to each other. For each zone, two observations are possible : *emptyRoom* and *notEmptyRoom*. The optimal policy was computed using different ratios to make the compromise between the Kullback-Leibler ratio and the entropy measure, as well as different probabilities of obtaining a false observation. The system was run 50 times with each set of parameters. To evaluate the policy, we measured the average number of messages sent by the robots, the average time needed to get a stable belief state and the number of false belief states at the end of exploration. We compared those measures with a multirobot system without communication and with a system in which agents communicate each observation they received immediately. The results are presented on figures 1a, 1b and 1c.

The linear part of the first two graphs represents the average time to reach a stable belief state, measured in number of iterations, an iteration being made up of the execution of an action, the reception of the associated observation and the reception of a communication message from another agent, if any. The bar part represents the average number of final states that were partially or totally incorrect. We notice that the number of false end states is reduced with partial 1a and complete 1c communication. In the worst case, that is to say a probability of 70% of receiving a correct observation after doing an explore action, the system with partial communication reduces the average number of false end states by 28%, and the system with complete communication

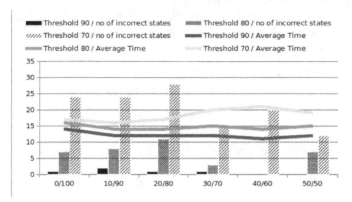

(a) Evaluation of the computed policy with relevant communication

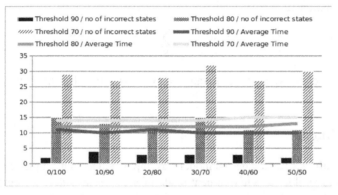

(b) Evaluation of the policy without communication

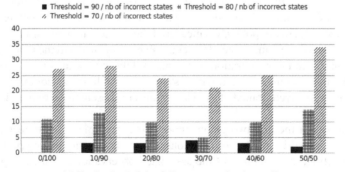

(c) Evaluation of the fully-communicative policy

Fig. 1. Evalutation of the three policies. The X-Axis represents the different ratios Kullback-Leibler / Entropy. The three different thresholds correspond to the probability to obtain a correct observation while doing an explore action

reduces it by 8%. In the average case, that is to say a probability of 80% of receiving a correct observation, the average number of false end states is reduced by 72% with our system and by 19% with a complete communication. We can conclude that choosing a relevant observation to communicate enable the system to be more robust to false observations coming from sensors. On top of that, Figure 2a shows that the average number of messages sent remains almost constant and, unsurprisingly, much lower than the system with complete communication (Figure 2b).

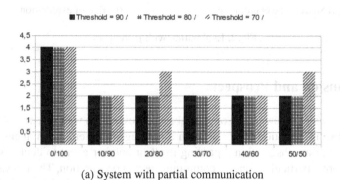

(a) System with partial communication

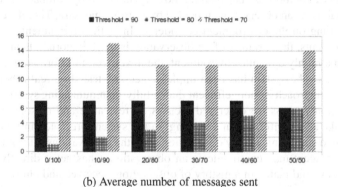

(b) Average number of messages sent

Those experiments seem to validate the hypothesis that choosing relevant information to communicate may improve system performances while reducing the number of communications.

Real Robots. We implemented the model on two μ-troopers in a simple scenario where two rooms must be explored. In the figure 2a, robot 1 decided to explore the room 1 and robot 2 decided to explore the room 2. Since beliefs of robot 2 about the environment are very accurate and it believes that robot 1 has incorrect beliefs, robot 2 decides to communicate the observation 0 to robot 1. Robot 1 receives this observation and updates its beliefs accordingly. This figure presents a case where an agent has approximated beliefs concerning the beliefs of the other agents. However this approximation does not prevent the robots to complete the mission and to reach a stable belief state, as presented on Figure 2b.

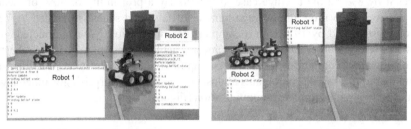

(a) Sharing the exploration   (b) End of exploration

Fig. 2. Exploration with μ-troopers

5   Conclusion and Prospects

We have introduced a new model of agent-based relevance as well as a decision process to make active information gathering with a multiagent system. Each agent computes the relevance of an observation regarding itself or another agent to decide whether it should explore a particular zone or communicate this observation. The relevance of an observation is a compromise between the novelty, modeled by Kullback-Leibler ratio, and the certainty of an observation, modeled by Entropy measure. Therefore it may be tuned depending on the environment considered. In static environment, as presented in the experiments, the certainty of an observation is more important than its novelty. However, in a highly dynamic environment, the novelty of an information may be the most important. The system has been implemented and tested on real robots. Results show that this approach is more efficient than a fully-communicating system.

The decision process we described focuses on relevance and reasoning on belief states to make active information gathering. In the system presented in this paper, an agent is able to communicate any observation from the observations set if it is relevant. Therefore, an agent may communicate an observation it has never directly received. Future works would maintain a history of observations received and allow an agent to communicate only observations it has previously received. Moreover, the beliefs about the beliefs of other agents are updated only when there is an explicit communication. We plan to work on a less naive method : since the same policy is used by all agents, we may update those beliefs more often by assuming the action taken by other agents. Finally, future works would consider the integration of the system presented in non-epistemic POMDPs.

References

1. Borlund, P.: The concept of relevance in IR. Journal of the American Society for information Science and Technology 54(10), 913–925 (2003)
2. Baeza-Yates, R., Ribeiro-Neto, B., et al.: Modern information retrieval, p. 463. ACM Press, New York (1999)
3. Floridi, L.: Understanding epistemic relevance. Erkenntnis 69(1), 69–92 (2008)

4. Roussel, S., Cholvy, L.: Cooperative interpersonal communication and Relevant information. In: ESSLLI Workshop on Logical Methods for Social Concepts (2009)
5. Ponzoni Carvalho Chanel, C., Teichteil-Königsbuch, F., Lesire, C.: POMDP-based online target detection and recognition for autonomous UAVs. In: ECAI, pp. 955–960 (2012)
6. Araya, M., Buffet, O., Thomas, V., Charpillet, F.: A POMDP extension with belief-dependent rewards. In: Advances in Neural Information Processing System, pp. 64–72 (2010)
7. Seuken, S., Zilberstein, S.: Formal models and algorithms for decentralized decision making under uncertainty. Autonomous Agents and Multi-Agent Systems 17(2), 190–250 (2008)
8. Bernstein, D., Givan, R., Immerman, N., Zilberstein, S.: The complexity of decentralized control of Markov decision processes. Mathematics of Operations Research 27(4), 819–840 (2002)
9. Roth, M., Simmons, R., Veloso, M.: Reasoning about joint beliefs for execution-time communication decisions. In: Proceedings of the Fourth International Joint Conference on Autonomous Agents and Multiagent Systems, pp. 786–793 (2005)
10. Cassandra, A., Kaelbling, L., Littman, M.: Acting optimally in partially observable stochastic domains. In: AAAI 1994, pp. 1023–1028 (1994)
11. Hansen, E., Feng, Z.: Dynamic Programming for POMDPs Using a Factored State Representation (2000)
12. Porta, J., Vlassis, N., Spaan, M., Poupart, P.: Point-based value iteration for continuous POMDPs. The Journal of Machine Learning Research 7, 2329–2367 (2006)
13. Hoey, J., Poupart, P.: Solving POMDPs with continuous or large discrete observation spaces. In: International Joint Conference on Artificial Intelligence, vol. 19, p. 1332 (2005)
14. Seuken, S., Zilberstein, S.: Memory-Bounded Dynamic Programming for DEC-POMDPs. In: IJCAI, pp. 2009–2015 (2007)
15. Spaan, M., Veiga, T., Lima, P.: Active cooperative perception in network robot systems using POMDPs. In: International Conference on Intelligent Robots and Systems, pp. 4800–4805 (2010)

Visual Consensus Feedback Mechanism for Group Decision Making with Complementary Linguistic Preference Relations

Francisco Chiclana[1], Jian Wu[2], and Enrique Herrera-Viedma[3]

[1] Centre for Computational Intelligence, Faculty of Technology,
De Montfort University, Leicester, UK
chiclana@dmu.ac.uk
[2] School of Economics and Management,
Zhejiang Normal University, Jinhua, Zhejiang, China
jyajian@163.com
[3] Department of Computer Science and Artificial Intelligence,
University of Granada, Granada, Spain
viedma@decsai.ugr.es

Abstract. A visual consensus feedback mechanism for group decision making (GDM) problems with complementary linguistic preference relations is presented. Linguistic preferences are modelled using triangular fuzzy membership functions, and the concepts of similarity degree (SD) between two experts as well as the proximity degree (PD) between an expert and the rest of experts in the group are defined and used to measure the consensus level (CL). A feedback mechanism is proposed to identify experts, alternatives and corresponding preference values that contribute less to consensus. The novelty of this feedback mechanism is that it provides experts with visual representations of their consensus status to easily 'see' their consensus position within the group as well as to identify the alternatives and preference values that should be reconsidered for changing in the subsequent consensus round. The feedback mechanism also includes individualised recommendations to those identified experts on changing their identified preference values and visual graphical simulation of future consensus status if the recommended values were to be implemented.

Keywords: Group decisions making, Consensus, linguistic preferences, Visual feedback mechanism.

1   Introduction

Subjectivity, imprecision and vagueness in the articulation of opinions pervade real world decision applications, and individuals usually find difficult to evaluate their preference using exact numbers. In these cases, individuals might feel more comfortable using words by means of linguistic labels or terms to articulate their preferences [1, 2].

V. Torra et al. (Eds.): MDAI 2014, LNAI 8825, pp. 72–83, 2014.

Let $\mathcal{L} = \{l_0, \ldots, l_s\}$ be a set of linguistic labels ($s \geq 2$), with semantics underlying a ranking relation that can be precisely captured with a linear order: $l_0 < l_1 < \cdots < l_s$. Assuming that the number of labels is odd and the central label ($l_{s/2}$) stands for the indifference state when comparing two alternatives, the remaining labels are usually located symmetrically around that central assessment, which guarantees that a kind of complementary or reciprocity property holds as in the case of numerical preferences [3]. Thus, if the linguistic assessment associated to the pair of alternatives (x_i, x_j) is $r_{ij} = l_h \in \mathcal{L}$, then the linguistic assessment corresponding to the pair of alternatives (x_j, x_i) would be $r_{ji} = l_{s-h}$. Therefore, the operator defined as $N(l_h) = l_g$ with $(g + h) = s$ is a negator operator because $N(N(l_h)) = N(l_g) = l_h$ [4].

The main two representation formats of linguistic information are [2]: the cardinal, which is based on the use of fuzzy sets characterised with membership functions and that are mathematically processed using Zadeh's *extension principle* [1]; and the ordinal, which is based on the use of the *symbolic computational model* [2]. Although the latter representation is able to capture some of the linguistic information to model, it is in fact processed using mathematical tools that are not appropriate for ordinal information but for information provided using a difference or ratio scale. Evidence of this is that the ordinal linguistic model is mathematically equivalent to the cardinal approach with fuzzy sets represented using a representative element of the corresponding membership functions, an example of which is the centroid [4]. Therefore, the uncertainty nature of the information is lost in the ordinal linguistic computational model. Furthermore, the linguistic cardinal approach is richer than the ordinal linguistic approach, not only because it has the latter one as a particular case but also because it provides a more flexible tool for GDM with LPRs because different types of fuzzy sets are possible to be used depending on the type and intensity of the imprecision and vagueness contained in the linguistic information to model.

In particular, convex normal fuzzy subsets of the real line, also known as fuzzy numbers, are commonly used to represent linguistic terms [5–7]. By doing this, each linguistic assessment is represented using a fuzzy number that is characterised by a membership function, with base variable the unit interval $[0, 1]$, describing its semantic meaning. The membership function maps each value in $[0, 1]$ to a degree of performance representing its compatibility with the linguistic assessment [1]. This paper focuses on the use of triangular fuzzy numbers to model linguistic information, which leads to the so-called triangular fuzzy complementary preference relations (TFCPRs) [8] because they extend both numeric preference relations and interval-valued preference relations.

GDM problems generally involve situations of conflict among its experts, and therefore it is preferable that the set of experts reach consensus before applying a selection process to derive the decision solution. There are two basic consensus models in GDM: the static consensus models [9] and the interactive consensus models [10]. The former does not implement any type of feedback mechanism to advice experts on how to change their preferences in order to achieve a higher consensus level while the latter does. Existing interactive consensus models methodology relies on the imposition to decision makers (DM) of changes in their opinion when consensus is below a threshold value. However, in practice, it is up to the decision maker to implement or not the recommendations given to him/her [11]. A more reasonable and suitable policy should rest on this premise and, consequently, it would allow the DM to revisit his/her evaluations using appropriate and meaningful consensus information representation. Therefore, the aim of this paper is to propose a visual consensus feedback mechanism for GDM to provide experts with visual representations of their consensus status to easily 'see' their consensus position within the group as well as to identify the alternatives and preference values that he/she should reconsider for changing in the subsequent consensus round. The feedback mechanism also includes individualised recommendations to those identified experts on changing their identified preference values as well as visual graphical simulation of future consensus status if the recommended values were to be implemented. To achieve this, we first define a TFCPRs similarity degree (SD) to measure, in the unit interval, how close two individual experts are. The proximity of an expert with respect to the whole group of experts is also measured, resulting in individual proximity degree (PD). Consensus level (CL) is defined as a linear combination of SD with PD, and all will be defined at the three different levels of a preference relation [12–14]: the pairs of alternatives, the alternatives and the whole set of alternatives.

The rest of paper is set out as follows: Section 2 focuses on the development of similarity and proximity degrees for TFCPRs. In Section 3, the level of consensus for TFCPRs is proposed, and a visual information feedback mechanism to increase the level of consensus is investigated. Finally, conclusions are drawn in Section 4.

2 Similarity and Proximity Degrees of Triangular Fuzzy Complementary Preference Relations

A fuzzy subset $\widetilde{A}$ of $\mathbb{R}$ is called a triangular fuzzy number (TFN) when its membership function $\mu_{\widetilde{A}}(x)\colon \mathbb{R} \to [0,1]$ is [15]:

$$\mu_{\widetilde{A}}(x) = \begin{cases} 0, & x < a \\ \frac{x-a}{b-a}, & a < x \leq b \\ \frac{c-x}{c-b}, & b \leq x < c \\ 0, & x > c \end{cases}$$

A TFN is shortly represented as $\widetilde{A} = (a, b, c)$, with a and c known as the lower and upper bounds, respectively, while b is known as its modal value.

A preference relation on a set of alternatives $X = \{x_1, x_2, \ldots, x_n\}$ with elements being TFNs, $\widetilde{P} = (\widetilde{p}_{ij})_{n \times n}$ and $\widetilde{p}_{ij} = (a_{ij}, b_{ij}, c_{ij})$, is called a triangular fuzzy complementary preference relation (TFCPR) if the following property holds [15]:

$$a_{ij} + c_{ji} = b_{ij} + b_{ji} = c_{ij} + a_{ji} = 1, \ \forall i, j = 1, 2, \ldots n, \tag{1}$$

2.1 Similarity Degrees

Given two TFNs, $\widetilde{A}_1 = (a_1, b_1, c_1)$ and $\widetilde{A}_2 = (a_2, b_2, c_2)$, their similarity $d(\widetilde{A}_1, \widetilde{A}_2)$ can be defined as follows [16, 12]:

$$s(\widetilde{A}_1, \widetilde{A}_2) = 1 - \frac{|a_1 - a_2| + |b_1 - b_2| + |c_1 - c_2|}{3}.$$

In the following, the similarity degree between two experts using TFCPRs is introduced:

Definition 1. *Let* $P^h = (p_{ik}^h)$ *and* $P^l = (p_{ik}^l)$ *be two TFCPRs on a set of alternatives* X *provided by two experts* e_h *and* e_l, *respectively. Then, the similarity degree between experts* e_h *and* e_l *on the pair of alternatives* (x_i, x_k), SD_{ik}^{hl}, *is :*

$$SD_{ik}^{hl} = SD(p_{ik}^h, p_{ik}^l) = 1 - d(p_{ik}^h, p_{ik}^l). \tag{2}$$

Notice that $SD_{ik}^{hl} = 1$ implies $|a_{ik}^h - a_{ik}^l| = |b_{ik}^h - b_{ik}^l| = |c_{ik}^h - c_{ik}^l| = 0$ and therefore $p_{ik}^h = p_{ik}^l$. Therefore, we have the following interpretation: the higher the value of SD_{ik}^{hl}, the more similar p_{ik}^h and p_{ik}^l are.

Definition 2. *The similarity degree between experts* e_h *and* e_l *on the alternative* x_i *is:* $SD_i^{hl} = SD(p_i^h, p_i^l) = \dfrac{\sum_{k=1}^{n} SD(p_{ik}^h, p_{ik}^l)}{n}.$

As above, when $SD_i^{hl} = 1$ experts e_h and e_l provide the same linguistic valuations for pairs of alternatives involving x_i. Thus, the higher the value of SD_i^{hl}, the more similar the experts' preferences are on the alternative x_i.

Definition 3. *The similarity degree between experts* e_h *and* e_l *on the whole set of alternatives* X *is:* $SD^{hl} = SD(P^h, P^l) = \dfrac{\sum_{i=1}^{n} \sum_{k=1}^{n} SD(p_{ik}^h, p_{ik}^l)}{n^2}.$

Clearly, $SD^{hl} = 1$ means that experts e_h and e_l provide identical TFCPRs, and we can interpret this similarity degree as follows: the higher the value SD^{hl}, the closer experts e_h and e_l are in their preferences on the set of alternatives.

The similarity degrees of an expert with the rest of the group of experts at the three different levels of a relation are defined as:

Level 1. *Similarity degree on the pair of alternatives* (x_i, x_k) *of expert* e_h *to the rest of experts in the group is* $SPA_{ik}^h = \dfrac{\sum_{l=1,\ l \neq h}^{m} SD_{ik}^{hl}}{m-1}.$

Level 2. *Similarity degree on the alternative* x_i *of expert* e_h *to the rest of experts in the group is* $SA_i^h = \dfrac{\sum_{k=1}^{n} SPA_{ik}^h}{n}.$

Level 3. *Similarity degree on the preference relation of expert* e_h *to the rest of experts in the group is* $SD^h = \dfrac{\sum_{i=1}^{n} SA_i^h}{n}.$

Finally, each expert in the GDM problem can be associated a relative (normalised) importance degree based on the similarity degrees at level 3 computed above, which we obviously refer to as the *relative similarity degree of an expert*: $RSD^h = \frac{SD^h}{\sum_{l=1}^{m} SD^l}$. These relative importance degrees could be different to particular importance weights the experts in the group are assigned before they provide their linguistic information on the set of alternatives. Our methodology is to implement both importance degrees in the computation of consensus to reflect the actual position of experts in the group as a collective [17, 18]. This will be developed in the following subsection. Next we provide a simple GDM example to illustrate the computation of the similarity degrees at the three levels of a relation and the final relative similarity degrees of the experts in the group.

Example 1. Suppose four experts $\{e_1, e_2, e_3, e_4\}$ with associated importance degrees $ID = (0.2, 0.1, 0.4, 0.3)^T$, are asked to provide their preference on a set of four alternatives $\{x_1, x_2, x_3, x_4\}$, being their linguistic preferences modelled via the following TFCPRs:

$$P^1 = \begin{pmatrix} - & (0.3,0.4,0.5) & (0.4,0.5,0.6) & (0.5,0.6,0.7) \\ (0.5,0.6,0.7) & - & (0.4,0.5,0.6) & (0.3,0.4,0.5) \\ (0.4,0.5,0.6) & (0.4,0.5,0.6) & - & (0.5,0.6,0.7) \\ (0.3,0.4,0.5) & (0.5,0.6,0.7) & (0.3,0.4,0.5) & - \end{pmatrix}$$

$$P^2 = \begin{pmatrix} - & (0.4,0.5,0.6) & (0.2,0.3,0.4) & (0.3,0.4,0.5) \\ (0.4,0.5,0.6) & - & (0.5,0.6,0.7) & (0.5,0.6,0.7) \\ (0.6,0.7,0.8) & (0.3,0.4,0.5) & - & (0.1,0.2,0.3) \\ (0.5,0.6,0.7) & (0.3,0.4,0.5) & (0.7,0.8,0.9) & - \end{pmatrix}$$

$$P^3 = \begin{pmatrix} - & (0.5,0.6,0.7) & (0.4,0.5,0.6) & (0.6,0.7,0.8) \\ (0.3,0.4,0.5) & - & (0.5,0.6,0.7) & (0.2,0.3,0.4) \\ (0.4,0.5,0.6) & (0.3,0.4,0.5) & - & (0.4,0.5,0.6) \\ (0.2,0.3,0.4) & (0.6,0.7,0.8) & (0.4,0.5,0.6) & - \end{pmatrix}$$

$$P^4 = \begin{pmatrix} - & (0.4,0.5,0.6) & (0.5,0.6,0.7) & (0.5,0.6,0.7) \\ (0.4,0.5,0.6) & - & (0.6,0.7,0.8) & (0.2,0.3,0.4) \\ (0.3,0.4,0.5) & (0.2,0.3,0.4) & - & (0.3,0.4,0.5) \\ (0.3,0.4,0.5) & (0.6,0.7,0.8) & (0.5,0.6,0.7) & - \end{pmatrix}$$

I) The similarity degree on pairs of alternatives for each expert are:

$$SPA^1 = \begin{pmatrix} 1.000 & 0.867 & 0.900 & 0.900 \\ 0.867 & 1.000 & 0.867 & 0.867 \\ 0.900 & 0.867 & 1.000 & 0.767 \\ 0.900 & 0.867 & 0.767 & 1.000 \end{pmatrix}; \quad SPA^2 = \begin{pmatrix} 1.000 & 0.933 & 0.767 & 0.767 \\ 0.933 & 1.000 & 0.933 & 0.733 \\ 0.767 & 0.933 & 1.000 & 0.700 \\ 0.767 & 0.733 & 0.700 & 1.000 \end{pmatrix}$$

$$SPA^3 = \begin{pmatrix} 1.000 & 0.867 & 0.900 & 0.833 \\ 0.867 & 1.000 & 0.933 & 0.867 \\ 0.900 & 0.933 & 1.000 & 0.833 \\ 0.833 & 0.867 & 0.833 & 1.000 \end{pmatrix}; \quad SPA^4 = \begin{pmatrix} 1.000 & 0.933 & 0.833 & 0.900 \\ 0.933 & 1.000 & 0.867 & 0.867 \\ 0.833 & 0.867 & 1.000 & 0.833 \\ 0.900 & 0.867 & 0.833 & 1.000 \end{pmatrix}$$

II) The similarity degrees on alternatives for each expert are:

$$SA^1 = (0.917, 0.900, 0.883, 0.883) ; \ SA^2 = (0.867, 0.900, 0.850, 0.800)$$

$$SA^3 = (0.900, 0.917, 0.917, 0.883) ; \ SA^4 = (0.917, 0.917, 0.883, 0.900)$$

III) The similarity degrees on the set of alternatives for each expert are:

$$SD^1 = 0.896 ; \ SD^2 = 0.854 ; \ SD^3 = 0.904 ; \ SD^4 = 0.904.$$

IV) The relative group similarity degrees for each expert are:

$$RSD^1 = 0.252 ; \ RSD^2 = 0.240 ; \ RSD^3 = 0.254 ; \ RSD^4 = 0.254.$$

2.2 Proximity Degrees

The proximity degrees measure the similarity between individual experts' opinions and the collective opinion for the group of experts. The aggregation of individual opinions will be weighted using a weight vector whose elements are a linear combination of the importance degree of individuals before the decision making process and the relative similarity degrees computed based on the information they provided as per the previous subsection. This is elaborated next:

(1) **Experts weighting vector:** $W = \eta \cdot ID + (1 - \eta) \cdot RSD$. If $\eta > 0.5$, then the group/moderator values higher the a priori importance degrees of the experts than their a posteriori relative similarity degrees. Obviously, for homogeneous GDM problems the value $\eta = 0$ applies.

(2) **The collective TFCPR**, $P = (p_{ik})_{n \times n}$, is computed as follows:

$$p_{ik} = w^1 \otimes p_{ik}^1 \oplus w^2 \otimes p_{ik}^2 \oplus \cdots \oplus w^m \otimes p_{ik}^m \tag{3}$$

Example 2 (Example 1 Continuation). Assuming a value of $\eta = 0.5$ we have the following weighting vector

$$W = 0.5 * ID + 0.5 * RSD = (0.22, 0.17, 0.33, 0.28)^T$$

and the collective TFCPR is

$$P = \begin{pmatrix} - & (0.41, 0.51, 0.61) & (0.39, 0.49, 0.59) & (0.50, 0.60, 0.70) \\ (0.39, 0.49, 0.59) & - & (0.51, 0.61, 0.71) & (0.27, 0.37, 0.47) \\ (0.41, 0.51, 0.61) & (0.29, 0.39, 0.49) & - & (0.34, 0.44, 0.54) \\ (0.30, 0.40, 0.50) & (0.53, 0.63, 0.73) & (0.46, 0.56, 0.66) & - \end{pmatrix}$$

Once the collective TFCPR is obtained, we compute the proximity degrees for each expert at the three different levels of a relation:

Level 1. *Proximity degree on pair of alternatives* (x_i, x_k) *of expert* e_h *to the group is* $PPA_{ik}^h = SD(p_{ik}^h, p_{ik})$.

Level 2. *Proximity degree on alternatives* x_i *of expert* e_h *to the group is* $PA_i^h = \frac{\sum_{k=1}^n PPA_{ik}^h}{n}$.

Level 3. *Proximity degree on the preference relation of expert* e_h *to the group is* $PD^h = \frac{\sum_{i=1}^n PA_i^h}{n}$.

Example 3 (Example 1 Continuation). Proximity degrees computation.

I) The proximity degree on pairs of alternatives for each expert are:

$$PPA^1 = \begin{pmatrix} 1.000 \ 0.889 \ 0.994 \ 0.999 \\ 0.889 \ 1.000 \ 0.894 \ 0.973 \\ 0.994 \ 0.894 \ 1.000 \ 0.843 \\ 0.999 \ 0.973 \ 0.843 \ 1.000 \end{pmatrix} ; \ PPA^2 = \begin{pmatrix} 1.000 \ 0.989 \ 0.806 \ 0.801 \\ 0.989 \ 1.000 \ 0.994 \ 0.773 \\ 0.806 \ 0.994 \ 1.000 \ 0.757 \\ 0.801 \ 0.773 \ 0.757 \ 1.000 \end{pmatrix}$$

$$PPA^3 = \begin{pmatrix} 1.000 \ 0.911 \ 0.994 \ 0.899 \\ 0.911 \ 1.000 \ 0.994 \ 0.927 \\ 0.994 \ 0.994 \ 1.000 \ 0.943 \\ 0.899 \ 0.927 \ 0.943 \ 1.000 \end{pmatrix} ; \ PPA^4 = \begin{pmatrix} 1.000 \ 0.989 \ 0.894 \ 0.999 \\ 0.989 \ 1.000 \ 0.906 \ 0.927 \\ 0.894 \ 0.906 \ 1.000 \ 0.957 \\ 0.999 \ 0.927 \ 0.957 \ 1.000 \end{pmatrix}$$

II) The proximity degrees on alternatives for each expert are:

$$PA^1 = (0.971, 0.939, 0.933, 0.954) ; \ PA^2 = (0.899, 0.939, 0.889, 0.833)$$

$$PA^3 = (0.951, 0.958, 0.983, 0.942) ; \ PA^4 = (0.971, 0.956, 0.940, 0.971)$$

III) The proximity degrees on the relation for each expert are:

$$PD^1 = 0.949 ; \ PD^2 = 0.890 ; \ PD^3 = 0.958 ; \ PD^4 = 0.959.$$

3   Consensus Model with Visual Information Feedback Mechanism for GDM with TFCPRs

Both similarity degree (SD) and proximity degree (PD) convey the concept of closeness of opinions between experts in a group: the first one between pairs of individual experts and the second one between an individual expert and the rest of experts in the group. Thus, both degrees could/should be used in measuring the level of consensus within a group of experts regarding the set of feasible alternatives in GDM. The simplest of the combinations is the linear one, and it is here used to propose the following definitions of the consensus level (CL) associated to each expert of the group at the three different levels of a relation:

Level 1. *Consensus level on the pair of alternatives (CLPA)* (x_i, x_k) *of expert* e_h *is* $CLPA_{ik}^h = \psi \cdot SPA_{ik}^h + (1 - \psi) \cdot PPA_{ik}^h$.

Level 2. *Consensus level on the alternatives(CLA)* x_i *of expert* e_h *is* $CLA_i^h = \psi \cdot SA_i^h + (1 - \psi) \cdot PA_i^h$.

Level 3. *Consensus level on the relation (CL) of expert* e_h *is* $CL^h = \psi \cdot SD^h + (1 - \psi) \cdot PD^h$.

The parameter $\psi \in [0,1]$ controls the weight of both similarity and proximity criteria. Unless there are specific reasons to prefer one index to the other one, the value to assume for the weighting parameter ψ should be 0.5, as it is assumed in the example below.

Example 4 (Example 1 Continuation). Consensus levels computation. Setting ψ at 0.5, the following consensus levels on the relation are obtained:

$$CL^1 = 0.922, \quad CL^2 = 0.872, \quad CL^3 = 0.932, \quad CL^4 = 0.932$$

The only expert with a consensus level below the threshold value is e_2 and therefore he/she will receive feedback advice on how to change his/her preferences to achieve a higher consensus level.

In practice, it is rare to achieve full and unanimous agreement of all the experts regarding all the feasible alternatives. As a consequence, the consensus threshold value (γ) to achieve is usually set to a value lower than 1. At the same time, the decision output should be acceptable for at least half of the experts, which means that the parameter γ should be set to a value no lower than 0.5. If the consensus level is not acceptable, that is, if it is lower than the specified threshold value, the experts are normally invited to discuss their opinions further in an effort to make them closer. To help experts in their discussion, in the following section a detailed description of a visual feedback methodology is provided.

3.1   Visual Information Feedback Mechanism

The visual information feedback mechanism consists of three stages: firstly, the *identification of the triangular fuzzy preference values* that should be subject to modification; secondly, the *generation of advice* on the direction–value of the required change; and, thirdly, the *automatic feedback process simulation* to show what would happen if experts are to accept the recommended preference values. These three stages are described in detail below:

(1) *Identification of the Triangular Fuzzy Preference Values:* The set of triangular fuzzy preference values that contribute less to reach an acceptable consensus level is identified and presented to the experts using visual graphs as illustrated in Fig. 1. Once consensus levels are computed, at the relation level, all experts will receive a visual representation of their consensus status in relation to the threshold value, which can be used to easily identify the experts furthest form the group. Following with Ex. 4, and using a threshold value $\gamma = 0.9$, Fig. 1(a) clearly identifies expert e_2 as the only expert contributing less to group consensus. If necessary, individual visual representations of consensus levels of alternatives and pair of alternatives are also provided to each expert to help them identify those alternatives and their associated preference values at the level of pairs of alternatives that contribute less to consensus and, consequently, potential to be reconsidered for changing in the next round of consensus. It is obvious from Ex. 4 that this is

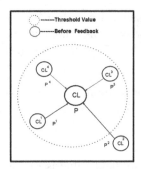

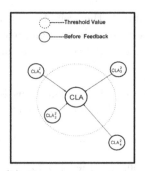

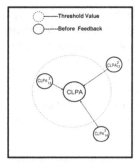

(a) Consensus levels on the relation: CL^h

(b) Consensus levels on the alternatives for e_2: CLA_i^2

(c) Consensus levels on the pairs of alternatives for A_1 and e_2: $CLPA_{1j}^2$

Fig. 1. Visual representation of consensus levels in relation to the consensus threshold value

necessary to be done for expert e_2, whom would receive visual representation at these levels as illustrated in Fig. 1(b) and Fig. 1(c), respectively.

Mathematically, these steps are modelled as follows:

Step 1. The set of experts with consensus levels below the threshold value γ is identified: $ECH = \{h \,|CL^h < \gamma\}$.

Step 2. For experts identified in step 1, those alternatives with a consensus level below γ are identified: $ACH = \{(h, i) \mid h \in ECH \land CLA_i^h < \gamma\}$.

Step 3. Finally, the triangular fuzzy preference values for the experts and alternatives identified in steps 1 and 2 that need to be changed are identified: $PACH = \{(h, i, k) \mid (h, i) \in ACH \land CLPA_{ik}^h < \gamma\}$.

Example 5 (Example 1 Continuation). The sets of 3-tuple identified as contributing less to consensus are:

$$PACH = \{(2,1,3), (2,1,4), (2,2,4), (2,3,1), (2,3,4), (2,4,1), (2,4,2), (2,4,3)\}$$

(2) *Generation of Advice:* The feedback mechanism also generates personalised recommendations rules, which will not only tell the experts which preference values they should change, but also provide them with the consensus advice to revisit their evaluation in the light of this extra information. For all $(h, i, k) \in PACH$, the following rule is feed backed to the corresponding experts:

"To increase your consensus level (CL), your preference value p_{ik}^h should be closer to $\overline{\overline{p}}_{ik}^h = \psi \cdot \overline{p}_{ik}^h + (1 - \psi) \cdot p_{ik}$," where $\overline{p}_{ik}^h = (\sum_{l=1,\, l \neq h}^m p_{ik}^l)/(m-1)$ and p_{ik} the collective preference value. The reciprocity property that the TFCPRs verify implies that when the pair of alternatives (i, k) is identified for change, the pair (k, i) has to be changed accordingly as well.

Example 6 ((Example 1 continuation)). The recommendations for expert e_2 are:

  - Preference value p_{13}^2 should be closer to (0.4,0.5,0.6).
  - Preference value p_{31}^2 should be closer to (0.4,0.5,0.6).
  - Preference value p_{14}^2 should be closer to (0.5,0.6,0.7).
  - Preference value p_{41}^2 should be closer to (0.3,0.4,0.5).
  - Preference value p_{24}^2 should be closer to (0.4,0.5,0.6).
  - Preference value p_{42}^2 should be closer to (0.4,0.5,0.6).
  - Preference value p_{34}^2 should be closer to (0.4,0.5,0.6).
  - Preference value p_{43}^2 should be closer to (0.4,0.5,0.6).

(3) *Automatic Feedback Process Simulation:* A what-if scenario analysis could be run to generate a visual graphical simulation of future consensus status if the recommended values were to be implemented, as shown in Fig. 2(a), Fig. 2(b) and Fig. 2(c). This will provide the decision makers with a clear picture of their actual position within the group, which they can then use to decide upon their actual position or subsequent action. If the advice is implemented, then the consensus level increases as example 7 shows. Not implementing these advices can lead to the consensus level to remain fixed or to increase at a very low rate, which would make the group consensus threshold value difficult to achieve. To avoid these situation, a maximum number of iterations maxIter can be incorporated in the visual information feedback mechanism following a similar approach of consensus models proposed in [19].

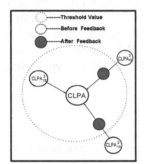 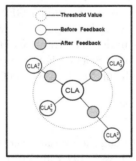 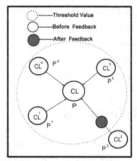

(a) $CLPA_{1j}^2$ before and after e_2 implements recommended values

(b) CLA_i^2 before and after e_2 implements recommended values

(c) CL^h before and after e_2 implements recommended values

Fig. 2. Simulation of consensus before and after recommended values are implemented by expert e_2

Example 7 (Finishing Example 1). After expert e_2 revisits his/her evaluation and implements the recommended TFNs, a new round of the consensus process is carried out, leading to the following new consensus levels:

$$CL^1 = 0.956, \ CL^2 = 0.976, \ CL^3 = 0.961, \ CL^4 = 0.957.$$

Because all experts are over the minimum consensus threshold value $\gamma = 0.9$, the consensual collective TFCPR is computed from which the final solution of consensus will be selected.

4 Conclusion

In this paper, a novel visual information feedback mechanism for GDM problems with TFCPRs has been presented. To achieve this, the concepts of similarity degree (SD) between two experts as well as the proximity degree (PD) between an expert and the rest of experts in the group are developed for TFCPRs. These degrees are used to compute both the aggregation weighting vector as well as the consensus level of the group of experts. The visual information feedback mechanism is investigated to identify experts, alternatives and corresponding preference values that contribute less to consensus. Recommendations to help experts the direction of the change required to increase their consensus are produced and an automatic visual feedback process simulation to show the experts what would happen if they were to follow recommendations by pictures is developed.

Acknowledgements. The authors would like to acknowledge FEDER financial support from the Project FUZZYLING-II Project TIN2010-17876; the financial support from the Andalusian Excellence Projects TIC-05299 and TIC-5991; and the University of Granada Excellence campus GENIL-BioTIC-UGR Research Visit programme. This work was also supported by National Natural Science Foundation of China (NSFC) under the Grant No.71101131 and No.713311002, and Zhejiang Provincial National Science Foundation for Distinguished Young Scholars of China under the Grant No. LR13G010001.

References

1. Zadeh, L.A.: The concept of a linguistic variable and its application to approximate reasoning-I. Information Sciences 8, 199–249 (1975)
2. Herrera, F., Alonso, S., Chiclana, F., Herrera-Viedma, E.: Computing with words in decision making: foundations, trends and prospects. Fuzzy Optimization and Decision Making 8(4), 337–364 (2009)
3. Chiclana, F., Herrera-Viedma, E., Alonso, S., Herrera, F.: Cardinal consistency of reciprocal preference relations: a characterization of multiplicative transitivity. IEEE Transactions on Fuzzy Systems 17(1), 14–23 (2009)
4. Pérez-Asurmendi, P., Chiclana, F.: Linguistic majorities with difference in support. Applied Soft Computing 18, 196–208 (2014)
5. Zhou, S.-M., Chiclana, F., John, R., Garibaldi, J.M.: Type-1 OWA operators for aggregating uncertain information with uncertain weights induced by type-2 linguistic quantifiers. Fuzzy Sets and Systems 159(24), 3281–3296 (2008)

6. Zhou, S.-M., Chiclana, F., John, R., Garibaldi, J.M.: Alpha-level aggregation: a practical approach to type-1 OWA operation for aggregating uncertain information with applications to breast cancer treatments. IEEE Transactions on Knowledge and Data Engineering 23(10), 1455–1468 (2011)
7. Chiclana, F., Zhou, S.-M.: Type-reduction of general type-2 fuzzy sets: the type-1 OWA approach. International Journal of Intelligent Systems 28(5), 505–522 (2013)
8. Xia, M.M., Xu, Z.S.: Methods for fuzzy complementary preference relations based on multiplicative consistency. Computers and Industrial Engineering 61, 930–935 (2011)
9. Xu, Z.S., Cai, X.Q.: Group consensus algorithms based on preference relations. Information Science 181, 150–162 (2011)
10. Alonso, S., Herrera-Viedma, E., Chiclana, F., Herrera, F.: A web based consensus support system for group decision making problems and incomplete preferences. Information Sciences 180, 4477–4495 (2010)
11. Eklund, P., Rusinowska, A., de Swart, H., Dong, Y.C., Xu, Y.F., Li, H., Feng, B.: A consensus model of political decision-making. Annals of Operations Research 158, 5–20 (2008)
12. Chiclana, F., Tapia-Garcia, J.M., del Moral, M.J., Herrera-Viedma, E.: A statistical comparative study of different similarity measures of consensus in group decision making. Information Sciences 221, 110–123 (2013)
13. Alonso, S., Pérez, I.J., Cabrerizo, F.J., Herrera-Viedma, E.: A linguistic consensus model for Web 2.0 communities. Applied Soft Computing 13(1), 149–157 (2013)
14. Wu, J., Chiclana, F.: A social network analysis trust-consensus based approach to group decision-making problems with interval-valued fuzzy reciprocal preference relations. Knowledge-Based Systems 59, 97–107 (2014)
15. Laarhoven, P.J.M., Pedrycz, W.: A fuzzy extension of Saaty's priority theory. Fuzzy Sets and Systems 11, 229–241 (1983)
16. Zwick, R., Carlstein, E., Budescu, D.V.: Measures of similarity among fuzzy concepts: A comparative analysis. International Journal of Approximate Reasoning 1, 221–242 (1987)
17. Chiclana, F., Herrera-Viedma, E., Herrera, F., Alonso, S.: Some induced ordered weighted averaging operators and their use for solving group decision-making problems based on fuzzy preference relations. European Journal of Operational Research 182, 383–399 (2007)
18. Mata, F., Perez, L.G., Zhou, S.-M., Chiclana, F.: Type-1 OWA methodology to consensus reaching processes in multi-granular linguistic contexts. Knowledge-Based Systems 58, 11–22 (2014)
19. Herrera-Viedma, E., Herrera, F., Chiclana, F.: A consensus model for multiperson decision making with different preference structures. IEEE Transactions on Systems, Man, and Cybernetics - Part A: Systems and Humans 32, 394–402 (2002)

Towards Armstrong-Style Inference System for Attribute Implications with Temporal Semantics*

Jan Triska and Vilem Vychodil

Dept. Computer Science, Palacky University, Olomouc
17. listopadu 12, CZ–77146 Olomouc, Czech Republic
jan.triska@upol.cz, vychodil@acm.org

Abstract. We show a complete axiomatization of a logic of attribute implications describing dependencies between attributes of objects which are observed in consecutive points in time. The attribute implications we consider are if-then formulas expressing presence of attributes of objects relatively in time. The semantics of the attribute implications is defined based on presence/absence of attributes of objects in consecutive points of time. The presented results extend the classic results on Armstrong-style completeness of the logic of attribute implications by using the time points as additional component. The ordinary results can be seen as special case of our results when only a single time point is considered.

Keywords: attribute implication, axiomatization, formal context, temporal semantics.

1 Introduction and Related Work

In this paper, we outline a complete axiomatic system for reasoning with particular if-then formulas describing dependencies in object/attribute data which are subject to change in consecutive points in time. The rules we investigate are related to attribute implications studied in formal concept analysis [11,22]. The principal difference between the attribute implications as they appear in formal concept analysis and the rules we consider in this paper is an additional component allowing to express the presence or absence of object attributes relatively in time. This may be viewed as though the objects can change their attributes in time and we are interested in describing dependencies that are preserved in all the time points. This paper is our initial study of this type of dependencies in which we outline the dependencies, their semantics, and show an Armstrong-style axiomatization of their semantic entailment.

In formal concept analysis (FCA), attribute implications play an important role since sets of attribute implications can be seen as alternative descriptions of concept lattices. Recall that the basic input data in formal concept analysis

---

* Supported by grant no. P202/14–11585S of the Czech Science Foundation.

V. Torra et al. (Eds.): MDAI 2014, LNAI 8825, pp. 84–95, 2014.

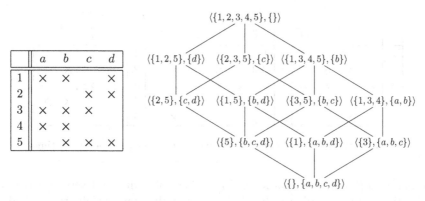

	a	b	c	d
1	×	×		×
2			×	×
3	×	×	×	
4	×	×		
5		×	×	×

Fig. 1. Formal context with objects $1, \ldots, 5$, attributes $a, \ldots, d$, and its concept lattice

is a binary relation $I \subseteq X \times Y$ between a set X of objects and a set Y of attributes (features) called a formal context. Formal contexts are often viewed as two dimensional tables with rows corresponding to objects, columns corresponding to attributes, and with crosses and blanks in the table, indicating whether the objects have/do not have the attributes. The primary output of FCA is a set of particular biclusters in the input data called formal concepts in I. Formal concepts [22] can be viewed as pairs $\langle A, B \rangle$ consisting of a subset $A \subseteq X$ of objects (called an extent) and a subset $B \subseteq Y$ of attributes (called an intent) such that B is the set of all attributes common to all the objects from A and, conversely, A is the set of all objects having all the attributes from B. In a table corresponding to I, formal concepts can be identified with maximal rectangles consisting of crosses. By the Basic Theorem of FCA [11], all formal concepts in I, when ordered by the set inclusion of their extents, form a structure called a concept lattice (which is a complete lattice). An example of a formal context with objects $X = \{1, 2, 3, 4, 5\}$, attributes $Y = \{a, b, c, d\}$, and the corresponding concept lattice can be found in Fig. 1.

An attribute implication over Y (the set of attributes of a given formal context) is an expression $A \Rightarrow B$ such that $A, B \subseteq Y$. The intended meaning of $A \Rightarrow B$ is to express a dependency "if an object has all the attributes in A, then it has all the attributes in B". Thus, in the narrow sense, $A \Rightarrow B$ can be seen as a (propositional) formula in the form of an implication between conjunctions of attributes in Y (which are considered as propositional variables). For $A, B, M \subseteq Y$, we call $A \Rightarrow B$ satisfied by M whenever $A \subseteq M$ implies $B \subseteq M$ and denote the fact by $M \models A \Rightarrow B$. Clearly, if M_x is considered to be the set of attributes of the object $x \in X$ in I, i.e., $M_x = \{y \in Y \mid \langle x, y \rangle \in I\}$, then $M_x \models A \Rightarrow B$ indeed formalizes the intended meaning of $A \Rightarrow B$ being satisfied by the attributes of the object x: "If x has all the attributes in A, then x has all the attributes in B."

One particular interest of FCA is to describe, in a concise way, the set of all attribute implications which are satisfied by all (or nearly all) objects in a given formal context (with X and Y being finite for computational reasons).

$t = 1$

	a	b	c	d
1				×
2	×	×	×	
3	×		×	×
4			×	×
5		×		

$t = 2$

	a	b	c	d
1		×	×	×
2		×		×
3	×	×	×	×
4	×		×	
5				×

$t = 3$

	a	b	c	d
1			×	×
2	×		×	
3	×	×		×
4		×	×	×
5				

..., ... , ... , ...

Fig. 2. Series of formal contexts observed in consecutive points in time

The major approaches [11,13] focus on finding non-redundant sets of attribute implications which entail exactly all attribute implications which are satisfied in a formal context. Therefore, the approaches rely on the notion of an entailment of attribute implications. As in other logical systems, we may consider two basic types of entailment—the semantic one based on models and the syntactic one based on provability. A set $M \subseteq Y$ is called a model of a set Σ of attribute implications if each $A \Rightarrow B \in \Sigma$ is satisfied by M. Furthermore, $A \Rightarrow B$ is semantically entailed by Σ, written $\Sigma \models A \Rightarrow B$, if $A \Rightarrow B$ is satisfied by all models of Σ. The semantic entailment of attribute implications has complete axiomatizations, i.e., for a suitably defined notion of provability, we have $\Sigma \models A \Rightarrow B$ iff $A \Rightarrow B$ is provable by Σ, denoted $\Sigma \vdash A \Rightarrow B$. The best known axiomatic system is based on the inference system of Armstrong [2] but there are other systems which recently received interest [4].

For a given formal context $I \subseteq X \times Y$, one may be interested in finding a set Σ of attribute implications such that Σ proves $A \Rightarrow B$ iff $A \Rightarrow B$ is satisfied by all objects in I. Such a Σ is called complete in I. In addition, Σ is called non-redundant if no proper subset of Σ does the job. A particular example of non-redundant complete sets are the so-called Guigues-Duquenne bases [13] which are in addition minimal in the number of formulas they contain. In addition to the fact that a complete set conveys information about the attribute implications which are satisfied in a formal context, it can be seen as an alternative description of the concept lattice since its models are exactly intents of all formal concepts of the formal context. In case of the example from Figure 1, a minimal complete set consists of a single implication $\{a\} \Rightarrow \{b\}$.

In this paper, we extend the considerations by assuming that the input data consists of a series of formal contexts which result by observing attributes of objects in consecutive points in time, see Figure 2 for an illustration. In this setting, it may be desirable to consider attribute implications which describe if-then dependencies between attributes relatively in time. We argue that such rules are often used in human reasoning, e.g., consider a rule "If x was unavailable yesterday and x is on sale today, then x will be demanded tomorrow." In this example, *yesterday*, *today*, and *tomorrow* refer to points in time relatively to the current point considered, x is an object, and "being on sale", "being unavailable", and "being demanded" are attributes from Y. By generalizing this example, we may consider rules of the form

$$\left\{y_1^{i_1}, \ldots, y_m^{i_m}\right\} \Rightarrow \left\{z_1^{j_1}, \ldots, z_n^{j_n}\right\}, \tag{1}$$

where $y_1, \ldots, y_m, z_1, \ldots, z_n$ are attributes from Y which have the same role as in the case of the classic attribute implications and $i_1, \ldots, i_m, j_1, \ldots, j_n$ are integers expressing points in time relatively to the current point. That is, 0 can be considered as the present point (today), -1 is the predecessor of 0 (yesterday), 1 is the successor of 0 (tomorrow), etc. Given a series of contexts, one may be interested in rules like (1) which are satisfied in all points of time or in all points in a predefined period of time. Such rules, when discovered from data, may be used for prediction based on the hypothesis that the rules may still hold in a "near future," i.e., beyond the time point of the last context observed. In case of the example with contexts in Figure 2, the implication $\{a^0\} \Rightarrow \{b^1\}$ saying that if the attribute a is present in the current time point then b is present in the next time point is satisfied by the three consecutive contexts in the figure. On the other hand, neither of them satisfies the ordinary attribute implication $\{a\} \Rightarrow \{b\}$.

In this paper, we propose a formalization of the rules of the form (1), propose the notion of semantic entailment which corresponds to the intuitive meaning we have just described, and focus on its axiomatization. We show that analogously as for the classic attribute implications, there is an Armstrong-like system of inference rules which is complete with respect to the considered semantic entailment. In the rest of this section, we make notes on related works.

Related Works The input data we consider in this paper can be viewed as particular example of temporal relational data [5] or triadic contexts [17] where the conditions correspond to time points. Note that in the triadic FCA, there has been approaches to attribute implications [12] which are, however, conceptually and technically different from the approach in our paper. In the context of association rule mining [1,23], analogous formulas have been proposed and studied as inter-transactional association rules, see [8,15,16,19,21], and applied in various problem domains [7,14]. The approaches focus on algorithms for mining dependencies from data with respect to additional parameters such that the confidence and support. Thus, the formulas we propose to study in this paper may be in the same relationship to the inter-transactional association rules as the attribute implications are related to the ordinary association rules. In a broader sense, our approach generalizes if-then rules which appear in other disciplines where reasoning with conjunctive if-then rules plays a role like the relational databases (and functional dependencies [6,20]) or logic programming (and particular definite programs [18]). Functional dependencies with a type of temporal interpretation which differs from ours are presented in [3].

2 Formulas and Semantic Entailment

In this section, we present a formalization of the formulas, their interpretation, and semantic entailment. Let us assume that Y is a non-empty and finite set of symbols called attributes. Furthermore, we use integers in order to denote time points. Our motivation is that the input data consists of several formal contexts which are observed in consecutive discrete time points and thus denoting the time point by integers is sufficient. We put

$$\mathcal{T}_Y = \{y^i \mid y \in Y \text{ and } i \in \mathbb{Z}\} \tag{2}$$

and interpret each $y^i \in \mathcal{T}_Y$ as "attribute y observed in time i" (technically, $\mathcal{T}_Y$ can be seen as the Cartesian product $Y \times \mathbb{Z}$). Under this notation, we may now formalize rules like (1) as follows:

Definition 1. An *attribute implication over Y annotated by time points* in $\mathbb{Z}$ is a formula of the form $A \Rightarrow B$, where A, B are finite subsets of $\mathcal{T}_Y$.

As we have outlined in the introduction, the purpose of the time points encoded by integers in antecedents and consequents of the considered formulas is to express points in time relatively to a current time point. Hence, the intended meaning of (1) abbreviated by $A \Rightarrow B$ is the following: "For all time points t, if an object has all the attributes from A considering t as the current time point, then it must have all the attributes from B considering t as the current time point". In what follows, we formalize the interpretation of $A \Rightarrow B$ in this sense.

Since we wish to define formulas being true in all time points (we are interested in formulas preserved over time), we need to shift relative times expressed in antecedents and consequents in formulas with respect to a changing time point. For that purpose, for each $M \subseteq \mathcal{T}_Y$ and $i \in \mathbb{Z}$, we may introduce a subset $M + j$ of $\mathcal{T}_Y$ by

$$M + j = \{y^{i+j} \mid y^i \in M\} \tag{3}$$

and call it a *time shift of M by j* (shortly, a j-shift of M). Obviously, we may consider more complex expression like $M + j + i$ meaning $(M + i) + j$ and denote $M + (-i)$ by $M - i$.

Definition 2. A formula $A \Rightarrow B$ *is true in* $M \subseteq \mathcal{T}_Y$ whenever, for each $i \in \mathbb{Z}$,

$$\text{if } A + i \subseteq M, \text{ then } B + i \subseteq M \tag{4}$$

and we denote the fact by $M \models A \Rightarrow B$.

Remark 1. (a) The value of i in the definition may be understood as a sliding time point. Moreover, $A + i$ and $B + i$ represent sets of attributes annotated by *absolute time points* considering i as the current time point. Note that using (3), the condition (4) can be equivalently restated as "$A \subseteq M - i$ implies $B \subseteq M - i$," i.e., instead of shifting the antecedents and consequents of the formula, we may shift the set M.

(b) Observe that $A \Rightarrow B$ is trivially true in M whenever $B \subseteq A$ because in that case (4) trivially holds for any i. By definition, $A \Rightarrow B$ is not true in M, written $M \not\models A \Rightarrow B$ iff there is i such $A + i \subseteq M$ and $B + i \not\subseteq M$. In words, in the time point i, M has all the attributes of A but does not have an attribute in B, i.e., the time point i serves as a counterexample.

(c) The classic attribute implications [11] can be seen as particular cases of the formulas considered here in case the only time point which appears in the formulas and M is 0. Indeed, in such a case, $M \models A \Rightarrow B$ iff $A \subseteq M$ implies $B \subseteq M$, i.e., iff $A \Rightarrow B$ is true in M in the classic sense.

(d) Let us comment on our initial motivation of evaluating the formulas in a series of contexts as in Figure 2. The set M in which we evaluate $A \Rightarrow B$ can be seen as encoding attributes of a single object changing in time, i.e., for contexts $I_i \subseteq X \times Y$ ($i \in \mathbb{Z}$), we may put $M_x = \{y^i \mid \langle x, y \rangle \in I_i\}$. Then, $M_x \models A \Rightarrow B$ is interpreted as the fact that "for all time points t, if the object x has all the attributes from A (in time t), then it has all the attributes from B (in time t)" which agrees with the desired meaning outlined in the introduction. Hence, $A \Rightarrow B$ being true in all the contexts (i.e., in a context changing over time) can be introduced as $M_x \models A \Rightarrow B$ for all $x \in X$.

We consider the following notions of a theory and a model:

Definition 3. Let Σ be a set of formulas (called a *theory*). A subset $M \subseteq \mathcal{T}_Y$ is called a *model of* Σ if $M \models A \Rightarrow B$ for all $A \Rightarrow B \in \Sigma$. The system of all models of Σ is denoted by $\mathrm{Mod}(\Sigma)$, i.e.,

$$\mathrm{Mod}(\Sigma) = \{M \subseteq \mathcal{T}_Y \mid M \models A \Rightarrow B \text{ for all } A \Rightarrow B \in \Sigma\}. \tag{5}$$

In general, $\mathrm{Mod}(\Sigma)$ is infinite and there may be theories that do not have any finite model. For instance, consider a theory containing $\{\} \Rightarrow \{y^0\}$. This is in contrast to the classic attribute implications in FCA where for finite Y, one always has a finite system of finite models. Interestingly, the systems of models in our case are exactly the systems of models which are closed under time shifts.

We call $\mathcal{S} \subseteq 2^{\mathcal{T}_Y}$ *closed under time shifts* if $M + i \in \mathcal{S}$ whenever $M \in \mathcal{S}$. We now have the following characterization of the systems of models as exactly the algebraic closure systems closed under time shifts:

Theorem 1. *Let Σ be theory. Then, $\mathrm{Mod}(\Sigma)$ is an algebraic closure system which is closed under time shifts. If $\mathcal{S} \subseteq 2^{\mathcal{T}_Y}$ is an algebraic closure system which is closed under time shifts then there is Σ such that $\mathcal{S} = \mathrm{Mod}(\Sigma)$.*

Proof (sketch). The fact that $\mathrm{Mod}(\Sigma)$ is an algebraic closure system can be checked as in the ordinary case. Suppose that $M \in \mathrm{Mod}(\Sigma)$ and take $A \Rightarrow B \in \Sigma$ and $j \in \mathbb{Z}$. If $A + i \subseteq M + j$, then $A + (i - j) \subseteq M$ and thus $B + (i - j) \subseteq M$ because $M \in \mathrm{Mod}(\Sigma)$ and $A \Rightarrow B \in \Sigma$. Therefore, $B + i \subseteq M + j$, i.e., $M + j \in \mathrm{Mod}(\Sigma)$. For the second claim, take $\Sigma = \{M \Rightarrow N \mid M \subseteq \mathcal{T}_Y, N \subseteq C_{\mathcal{S}}(M), \text{ and } M, N \text{ are finite}\}$ where $C_{\mathcal{S}}$ is the algebraic closure operator induced by $\mathcal{S}$ and observe that $M = C_{\mathcal{S}}(M)$ iff M is a model of Σ. □

According to the following observation, for each Σ, we may consider the induced closure operator $[\cdots]_{\Sigma}$ of a *semantic closure*, i.e.,

$$[M]_{\Sigma} = \bigcap \{N \in \mathrm{Mod}(\Sigma) \mid M \subseteq N\}. \tag{6}$$

In general, $[M]_\Sigma$ is infinite. We now turn our attention to the semantic entailment and its basic properties.

Definition 4. Let Σ be a theory. Formula $A \Rightarrow B$ *is semantically entailed by* Σ if $M \models A \Rightarrow B$ for each $M \in \mathrm{Mod}(\Sigma)$.

The following lemma justifies the description of time points in attribute implications as *relative* time points. Namely, it states that each $A \Rightarrow B$ semantically entails all formulas resulting by shifting the antecedent and consequent of $A \Rightarrow B$ by a constant factor.

Lemma 1. $\{A \Rightarrow B\} \models \{A + i \Rightarrow B + i\}$.

Proof. Take $M \in \mathrm{Mod}(\{A \Rightarrow B\})$ and let $(A + i) + j \subseteq M$. Then, $A + i \subseteq M - j$ and by Theorem 1, we get $M - j \in \mathrm{Mod}(\{A \Rightarrow B\})$ which yields $B + i \subseteq M - j$ and thus $(B + i) + j \subseteq M$, proving $M \models A + i \Rightarrow B + i$ □

Analogously as for the classic attribute implications, the semantic entailment of $A \Rightarrow B$ by a theory Σ can be checked by means of the least model of Σ generated by A:

Theorem 2. *For any Σ and $A \Rightarrow B$, the following conditions are equivalent:*

   (i) $\Sigma \models A \Rightarrow B$,

   (ii) $[A]_\Sigma \models A \Rightarrow B$,

   (iii) $B \subseteq [A]_\Sigma$.

Proof (sketch). Clearly, (i) implies (ii) since $[A]_\Sigma \in \mathrm{Mod}(\Sigma)$; (ii) implies (iii) because $A + 0 \subseteq [A]_\Sigma$. Assume that (iii) holds and take $M \in \mathrm{Mod}(\Sigma)$ and $i \in \mathbb{Z}$ such that $A + i \subseteq M$. Then, $A \subseteq M - i$ and thus $B \subseteq [A]_\Sigma \subseteq [M - i]_\Sigma \subseteq [M]_\Sigma - i$ from which it follows that $B + i \subseteq [M]_\Sigma = M$, proving (i).   □

Remark 2. Let us note that subsets $M \subseteq \mathcal{T}_Y$ we use to evaluate our formulas can be seen as particular Kripke models [9,10] of a propositional language which contains the attributes from Y as propositional variables. Namely, for $M \subseteq \mathcal{T}_Y$, we may consider $\mathbf{K}_M = \langle W, e, r \rangle$, where the set of worlds $W = \mathbb{Z}$, the accessibility relation $r \subseteq W \times W$ is defined by $\langle w, w + 1 \rangle \in r$ for all $w \in \mathbb{Z}$, and $e(w, y) = 1$ if $y^w \in M$ and $e(w, y) = 0$ otherwise. The formulas introduced in our paper use time points (worlds in sense of the Kripke model $\mathbf{K}_M$) as annotations in their antecedents and consequents. Another approach is to introduce modalities G (meaning "in the world after the current one") and H (meaning "in the world before the current one") and represent each formula of the form (1) by

$$\left(\triangle^{i_1} y_1 \,\&\, \cdots \,\&\, \triangle^{i_m} y_m \right) \Rightarrow \left(\triangle^{j_1} z_1 \,\&\, \cdots \,\&\, \triangle^{j_n} z_n \right), \tag{7}$$

where $\&$ stands for conjunction and

$$\triangle^i y = \begin{cases} y, & \text{if } i = 0, \\ G\triangle^{i-1} y, & \text{if } i > 0, \\ H\triangle^{i+1} y, & \text{if } i < 0. \end{cases} \tag{8}$$

Under this notation, (1) is true in M in our sense iff the corresponding (7) is true in $\mathbf{K}_M$ in all worlds $w \in W$ provided that the value of $G\varphi$ in $\mathbf{K}$ and w is defined as the value of φ in $\mathbf{K}$ and w' such that $\langle w, w' \rangle \in r$ (and analogously for H). Thus, up to a different formalization, the interpretation of attribute implications annotated by time points can be seen as the interpretation of particular modal formulas in propositional Kripke models.

Let us conclude this section by making notes on the relationship of our formulas to important structures and operators which appear in the formal concept analysis [22]. As we have mentioned in the introduction, a series of contexts $I_j \subseteq X \times Y$ $(j \in \mathbb{Z})$ can be seen as a triadic context [17], i.e., a structure $\mathbf{T} = \langle X, Y, \mathbb{Z}, I \rangle$ such that $I \subseteq X \times Y \times \mathbb{Z}$. Namely, we may put $\langle x, y, j \rangle \in I$ iff $\langle x, y \rangle \in I_j$. Attribute implications annotated by time points can be interpreted in such triadic contexts:

Definition 5. A formula $A \Rightarrow B$ *is true in a triadic context* $\mathbf{T} = \langle X, Y, \mathbb{Z}, I \rangle$, written $\mathbf{T} \models A \Rightarrow B$, if for each $x \in X$ and $M_x = \{y^i \mid \langle x, y, i \rangle \in I\}$, we have that $M_x \models A \Rightarrow B$.

A classic result on attribute implications says that an attribute implication is true in a context if and only if it is true in all its intents. In addition, $A \Rightarrow B$ is true in a context if B is included in the intent generated by A. In our case, we may establish an analogous characterization.

Given $\mathbf{T} = \langle X, Y, \mathbb{Z}, I \rangle$, we may define operators $^\uparrow$ and $^\downarrow$ assigning subsets of $\mathcal{T}_Y$ to subsets of $\mathcal{T}_X$ and *vice versa* as follows: For any $A \subseteq \mathcal{T}_X$ (defined as in (2) with Y replaced by X) and $B \subseteq \mathcal{T}_Y$, we put

$$A^\uparrow = \{y^j \in \mathcal{T}_Y \mid \langle x, y, i+j \rangle \in I \text{ for all } x^i \in A\}, \tag{9}$$

$$B^\downarrow = \{x^i \in \mathcal{T}_X \mid \langle x, y, i+j \rangle \in I \text{ for all } y^j \in B\}. \tag{10}$$

The pair of operators $^\uparrow : 2^{\mathcal{T}_X} \to 2^{\mathcal{T}_Y}$ and $^\downarrow : 2^{\mathcal{T}_Y} \to 2^{\mathcal{T}_X}$ defined by (9) and (10) forms an antitone Galois connection and thus the composed operator $^{\downarrow\uparrow} : 2^{\mathcal{T}_Y} \to 2^{\mathcal{T}_Y}$ is a closure operator.

Theorem 3. *For any triadic context* $\mathbf{T} = \langle X, Y, \mathbb{Z}, I \rangle$ *and formula* $A \Rightarrow B$, *the following conditions are equivalent:*

(i) $\mathbf{T} \models A \Rightarrow B$,

(ii) $A^\downarrow \subseteq B^\downarrow$,

(iii) $B \subseteq A^{\downarrow\uparrow}$.

Proof (sketch). In order to see (i), prove that $x^i \in A^\downarrow$ iff, for each $y^j \in A$, we have $\langle x, y, i+j \rangle \in I$ which is according to Definition 5 iff $y^{i+j} \in M_x$ for all $y^j \in A$. The latter observation is true iff $A + i \subseteq M_x$. An analogous observation can be made for B. Hence, (i) is equivalent to (ii). The equivalence of (ii) and (iii) follows by properties of antitone Galois connections. $\square$

The following corollary of the previous assertion gives analogy to the classic result on validity of attribute implications in contexts as implications satisfied by all concept intents:

Corollary 1. *For any triadic context* $\mathbf{T} = \langle X, Y, Z, I \rangle$ *and formula* $A \Rightarrow B$, $\mathbf{T} \models A \Rightarrow B$ *iff* $M^{\downarrow\uparrow} \models A \Rightarrow B$ *for each* $M \subseteq \mathcal{T}_Y$.

Proof (sketch). The if-part follows by Theorem 3 (iii) for $M = A$. Conversely, if $A + i \subseteq M^{\downarrow\uparrow}$, then $M^{\downarrow} \subseteq (A + i)^{\downarrow} = A^{\downarrow} - i \subseteq B^{\downarrow} - i$ by Theorem 3 (ii). Thus, $B + i \subseteq B^{\downarrow\uparrow} + i = (B^{\downarrow} - i)^{\uparrow} \subseteq M^{\downarrow\uparrow}$, proving $M^{\downarrow\uparrow} \models A \Rightarrow B$.   □

3   Axiomatization

In this section, we present a deduction system for our formulas and the notion of syntactic entailment. The syntactic entailment of formulas is based on an extension of the Armstrong axiomatic system [2] which is well known mainly in database systems [20]. The extension we propose accommodates the fact that time points in formulas are relative.

Each formula of the form $A \cup B \Rightarrow A$ (A, B are finite subsets of $\mathcal{T}_Y$) is considered as an *axiom*. Furthermore, we consider the following *deduction rules*:

(Cut) from $A \Rightarrow B$ and $B \cup C \Rightarrow D$ infer $A \cup C \Rightarrow D$,

(Shf) from $A \Rightarrow B$ infer $A + i \Rightarrow B + i$,

where $i \in \mathbb{Z}$. A proof of $A \Rightarrow B$ from Σ is a sequence $\varphi_1, \ldots, \varphi_n$ such that φ_n equals $A \Rightarrow B$ and for each φ_i we either have $\varphi_i \in \Sigma$, or φ_i is an axiom, or φ_i is derived by (Cut) or (Shf) from $\varphi_1, \ldots, \varphi_{i-1}$.

Remark 3. Note that there are several equivalent systems which are called the Armstrong systems [20]. In our presentation, the axioms can be seen as nullary deduction rules and (Cut) is a binary deduction rule. Together, they form a system which is equivalent to that from [2]. We call the additional rule (Shf) the rule of "time shifts." Also note that in the database literature, (Cut) is also referred to as the rule of pseudo-transitivity [20].

We say that $A \Rightarrow B$ *is provable from* Σ, denoted $\Sigma \vdash A \Rightarrow B$ if there is a proof of $A \Rightarrow B$ from Σ. Since we in fact extend the Armstrong system, we immediately get the following properties:

(Add) $\{A \Rightarrow B, A \Rightarrow C\} \vdash A \Rightarrow B \cup C$,

(Aug) $\{B \Rightarrow C\} \vdash A \cup B \Rightarrow A \cup C$,

(Pro) $\{A \Rightarrow B \cup C\} \vdash A \Rightarrow B$,

(Tra) $\{A \Rightarrow B, B \Rightarrow C\} \vdash A \Rightarrow C$.

Our inference system is sound in the usual sense:

Theorem 4. *If* $\Sigma \vdash A \Rightarrow B$ *then* $\Sigma \models A \Rightarrow B$.

Proof. The proof goes by induction on the length of a proof, considering the facts that each axiom is true in all models, (Cut) is a sound deduction rule [20], and (Shf) is sound on account of Theorem 1. □

Lemma 2. *If* $\Sigma \nvdash A \Rightarrow B$, *then there is* $M \in \mathrm{Mod}(\Sigma)$ *such that* $M \nvDash A \Rightarrow B$.

Proof (sketch). Consider Σ and for each non-negative integer n, put

$$A_\Sigma^0 = A,$$
$$A_\Sigma^{n+1} = A_\Sigma^n \cup \bigcup\{F + i \mid E \Rightarrow F \in \Sigma \text{ and } E + i \subseteq A_\Sigma^n\},$$
$$A_\Sigma^\omega = \bigcup_{n=0}^\infty A_\Sigma^n.$$

It suffices to show that one can take A_Σ^ω for M. Take $E \Rightarrow F \in \Sigma$, $i \in \mathbb{Z}$ and let $E + i \subseteq A_\Sigma^\omega$. Since $E + i$ is finite, there must be n such that $E + i \subseteq A_\Sigma^n$ and thus $F + i \subseteq A_\Sigma^{n+1} \subseteq A_\Sigma^\omega$, proving that $A_\Sigma^\omega \in \mathrm{Mod}(\Sigma)$. Next, prove that $B \subseteq A_\Sigma^\omega$ implies $\Sigma \vdash A \Rightarrow B$. To see that, it suffices to check that for every n and every finite $D \subseteq A_\Sigma^n$, we have $\Sigma \vdash A \Rightarrow D$ since then the claim readily follows for $B = D$. Assume the claim holds for n and all finite $D \subseteq A_\Sigma^n$. Consider $n + 1$ and take a finite $D \subseteq A_\Sigma^{n+1}$. Now, consider a finite

$$D' = \{\langle E \Rightarrow F, i\rangle \mid E \Rightarrow F \in \Sigma \text{ and } E + i \subseteq A_\Sigma^n\}$$

such that $D \subseteq \bigcup\{F + i \mid \langle E \Rightarrow F, i\rangle \in D'\} \cup A_\Sigma^n$. By induction hypothesis, for each $\langle E \Rightarrow F, i\rangle \in D'$, we have $\Sigma \vdash A \Rightarrow E + i$ and for $E \Rightarrow F \in \Sigma$, we have $\Sigma \vdash E + i \Rightarrow F + i$ using (Shf). Thus, (Tra) gives $\Sigma \vdash A \Rightarrow F + i$. Since D' is finite, $\Sigma \vdash A \Rightarrow D$ follows by finitely many applications of (Add) and (Pro). Therefore, our initial assumption $\Sigma \nvdash A \Rightarrow B$ yields $B \nsubseteq A_\Sigma^\omega$. Now, for $i = 0$, we have that $A + i = A \subseteq A_\Sigma^\omega$ and $B + i = B \nsubseteq A_\Sigma^\omega$, i.e., $A_\Sigma^\omega \nvDash A \Rightarrow B$. □

Theorem 5 (completeness). $\Sigma \vdash A \Rightarrow B$ *iff* $\Sigma \vDash A \Rightarrow B$.

Proof. Consequence of Theorem 4 and Lemma 2. □

The set A_Σ^ω introduced in the proof of Lemma 2 can be seen as a constructive description of $[A]_\Sigma$ since both the sets coincide:

Corollary 2. *For every* $A \subseteq \mathcal{T}_Y$, *we have* $[A]_\Sigma = A_\Sigma^\omega$.

Proof. We get $[A]_\Sigma \subseteq A_\Sigma^\omega$ since $[A]_\Sigma$ is the least model of Σ containing A and the converse inclusion follows by the monotony of the operator ${}_\Sigma^\omega$ used in the proof of Lemma 2 and the fact that $([A]_\Sigma)_\Sigma^\omega = [A]_\Sigma$. □

Remark 4. Let us stress that the notions of semantic and syntactic entailment we have considered in our paper are different from their classic counterparts. Indeed, each attribute implication annotated by time points can also be seen as a classic attribute implication *per se* because the sets A and B in $A \Rightarrow B$ are subsets of $\mathcal{T}_Y$. Therefore, in addition to the semantic entailment from Definition 4, we may consider the ordinary one which disregards the special role of time points. The same applies to the provability—the classic notion is obtained by omitting the rule (Shf). For instance, $\Sigma = \{\{x^1\} \Rightarrow \{y^2\}, \{y^5\} \Rightarrow \{z^2\}\}$ proves $\{x^4\} \Rightarrow \{y^5\}$ by (Shf) and thus $\{x^4\} \Rightarrow \{z^2\}$ by (Tra). On the other hand, Σ does not prove $\{x^4\} \Rightarrow \{z^2\}$ without (Shf).

4 Conclusion and Future Research

We have introduced attribute implications annotated by time points and their semantics based on evaluating the implications in models using a sliding time point. The formulas are considered true if they are preserved in all time points. We have defined semantic entailment, showed closure properties of systems of models and provide a characterization based on least models. Furthermore, we have shown an axiomatization which extends the classic axiomatization by considering an additional rule of time shifts.

Our future research in the area will focus on the following issues:

- Relationship to other logical systems for rule-based reasoning;
- connections to concept lattices, Galois connections, and related structures;
- non-redundant descriptions of dependencies which hold in data;
- relationship to modal and temporal (tense) logics;
- generalization of the approach by extensions of Armstrong-like systems;
- generalization for similarity-based and graded rules;
- applications in prediction.

References

1. Agrawal, R., Imieliński, T., Swami, A.: Mining association rules between sets of items in large databases. In: Proceedings of the 1993 ACM SIGMOD International Conference on Management of Data, SIGMOD 1993, pp. 207–216. ACM, New York (1993)
2. Armstrong, W.W.: Dependency structures of data base relationships. In: Rosenfeld, J.L., Freeman, H. (eds.) Information Processing 1974: Proceedings of IFIP Congress, pp. 580–583. North Holland, Amsterdam (1974)
3. Bettini, C., Jajodia, S., Wang, X.S.: Time Granularities in Databases, Data Mining, and Temporal Reasoning. Springer (2000)
4. Cordero, P., Mora, A., de Guzmán, I.P., Enciso, M.: Non-deterministic ideal operators: An adequate tool for formalization in data bases. Discrete Applied Mathematics 156(6), 911–923 (2008)
5. Date, C.J., Darwen, H., Lorentzos, N.A.: Temporal Data and the Relational Model. Elsevier (2002)
6. Fagin, R.: Functional dependencies in a relational database and propositional logic. IBM Journal of Research and Development 21(6), 534–544 (1977)
7. Feng, L., Dillon, T., Liu, J.: Inter-transactional association rules for multi-dimensional contexts for prediction and their application to studying meterological data. Data Knowl. Eng. 37(1), 85–115 (2001)
8. Feng, L., Yu, J.X., Lu, H., Han, J.: A template model for multidimensional inter-transactional association rules. The VLDB Journal 11(2), 153–175 (2002)
9. Gabbay, D.M.: Tense systems with discrete moments of time, part I. Journal of Philosophical Logic 1(1), 35–44 (1972)
10. Gabbay, D.M.: Model theory for tense logics. Annals of Mathematical Logic 8(1-2), 185–236 (1975)
11. Ganter, B., Wille, R.: Formal Concept Analysis: Mathematical Foundations, 1st edn. Springer-Verlag New York, Inc., Secaucus (1997)

12. Ganter, B., Obiedkov, S.: Implications in triadic formal contexts. In: Wolff, K.E., Pfeiffer, H.D., Delugach, H.S. (eds.) ICCS 2004. LNCS (LNAI), vol. 3127, pp. 186–195. Springer, Heidelberg (2004)
13. Guigues, J.L., Duquenne, V.: Familles minimales d'implications informatives resultant d'un tableau de données binaires. Math. Sci. Humaines 95, 5–18 (1986)
14. Huang, Y.P., Kao, L.J., Sandnes, F.E.: Efficient mining of salinity and temperature association rules from argo data. Expert Syst. Appl. 35(1-2), 59–68 (2008)
15. Lee, A.J.T., Wang, C.S., Weng, W.Y., Chen, Y.A., Wu, H.W.: An efficient algorithm for mining closed inter-transaction itemsets. Data Knowl. Eng. 66(1), 68–91 (2008)
16. Lee, A.J.T., Wu, H.W., Lee, T.Y., Liu, Y.H., Chen, K.T.: Mining closed patterns in multi-sequence time-series databases. Data Knowl. Eng. 68(10), 1071–1090 (2009)
17. Lehmann, F., Wille, R.: A triadic approach to formal concept analysis. In: Ellis, G., Rich, W., Levinson, R., Rich, W., Sowa, J.F. (eds.) ICCS 1995. LNCS, vol. 954, pp. 32–43. Springer, Heidelberg (1995)
18. Lloyd, J.W.: Foundations of Logic Programming. Springer-Verlag New York, Inc., New York (1984)
19. Lu, H., Feng, L., Han, J.: Beyond intratransaction association analysis: Mining multidimensional intertransaction association rules. ACM Trans. Inf. Syst. 18(4), 423–454 (2000)
20. Maier, D.: Theory of Relational Databases. Computer Science Pr., Rockville (1983)
21. Tung, A.K., Lu, H., Han, J., Feng, L.: Breaking the barrier of transactions: Mining inter-transaction association rules. In: Proceedings of the Fifth ACM SIGKDD International Conference on Knowledge Discovery and Data Mining, KDD 1999, pp. 297–301. ACM, New York (1999)
22. Wille, R.: Restructuring lattice theory: An approach based on hierarchies of concepts. In: Rival, I. (ed.) Ordered Sets. NATO Advanced Study Institutes Series, vol. 83, pp. 445–470. Springer, Netherlands (2009)
23. Zaki, M.J.: Mining non-redundant association rules. Data Mining and Knowledge Discovery 9, 223–248 (2004)

A Dynamic Approach for the Online Knapsack Problem

Hajer Ben-Romdhane[1], Sihem Ben Jouida[1], and Saoussen Krichen[2]

[1] LARODEC Laboratory, ISG of Tunis, 41 Rue de la Liberté, Le Bardo, Tunisia
[2] FSJEG de Jendouba, Avenue de l'U.M.A, 8189 Jendouba, Tunisia

Abstract. The online knapsack problem (OKP) is a generalized version of the 0-1 knapsack problem (0-1KP) to a setting in which the problem inputs are revealed over time. Whereas the 0-1KP involves the maximization of the value of the knapsack contents without exceeding its capacity, the OKP involves the following additional requirements: items are presented one at a time, their features are only revealed at their arrival, and an immediate and irrevocable decision on the current item is required before observing the next one. This problem is known to be non-approximable in its general case. Accordingly, we study a relaxed variant of the OKP in which items delay is allowed: we assume that the decision maker is allowed to retain the observed items until a given deadline before deciding definitively on them. The main objective in this problem is to load the best subset of items that maximizes the expected value of the knapsack without exceeding its capacity. We propose an online algorithm based on dynamic programming, that builds-up the solution in several stages. Our approach incorporates a decision rule that identifies the most desirable items at each stage, then places the fittest ones in the knapsack. Our experimental study shows that the proposed algorithm is able to approach the optimal solution by a small error margin.

Keywords: Knapsack problem, Online algorithms, Optimal stopping, Dynamic programming.

1 Introduction

Decision making under dynamic environments is a challenging task as information on the problem being optimized are time-dependent, not completely known a priori, or uncertain. Typically, the problem inputs are revealed over time and/or get updated as the optimization process proceeds. Besides, the solution to these problems is constructed sequentially over time based only on the previously revealed data, and with partial or imperfect knowledge about the future [1]. Such problems, termed as "online optimization problems", are receiving increasing interest due to the wide range of applications it can be applied to. Real-world applications involving online scenarios include [2][3]: ad allocation problems, selection of investment projects, and freight transportation.

In this work, we are particularly interested in an online resource allocation problem that involves the sequential arrival of items for a limited resource. The so-called "online knapsack problem" (OKP) is defined as follows. A decision maker (DM) observing a sequence of items, arriving one by one, is required to fill his knapsack with the most valuable ones. Each item differs by its weight and its reward, and both values remain

V. Torra et al. (Eds.): MDAI 2014, LNAI 8825, pp. 96–107, 2014.

unknown until the item is received. The knapsack is of a limited capacity and the selection is made in an online fashion: the DM have to decide irrevocably whether to load the current item or to discard it before observing the next one. Therefore, the OKP requires an online approach that builds the decision concurrently with the arrival of new items.

This problem was firstly addressed in [4]. In that study, it was shown that there is no algorithm able to approximate the optimal solution within a constant competitive ratio (is the ratio of the payoff of an online algorithm to the payoff of an offline algorithm that have a complete knowledge of items features), and to bypass this difficulty, the authors considered the stochastic case. They proposed a linear time online algorithm that approximates the optimum to within a factor of $O(log^{3/2}n)$, and defined a lower bound of $\Omega(1)$ in the special case when a buffer storage to defer the decision about some items is used. This factor was improved in [5] to $O(logn)$. Iwama and Taketomi [6] introduced the removable OKP where the DM is allowed to replace some of the loaded items when he finds better ones. The same problem was equally investigated in [7] while authorizing resource augmentation, and in [8] where items are permitted to be fractionalized. A more recent work [9] addressed the OKP with removal cost where items removed from the knapsack, due to capacity surplus, incur cancelation charge. Auctions are also a potential application of the OKP [10]: a buyer with a limited budget is looking to purchase bids from a given set. Each bid is the property of a single bidder wishing to place his object in the knapsack. The bid value and weight correspond respectively to the bidder valuation of having its item in the knapsack and the required place in the knapsack, and the objective is to purchase the best subset of bids.

Babaioff et al. [11] showed that the OKP can be viewed as a weighted form of the multiple-choice secretary problem: a special case of the optimal stopping problem where a manager is receiving requests for vacant secretarial posts, and would like to select the most competitive candidates. Knowing nothing about the future applicants, he is required to decide irrevocably about the current candidate: to hire him immediately, or to send him away. If each candidate is symbolized by an item and the number of available posts is considered as the capacity of the knapsack, the problem would be equivalent to an OKP. Conversely speaking, if all item's weight are set to 1 and the knapsack capacity is fixed to a predefined value x, the OKP is reduced to an -unweighted- multiple choice secretary problem. This problem was solved via a $10e$-competitive algorithm while assuming arbitrary weights, and a e-competitive algorithm was proposed for the equal weights case. Besides, several studies have considered stochastic OKPs. The dynamic and stochastic knapsack problem was introduced in [2], and two special cases were examined later while assuming equal [12] and random weights [13]. These studies showed that the optimal policy for OKPs is a threshold type policy.

The common in the aforesaid works is that, in order to overcome the difficulty of such problems, the authors relaxed the original problem (e.g. by allowing items removal, through capacity augmentation) or considered their stochastic counterparts. Specifically, the removability assumption was the most appealing and the most used approach to achieve a competitive ratio. Indeed, we think that the fact of loading items at their arrival time and allowing their removal when better items are encountered is none other than a special case of deferring the decision about these items to advanced stages of the searching process. And by referring to real-world, we can think of many applications of OKPs that involve the maintain of -at least- some items until later stages, before abandoning them. This is the case when one needs to a hire a secretary: as he seeks the most qualified applicant, he asks the best interviewed ones to leave their contact details so that they get a response, and make his selection once the candidate list is formed. The problem of cargo loading is another potential application, where a logistic company takes care of loading items in a cargo before being sent to their final destination. Because the available cargo cannot accommodate all items at once, the problem is to select some of the online loading requests, while holding on the rest for future shipments.Other examples of real-world applications in the resource allocation field could include: web advertising, selling real- estate, and auctions [10]. From there comes the need to address the OKP with items delay.

This study introduces a new version of the OKP in which the DM is allowed to delay items. We define the delay of an item as being the defer of the decision about that item until observing the next one(s). That is, an item i received at t will still available for the selection in subsequent stages until it is loaded in the knapsack (at any stage $g >= t$) or the delay deadline occurs. However, delayed items are penalized in terms of utility. We note that the concept of delay has already been considered within the context of online decision making, notably with the optimal stopping problem [14].

The OKP with delay is defined as the decision process aiming to maximize the expected reward of the knapsack contents. The DM, receiving items sequentially, is allowed to select immediately the current item or to defer his decision about that item to next stages. The selection and the delay of a given item depends on the previously received items and on its expected utilities. When an item is delayed, its utility incurs a penalty which means that this item loses some of its desirability over time. Two stopping criteria are considered: the knapsack is full, or all the potential items have been received. The main challenge of this problem is to determine which items to select and in the opportune moment especially that items selection is irrevocable.

In order to solve this problem, we propose a dynamic programming approach managed by a stopping rule. Our approach reduces the online problem to a number of static knapsack sub-problems equal to the number of stages. The proposed algorithm operates as follows. Every time that a new item appears, it is ranked among the foregoing ones and on the basis of these ranks the dynamic equations are computed to identify a subset of candidate items. The chosen items serve as inputs for a 0-1KP, the solution of which are items to be inserted irrevocably in the knapsack. These steps are repeated until the process ends. The collection of items contained into the knapsack constitutes the solution of the OKP. A comparative study is carried out to evaluate the performance of the proposed approach with respect to the results of an offline algorithm, and in terms of several performance measures. Our results indicate that the proposed approach is able to approximate the optimal solution with a small gap.

The remainder of this paper is organized as follows. In section 2, we state the problem and present our dynamic formulation. Section 3 describes the proposed approach and draws its algorithm. The experimental evaluation of our algorithm and results analysis are presented in section 4. The effect of using different utility functions is also discussed in the same section. The last section contains our concluding remarks and possible directions for future work.

2 Problem Statement

We consider a knapsack with a limited capacity C and a sequence of n items arriving one by one over time. Each item is characterized by a value and a weight, which become known when the item is received. Each new stage (or a time step), exactly one item is presented. The DM is allowed to delay items to subsequent stages (while incurring a penalty), but once inserted in the knapsack, they can no longer be removed. Items delayed at previous stages remain available until the stopping criteria is met, or the DM selects them. Two stopping criteria are considered in this work: when the capacity of the knapsack is exhausted, or all the potential items are already received. The DM aims to select -progressively over time- the fittest of all items in order to maximize his profit without exceeding the capacity of the knapsack.

2.1 Notations

For the better understanding of our problem formulation, we start by defining all the notations that will be used in the rest of this paper.

Symbols	Explanations
n	the total number of potential items
C	the knapsack weight capacity
i	item's number (refers to the ith received item)
j	stage's number (i.e. j items were so far revealed)
v_i	value (or reward) of item i
w_i	weight of item i
d_i	density of item i, computed as its value per unit weight $d_i = \frac{v_i}{w_i}$
c_j	the remaining capacity of the knapsack at j
r	the relative rank of the current item (its rank among the so far observed items)
k	the absolute rank (the rank of the current item among the whole sequence)
$U^i(k,j)$	the utility function of item i at j
$EU^{i*}(j,r)$	the expected utility of the item i
$EU_s^i(j,r)$	the expected utility when stopping i (when selecting it)
$EU_c^i(j)$	the expected utility when continuing (when delaying i)
$P(k|r,j)$	the probability of having k given r at j
S	the set of candidate items
x_i	if item i is selected $x_i = 1$, otherwise $x_i = 0$

2.2 Dynamic Formulation of the OKP

The OKP can be regarded as a decision process aiming to identify the best offers from a sequence of offers arriving successively. Therefore, we use the dynamic equations of the optimal stopping problem as a base for our dynamic formulation.

We are given n items, arriving over n discrete periods, and a knapsack of capacity C. Each item has an associated value v_i and a weight w_i. The problem is to fill the knapsack by the fittest of all items in an online fashion, and without violating the capacity constraint. Each new stage, the DM ranks the available items (the one received at the current stage and the delayed ones). In this framework, it is essential to distinguish between two types of rank: relative and absolute. The relative rank r of an item i is attributed by the DM and indicates its desirability (measured in terms of density: the greater is the density, the more desirable is the item) among the so far received items.

However, the absolute rank of i is its rank among the n potential items. As no prior information is available, the absolute rank (k) of an item can only be determined when all the items are received. Therefore, the DM decides to select or to delay each available item based only on the relative ranks. Our decision strategy reposes on two steps: identifying candidates and loading the best of them.

Identifying candidates. A first step consists in identifying the set of candidates by computing items expected utilities. This is accommodated by the following formulation.

The DM's utility is a measure of its desirability of the consequences to which can lead his decision. In our case, $U^i(k, j)$ denotes the DM's utility of selecting item i whose absolute rank is equal to k at the jth stage. The utility is a non-increasing function of the absolute rank: $U^i = f(k)$. As we are looking for the best subset of items to be packed in the knapsack, we adopted a utility function which attributes decreasing values in terms of the absolute rank. Therefore, our utility function is expressed in terms of the absolute rank and the stage's number: $U^i = f(k, j)$. We consider, in this work, two different utility functions in order to study the influence of each utility function on the final decision. Their formulas are given by:

$$U_1^i(k, j) = \frac{1}{k} \qquad (1) \qquad U_2^i(k, j) = \frac{n - k + 1}{n} \qquad (2)$$

where *the inverse-rank utility* $U_1^i(k, j)$, is computed as the inverse number of the absolute rank, while *the regressive fraction utility* $U_2^i(k, j)$, is computed in terms of the the total number of items. As to delayed items (reconsidered in subsequent stages, $j > i$), their utilities are discounted to the utility of the next rank.

Because the decision about a given item i is between two alternatives (to select the item or to delay it), it is reasonable to consider the expected utility of each alternative as base to make the decision. To determine the expected utility in each case, we refer to the optimal stopping context. Indeed, the OKP can be viewed as a constrained multiple choice optimal stopping problem where a DM is required to select the best offers of the sequence of offers arriving over time. The stopping rule in optimal stopping problems consists in stopping the process when the DM believes that the current offer is the best of the sequence. This is translated in the OKP by stopping at the items that are expected to be the more desirable of the sequence, so to load them in the knapsack. However, continuing to the next item means in our context to delay the current item.

We denote by $EU_s^i(j, r)$ the expected utility of selecting item i at j with a relative rank r. The expected utility when continuing, denoted by $EU_c^i(j)$, is the expected utility of delaying item i at j and continuing to the next stage.

The decision at any stage of the selection process depends on the values of these two components, and the DM will react in accordance with the decision that maximizes his expected utility. That is, if $EU_s^i(j, r) \geq EU_c^i(j)$, item i will be considered as a candidate at j, otherwise it is delayed to next stages. Therefore, the expected utility of item i at j can be stated as:

$$EU^{i*}(j, r) = \max[EU_s^i(j, r), EU_c^i(j)] \qquad (3)$$

where the expected utility of selecting i is given by:

$$EU_s^i(j,r) = \sum_{k=r}^{n-j+r} U^i(k,j)P(k|r,j) \quad (4) \qquad \text{with} \quad P(k|r,j) = \frac{\binom{k-1}{r-1}\binom{n-k}{j-r}}{\binom{n}{j}} \quad (5)$$

This expected utility of selecting an item is expressed in terms of the absolute rank of that item. As absolute ranks become only known when all items are received, we compute the probability of having the current item as the kth best item of the sequence for each possible rank $k \in \{r,...,n-j+r\}$. Subsequently, $EU_s^i(j,r)$ is given by summing up each of these probabilities weighted by its corresponding utility. However, the expected utility when continuing with delaying item i at j, is computed as the average sum of the expected utilities of item i until stage $j+1$, and this to measure the effect of delaying i for the next stage. It can be written as:

$$EU_c^i(j) = \frac{1}{j+1} \sum_{r=1}^{j+1} EU^i(j+1,r) \quad (6)$$

We note that Eq. (6) is only available when $j < n$, otherwise ($j = n$) no item can be delayed anymore and all the available items should be nominated for the final selection.

Thereby, the DM can identify which items to delay and the ones to be inserted in the knapsack. However, if the the knapsack cannot carry all items considered for the selection, then only the best ones will be inserted in the knapsack. We denote by S the subset of items verifying the inequality $EU_s^i(j,r) \geq EU_c^i(j)$ at j, Hence, S is the subset of candidates for selection.

Select the fittest items. If the set of candidate items cannot fit in the knapsack, the best items in this set are chosen. To insure selecting the best of all items in S, we solve a 0-1KP having as inputs the set of items S and as a capacity constraint the remaining capacity at j. The knapsack subproblem at stage j, $KP_j(S,c_j)$, can be stated as:

$$\text{Maximize } Z(x) = \sum_{i \in S} v_i x_i \quad \text{Subject to } \sum_{i \in S} w_i x_i \leq c_j \quad (7)$$

The solution of $KP_j(S,c_j)$ is the subset of items to be loaded in the knapsack at stage j. Hence, at any stage j ($j \in [1,n]$), the knapsack will contain items selected during the previous $j-1$ stages in addition to items selected at the current stage.

3 Online Algorithm for the OKP with Items Delay

This section deals with the description of our online algorithm. We present, in a first part of this section, the proposed approach and we draw up its pseudocode. The second subsection details the solution steps of a small size problem for demonstration purpose.

3.1 The Algorithm

The algorithm inputs are the total number of items n and the capacity C. Items are then presented one per stage, and the next one is only revealed when the current stage is achieved. Each new stage, the algorithm assigns relative ranks to the observed items by density. Then, the expected utilities of available items are computed based on the attributed ranks and the stage index. The candidate set is then formed by items whose expected utility when stopping is greater than their expected utility when continuing. If the sum of the weights of items in the candidate set S is greater than the remaining capacity of the knapsack, the fittest items of the candidate set are selected and placed in the knapsack by solving a 0-1 knapsack subproblem. There, our algorithm makes appeal to the offline algorithm proposed in [15] to get the optimal subset of items from items in the candidate set. The selected items are inserted in the knapsack, while the remaining ones are definitively rejected and will not be considered in the subsequent stages. Afterwards, the capacity of the knapsack is updated and the algorithm moves to the next stage. These steps are repeated until the capacity is exhausted or the last stage is achieved. The pseudocode of our algorithm is given in Fig. 1.

```
j ← 1;   i ← 1;      /* j and i are, respectively, the stage's index and item's index */
while j ≤ n do
    Receive item i
    Rank the observed items from 1 to j    /* Attribute relative ranks to available items */
    for each i ∈ [1, j] do
        Compute the expected utilities of item i
        if (EU_s^i ≥ EU_c^i) then
            S ∪ i          /* Item i is selected as a candidate */
        end if
    end for
         |S|
    W ← ∑ w_s     /* Compute the weight sum of candidates in S */
        s=1
    if (c_j ≤ W) then
        Load all candidates in the knapsack
    else
        Solve KP_j(S, c_j)
        Load the selected items in the knapsack
        S ← ∅         /* All remaining items are rejected */
    end if
    Update(c_j)    /* Compute the remaining capacity in the knapsack */
    if (c_j = 0) then
        Quit the procedure
    else
        j ← j + 1
    end if
end while
```

Fig. 1. Pseudocode of the proposed dynamic approach for the OKP

3.2   Example

For the sake of thoroughness and full understanding of the proposed approach, we detail the solution steps through an example of a small size. We consider a knapsack of capacity with a capacity $C = 904$, and a set of items with the following values and weights: $v_i = \{94, 506, 992, 416\}$ and $w_i = \{485, 300, 421, 240\}$, where $i \in \{1, 2, 3, 4, 5\}$.

These values are provided to the algorithm as soon as the item in question becomes available (i.e, at stage i, the features of item i becomes known). We note that, in this example, we compute the expected utilities using the utility function U_2^i (stated in eq. (2)). In what follows, we analyze stage by stage the solution.

Stage 1: the first item I_1 appears. We compute its expected utilities (with a relative rank equal to 1): $EU_s^1(1,1)=0.6$ and $EU_c^1(1)=0.79$. As the expected utility when continuing of I_1 is greater than its expected utility when stopping, the decision will be to delay I_1 and to move to the next stage.

Stage 2: the second item I_2 is received. Available items are I_1 (delayed in stage 1) and I_2. By means of the density function, we can assign relative ranks to the available items: $d_1=0.19$ and $d_2=1.68$. Hence, I_2 is ranked first and its expected utilities are computed with a relative rank 1: $EU_s^2(2,1) = 0.8$ and $EU_c^2(2) = 0.78$. As its expected utility when stopping is greater than its expected utility when continuing, I_2 is a candidate. However I_1 is not, because $EU_c^1(2) = 0.78 > EU_s^1(2,2) = 0.4$. Since the knapsack is empty and I_2 at this stage is the only candidate, we can load I_2 without going through the solution of $KP_2(S,904)$. Therefore, the knapsack contains at the end of the second stage item I_2 and the remaining capacity in the knapsack is $c_2 = 904 - 300 = 604$.

Stage 3: Available items are I_1 and I_3. The relative ranks are $\{r_1,r_2,r_3\}=\{3,2,1\}$, where r_i denotes the relative rank of item i. By computing the expected utilities, we found that I_3 is the unique candidate ($EU_s^3(3,1) = 0.9 > EU_c^3(3) = 0.72$). Hence, I_3 is loaded in the knapsack and the knapsack capacity is updated $c_3= 183$.

Stage 4: Available items at stage 4 are I_1 and I_4. The relative ranks are $\{r_1,r_2,r_3,r_4\} = \{4,3,1,2\}$, and I_4 is a candidate. As the weight of item I_4 is greater than the remaining capacity in the knapsack, I_4 is rejected and we continue to the next stage.

Stage 5: At the last stage, all potential items are already received. There is no need to compute the expected utilities as all the available items are candidates for the selection. Therefore, we solve the $KP_5(S,183)$ with $S=\{I_1, I_5\}$. The remaining capacity cannot carry any of the available items, so the processes is stopped.

Accordingly, the solution of this online problem is the subset of items: $\{I_2,I_3\}$ and the total value of the knapsack contents is $Z=1498$. Knowing that the optimal subset of items obtained by an offline algorithm is $\{I_3,I_5\}$ with a total reward of 1641, we can induce that our algorithm closely approached the optimal solution.

4 Experimental Study

This section is concerned with the evaluation of the proposed algorithm. First, we present the settings of our experimentations. Then, we describe the evaluation measures being used in the experimental study. Finally, we report and discuss the computational results for the performance of the proposed algorithm.

4.1 Experimentation Settings

Our algorithm is run on several instances ranging in size from 10 to 1000. To the best of our knowledge, there is no benchmark tool to evaluate OKPs. Therefore, we use the test instances generator of Pisinger [16] to generate the parameters of 0-1KPs. We set the capacity value for each instance to 50% of the sum of all items weight. For comparison purposes, we generate the optimal solution for the tested instances using the exact algorithm proposed in [15].

In our experimentations, we run each test instance over 30 random permutations of features. We define a random permutation as a random ordering of the items arrival times over the n stages. Results reported in this work are the average of 30 independent runs. Moreover, and in order to analyze the effect of the utility function in the final solution, we present results obtained by two different utility functions (eq. (1-2)).

4.2   Performance Measures

Performance measures are quantitative tools for assessing the efficiency of an algorithm with regards to the achieved fitness value or a specific behavioral aspect. We define, in this section, several performance measures used to evaluate the proposed algorithm.

Average Reward. It is a fitness-based measure that indicates how well the algorithm optimizes the objective function. The average reward (AR) is average of the fitness values obtained in each of the independent runs.

Fitness Ratio. The fitness ratio (FR) assesses the quality of the obtained solution with regards to the optimal fitness value Z^*. It is computed by dividing the *Average Reward* by the approximate algorithm by the optimal fitness. The greater is FR, the closer the algorithm approach the optimal solution.

Fitness Error. Fitness defect is one of the most informative measures. It is computed as the difference between the optimal solution and the fitness value achieved by our algorithm. The smaller is the error, the better the algorithm performs.

First Loading Stage. Although we allow the DM to defer decision-making, it is crucial to know when the DM starts to fill his knapsack. A particular interest should be given to the first selection stage as it is the stage in which the DM decided that he has observed enough items to proceed to the selection. Hence, we record the first loading stage (FLS) index for each of the tested instances and report the average value.

Process Ending Stage. This measure aims to determine the stage in which the stopping criteria is met. Indeed, the process ends when the capacity of the knapsack is exhausted or when all items are observed. By tracking down the stopping criteria, the process ending stage (PES) will help us to better understand the behavior of the online algorithm and to know if the last stage is of importance to achieve good solutions. The PES is computed by averaging the recorded indexes of final stages.

Ratio of Loads Before the Last Stage. Online algorithms are required to make decision sequentially over time. However, if the decision is delayed until observing the last item, the problem will be reduced to a 0-1KP (with full knowledge about all items). Accordingly, an evaluation measure is required to assess whether the proposed algorithm is performing in an online manner (with partial or imperfect knowledge) or not. We propose a new measure that we call "ratio of loads before the last stage" (LBLS). This measure is intuitive and easy to compute. It is given by dividing the number of items contained in the knapsack at the penultimate stage by the total number of loaded items. The LBLS measure can be viewed as a sanity check of the online aspect of the solution, but it does not rely on the ability of the algorithm to produce good solutions.

4.3 Experimental Results

We report, in Table 1, the experimental results obtained by the proposed algorithm in terms of several performance measures. Besides, we draw the algorithm behavior according to the FR and LBLS measures in respectively Fig.2a and Fig.2b.

Table 1. Experimental Results in terms of numerous performance measures

n	AR		FR		Error		FLS		PES		LBLS (%)		CPU
	U_1	U_2	U_1	U_2	U_1	U_2	U_1	U_2	U_1	U_2	U_1	U_2	
10	3646	3440	0.95	0.89	194	400	4	4	9	8	0.97	0.98	3.7e-4
50	19428	16099	0.85	0.76	1726	5057	17	14	47	40	0.99	0.99	8.2e-4
100	36248	30697	0.89	0.76	4142	9693	32	28	88	78	0.99	1	10e-3
200	70042	59636	0.88	0.75	9577	19984	64	54	179	148	0.99	0.99	12e-2
300	114644	102815	0.94	0.85	6649	18478	170	170	290	267	0.99	1	79e-2
400	152953	142743	0.95	0.89	6400	16610	279	279	378	364	1	1	4.1
500	192580	182062	0.96	0.91	7109	17628	384	384	482	429	0.99	1	7.1
600	232821	221073	0.98	0.93	3518	15266	488	488	578	558	1	1	18.4
700	271195	259510	0.98	0.94	4011	15696	591	591	678	633	1	1	36.7
800	310225	299511	0.98	0.95	3854	14568	693	693	780	725	1	1	92.9
900	349829	339368	0.99	0.96	3040	13501	795	795	881	826	1	1	117.6
1000	388655	376492	0.99	0.95	3649	15812	897	897	982	927	1	1	142.4

As it can be seen from the table, AR values obtained by the algorithm are very close to the optimal values obtained by the offline algorithm. Additionally, the results of FR measure are too close to 1. This indicates that the algorithm meets the objective of maximizing the overall profit. The results of the fitness error measure confirm the previous interpretations and showed that the error margin is small. One further observation that deserve to be pointed out, from these measures, relates to the utility functions being used. It is apparent from the obtained values that the utility function can influence the fitness value, since when using U_1, the algorithm provides better results than with U_2. Also, Fig.2a shows that the curve of U_1 is higher than the curve of U_2 which leads to say that the algorithm performs better when using U_1. So as an illustration of the fitness-related measures, we can say that our algorithm fulfilled a very good performance.

The rest of measures used in this paper are behavioral-based measures. We have considered the FLS to measure the latency between the initiation of the process and the first selection. Indeed, FLS values showed that also this goal is respected and the algorithm proceeds to the selection before all items are revealed. However, this measure indicates that the two utility functions gave the same values in almost all cases. Hence, we can say that the FLS does not relate to the value of the final solution since with the two utility functions the algorithm starts the selection at the same stage. However, the PES measure is more informative. The values of PES obtained when using U_1 are always greater than those of U_2, which implies that by using U_1 the process ends later to the process using U_2. This explains why the algorithm obtains higher fitness values when using U_1. Indeed, with U_1 the algorithm is more prudent when making decisions and defer the selection decision more than do it with U_2. From the LBLS values and curve (see

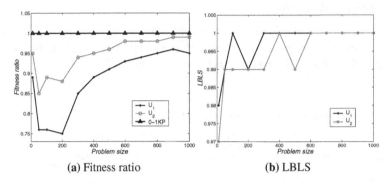

(a) Fitness ratio   (b) LBLS

Fig. 2. A comparison of the results using U_1 and U_2 in terms of two performance measures

Fig.2b), we can induce that our algorithm is able to solve the problem effectively and in an online manner. Almost 99% of the loads has been made before reaching the last stage. This says that our approach is able to achieve near-optimal solutions without having full knowledge about the problem data, and this regardless of the used function of utility. Our last evaluation tool is the CPU time. With small and medium instances, the consumed CPU time is reasonable. But it becomes prohibitively large as the problem size grows up. This can be deemed as the weak point of our algorithm.

To sum up, we can say that our dynamic approach has proved to be efficient in solving the OKP. By comparing our results to those provided by offline algorithms, we showed that we reached near-optimal solutions with small error margin. The comparative analysis of the results provided by two different utility functions allowed us to see when they influence the final solution and when they does not affect it at all. Indeed, we can say that the utility function U_2 is more convenient for DMs who desire to make decisions in a close time horizon, while U_1 is more suitable for DMs who prefer to delay their decisions until a considerable number of items appears. In other words, U_2 quantifies the attitude of flexible DMs who accept moderate near-term profit, however U_1 is for more demanding DMs.

5   Conclusion

This article deals with a special case of the OKP where the DM is allowed to delay items until a predefined deadline is reached. However, delayed items incurs penalization in terms of utility. We presented a dynamic approach to solve the problem by decomposing the online problem to a series of static knapsack subproblems. Each stage, the DM can identify his most desirable items among the available ones by computing their expected utilities, then selects the fittest ones while respecting the remaining capacity. Our experimental investigation showed that we were able to reach a very good performance using our online approach. Besides, the use of two different utility functions allowed us to come up to near-optimal solutions while involving two different attitudes to risk.

Future work may focus on generalizing the OKP with delay to a more natural setting where items availability is constrained and independent from the stopping criteria.

A comparative study can also be conducted to determine the more appropriate utility function to use in such problems.

References

1. Albers, S.: Online algorithms: a survey. Mathematical Programming 97(1-2), 3–26 (2003)
2. Papastavrou, J.D., Rajagopalan, S., Kleywegt, A.J.: The dynamic and stochastic knapsack problem with deadlines. Management Science 42, 1706–1718 (1996)
3. Mahdian, M., Nazerzadeh, H., Saberi, A.: Online optimization with uncertain information. ACM Trans. Algorithms 8, 2:1–2:29 (2012)
4. Marchetti-Spaccamela, A., Vercellis, C.: Stochastic on-line knapsack problems. Mathematical Programming 68, 73–104 (1995)
5. Lueker, G.S.: Average-case analysis of off-line and on-line knapsack problems. In: SODA 1995: Proceedings of the Sixth Annual ACM-SIAM Symposium on Discrete Algorithms, pp. 179–188. Society for Industrial and Applied Mathematics (1995)
6. Iwama, K., Taketomi, S.: Removable online knapsack problems. In: Widmayer, P., Triguero, F., Morales, R., Hennessy, M., Eidenbenz, S., Conejo, R. (eds.) ICALP 2002. LNCS, vol. 2380, pp. 293–305. Springer, Heidelberg (2002)
7. Iwama, K., Zhang, G.: Optimal resource augmentations for online knapsack. In: Charikar, M., Jansen, K., Reingold, O., Rolim, J.D.P. (eds.) APPROX and RANDOM 2007. LNCS, vol. 4627, pp. 180–188. Springer, Heidelberg (2007)
8. Han, X., Makino, K.: Online minimization knapsack problem. In: Bampis, E., Jansen, K. (eds.) WAOA 2009. LNCS, vol. 5893, pp. 182–193. Springer, Heidelberg (2010)
9. Han, X., Kawase, Y., Makino, K.: Online knapsack problem with removal cost. In: Gudmundsson, J., Mestre, J., Viglas, T. (eds.) COCOON 2012. LNCS, vol. 7434, pp. 61–73. Springer, Heidelberg (2012)
10. Zhou, Y., Chakrabarty, D., Lukose, R.: Budget constrained bidding in keyword auctions and online knapsack problems. In: Papadimitriou, C., Zhang, S. (eds.) WINE 2008. LNCS, vol. 5385, pp. 566–576. Springer, Heidelberg (2008)
11. Babaioff, M., Immorlica, N., Kempe, D., Kleinberg, R.D.: A knapsack secretary problem with applications. In: Charikar, M., Jansen, K., Reingold, O., Rolim, J.D.P. (eds.) APPROX and RANDOM 2007. LNCS, vol. 4627, pp. 16–28. Springer, Heidelberg (2007)
12. Papastavrou, J.D., Kleywegt, A.J.: The dynamic and stochastic knapsack problem. Operations Research 46, 17–35 (1998)
13. Papastavrou, J.D., Kleywegt, A.J.: The dynamic and stochastic knapsack problem with random sized items. Operations Research 49, 26–41 (2001)
14. Kramer, A.D.I.: Delaying decisions in order to learn the distribution of options. PhD thesis (2010)
15. Pisinger, D.: A minimal algorithm for the 0-1 knapsack problem. Operations Research 45(5), 758–767 (1997)
16. Pisinger, D.: Core problems in knapsack algorithms. Operations Research 47, 570–575 (1999)

On Kernelization for a Maximizing Model
of Bezdek-Like Spherical Fuzzy c-Means Clustering

Yuchi Kanzawa

Shibaura Institute of Technology, Koto, Tokyo 135-8548, Japan
kanzawa@sic.shibaura-it.ac.jp

Abstract. In this study, we propose three modifications for a maximizing model of spherical Bezdek-type fuzzy c-means clustering (msbFCM). First, we kernelize msbFCM (K-msbFCM). The original msbFCM can only be applied to objects on the first quadrant of the unit hypersphere, whereas its kernelized form can be applied to a wider class of objects. The second modification is a spectral clustering approach to K-msbFCM using a certain assumption. This approach solves the local convergence problem in the original algorithm. The third modification is to construct a model providing the exact solution of the spectral clustering approach. Numerical examples demonstrate that the proposed methods can produce good results for clusters with nonlinear borders when an adequate parameter value is selected.

Keywords: fuzzy c-means clustering, kernelization, spectral clustering approach.

1 Introduction

The hard c-means (HCM) clustering algorithm [1], also known as K-means, splits objects into well-separated clusters by minimizing the sum of the squared distances between the objects and cluster centers. This concept has been extended to fuzzy clustering, where object membership is shared among all of the cluster centers, rather than being restricted to a single cluster. To attain fuzzy clustering, Bezdek's algorithm replaces linear membership weights with a power function, and creates cluster centers based on weighted means [2]. This produces what is commonly known as the fuzzy c-means (FCM) algorithm. To distinguish this algorithm from the many variants that have subsequently been proposed, we refer to this as Bezdek-type FCM (bFCM). Another fuzzy approach used for cluster analysis is the regularization of the HCM objective function. Recognizing that HCM is singular, and that an appropriate cluster cannot be obtained using the Lagrangian multiplier method, Miyamoto and Mukaidono introduced a regularization term into the objective function via the negative entropy of membership [3] with a positive parameter, thereby producing entropy-regularized FCM (eFCM).

In HCM, bFCM, and eFCM, the squared Euclidean distance is assumed to measure the dissimilarity between an object and a cluster center. However, there are many other (dis)similarity measures. In particular, spherical K-means [4] and its fuzzified variants [5],[6] calculate the cosine correlation between an object and a cluster center, and effectively employ this as the dissimilarity measure. These methods are referred to

V. Torra et al. (Eds.): MDAI 2014, LNAI 8825, pp. 108–121, 2014.

as spherical because the cosine correlation ignores the magnitude of the objects, thus assuming that all objects lie on the unit hypersphere. The spherical clustering methods that correspond to HCM, bFCM, and eFCM are denoted as sHCM, sbFCM, and seFCM in the present study.

All of the aforementioned clustering methods are minimizing models, i.e., the algorithms solve the corresponding minimization problems based on the dissimilarities between objects and clusters. Maximization models can also be considered, i.e., maximization problems based on the similarities between objects and clusters. The maximizing models of sHCM and seFCM (msHCM and mseFCM) are equivalent to the corresponding minimizing models (sHCM and seFCM), whereas a maximizing model of sbFCM has not been well defined [7]. In a previous study, a maximizing model of sbFCM (msbFCM) was proposed by the author [8].

To cluster objects with nonlinear borders, an algorithm [9] that uses a nonlinear transformation from the original pattern space into a higher-dimensional feature space with kernel functions [10] has been decribed. The explicit mapping is generally unknown for kernel data analysis, but the inner products between pairs of objects in feature space should be known. However, an explicit mapping has been introduced by Miyamoto, and this enabled the appearance of kernelinzed clustering in a higher-dimensional space to be described via kernel principal component analysis [11], [12]. With the exception of msbFCM, the aforementioned spherical clustering methods are kernelized (e.g., [5], [6]), and the kernelized algorithms produce good results for clusters with nonlinear borders.

In this study, we propose three modifications to msbFCM. The first modification is to kernelize msbFCM (K-msbFCM). The original msbFCM can only be applied to objects on the first quadrant of the unit hypersphere, whereas kernelized algorithms can be applied to wider classes of objects, because the Gaussian kernel and its variants [13] map the original data into the first quadrant of the unit hypersphere. The second modification is a spectral clustering approach to K-msbFCM undera certain assumption. This approach solves the local convergence problem of the original algorithm. The third modification constructs a model that gives an exact solution to the spectral clustering approach. The abbreviations and names of methods in this paper are summarized in Table 1.

The rest of this paper is organized as follows. In section 2, the notation and msbFCM that forms the basis of the proposed methods are introduced. Section 3 presents the basic concepts and proposed methods, and section 4 provides some illustrative examples. Section 5 contains our concluding remarks.

2 Preliminaries

Let $X = \{x_k \in \mathbb{R}^p \mid k \in \{1, \cdots, N\}\}$ be a dataset of p-dimensional points, and assume that every datum lies on the first quadrant of the unit hypersphere. The membership of x_k that belongs to the i-th cluster is denoted by $u_{i,k}$ ($i \in \{1, \cdots, C\}, k \in \{1, \cdots, N\}$), and the set of $u_{i,k}$ is denoted by u, which is also known as the partition matrix. The cluster center set is denoted by $v = \{v_i \mid v_i \in \mathbb{R}^p, i \in \{1, \cdots, C\}\}$. The inner product between the k-th datum and the i-th cluster center is denoted by $s_{i,k} = x_k^\mathsf{T} v_i$.

Table 1. Methods discussed in this paper

Abbreviation	Methods
HCM	hard c-means
bFCM	Bezdek-type fuzzy c-means
eFCM	entropy-regularized fuzzy c-means
sHCM	spherical hard c-means
sbFCM	spherical Bezdek-type fuzzy c-means
seFCM	spherical entropy-regularized fuzzy c-means
msHCM	maximizing model of spherical hard c-means, which is equivalent to sHCM
mseFCM	maximizing model of spherical entropy-regularized fuzzy c-means, which is equivalent to seFCM
msbFCM	maximizing model of spherical Bezdek-type fuzzy c-means, the basis of the proposed method
K-msbFCM	kernelized msbFCM, the proposed method
sK-msbFCM	spectral clustering approach to K-msbFCM, the proposed method
msK-msbFCM	model giving the exact solution of sK-msbFCM, the proposed method

msbFCM [8] aims to solve the optimization problem:

$$\underset{u,v}{\text{maximize}} \sum_{i=1}^{C} \sum_{k=1}^{N} u_{i,k}^{\frac{1}{m}} s_{i,k}, \tag{1}$$

$$\text{subject to } \sum_{i=1}^{C} u_{i,k} = 1, \tag{2}$$

$$\|v_i\|_2 = 1 \tag{3}$$

where $m > 1$ is a fuzzification parameter. The msbFCM algorithm is as follows:

Algorithm 1 (msbFCM [8])

STEP 1. Give the number of clusters C and the fuzzification parameter m, and set the initial cluster centers as v.

STEP 2. Calculate s by

$$s_{i,k} = x_k^\mathsf{T} v_i. \tag{4}$$

STEP 3. Calculate u by

$$u_{i,k} = \frac{s_{i,k}^{\frac{m}{m-1}}}{\sum_{j=1}^{C} s_{j,k}^{\frac{m}{m-1}}}. \tag{5}$$

STEP 4. Calculate v by

$$v_i = \frac{\sum_{k=1}^{N} u_{i,k}^{\frac{1}{m}} x_k}{\|\sum_{k=1}^{N} u_{i,k}^{\frac{1}{m}} x_k\|_2}. \tag{6}$$

STEP 5. Check the stopping criterion for (u, v). If the criterion is not satisfied, go to STEP 2. □

Note that msbFCM can only be applied to objects on the first quadrant of the unit hypersphere, because the concavity of the objective function, a necessary condition for optimality, is violated for the membership of objects outside this domain. This is in contrast to conventional spherical clustering methods, which are valid for objects on the whole unit hypersphere. However, this constraint does not greatly restrict the applicability of msbFCM. Document clustering is an application for msbFCM, where each document was described as a normalized term-frequency vector or term frequency-inverse document frequency (tf-idf) weighted vector with positive elements [8]. Regardless, it is desirable that msbFCM can be applied to a wider class of objects.

3 Proposed Method

3.1 Basic Concept

We first kernelize msbFCM (K-msbFCM). As stated in section 2, the original msbFCM can only be applied to objects on the first quadrant of the unit hypersphere, whereas its kernelized algorithm can be applied to wider classes of objects. Here, we consider objects $\{x_k\}_{k=1}^N$ not on the first quadrant of the unit hypersphere. The elements of the kernel matrix obtained from the Gaussian kernel are described as

$$K_{k,\ell} = \exp(-\sigma^2 \|x_k - x_\ell\|_2^2). \tag{7}$$

Note that

$$K_{k,k} = 1 \text{ and } K_{k,\ell} \in [0, 1], \tag{8}$$

that is, the norm induced by the inner product of each feature vector is 1, and the inner product of a pair of feature vectors ranges from zero to one. This implies that the feature vectors corresponding to the Gaussian kernel are on the first quadrant of the unit hypersphere. Therefore, using an adequate kernel does not restrict the dataset that can be applied to K-msbFCM.

Second, we derive a spectral clustering approach to K-msbFCM. We can obtain an equivalent maximization problem for the cluster centers by substituting membership update equation into the original problem. Next, assuming an extremely well-separated situation, an eigenproblem is obtained from the optimization problem with the fuzzification parameter $m = 2$, using the coefficients by which cluster centers are expressed as linear combinations of objects. This implies that the globally optimal solution can be obtained by solving that eigenproblem. Thus, no specific initial value setting is needed, so we overcome the local convergence problem of the original algorithm. Because the actual situation is not well-separated, some iterations are executed to update the membership values. Therefore, similar to spectral clustering, this algorithm consists of two stages: (1)solving the eigenproblem, and (2)updating the optimal solutions.

In experiments using the spectral clustering approach of msbFCM, we found that one iteration in the second stage was sufficient to produce good clustering results. This implies that there is an algorithm for the clustering model that consists of only the first stage and one iteration of the second stage in the spectral clustering approach of K-msbFCM. This is the motivation for the third modification to msbFCM, which provides an exact solution to the spectral clustering approach. For slightly generalized objective function with additional parameter, when the fuzzification parameter value is set to

2.0 and another parameter is set to 2.0, the derived algorithm consists of solving the eigenproblem and one iteration to update the optimal solutions.

3.2 K-msbFCM

For a given set of objects $X = \{x_k \mid k \in \{1,\ldots,N\}\}$, K-msbFCM assumes that the kernel matrix $K \in \mathbb{R}^{N \times N}$ is known. Let $\mathbb{H}$ be a higher-dimensional feature space, $\Phi : X \to \mathbb{H}$ be a map from the data set X to the feature space $\mathbb{H}$, and $W = \{W_i \in \mathbb{H} \mid i \in \{1,\cdots,C\}\}$ be a set of cluster centers in the feature space.

K-msbFCM solves the following optimization problem:

$$\operatorname*{maximize}_{u,W} \sum_{i=1}^{C} \sum_{k=1}^{N} u_{i,k}^{\frac{1}{m}} s_{i,k} \tag{9}$$

$$\text{subject to } \langle W_i, W_i \rangle = 1, \tag{10}$$

and Eq. (2), where $s_{i,k} = \langle \Phi(x_k), W_i \rangle$. The Lagrangian $L(u,v)$ is described as

$$L(u,v) = \sum_{i=1}^{C} \sum_{k=1}^{N} u_{i,k}^{\frac{1}{m}} s_{i,k} + \sum_{k=1}^{C} \gamma_k \left(1 - \sum_{i=1}^{C} u_{i,k} \right) + \sum_{i=1}^{C} \nu_i \left(1 - \|W_i\|_{\mathbb{H}}^2 \right) \tag{11}$$

with Lagrange multipliers (γ, ν). The necessary conditions for optimality can be written as

$$\frac{\partial L(u,W)}{\partial u_{i,k}} = 0, \tag{12}$$

$$\frac{\partial L(u,W)}{\partial W_i} = 0, \tag{13}$$

$$\frac{\partial L(u,W)}{\partial \gamma_k} = 0, \tag{14}$$

$$\frac{\partial L(u,W)}{\partial \nu_i} = 0. \tag{15}$$

Optimal membership is given by Eq. (12) in the form

$$u_{j,k} = \left(\frac{1}{m\gamma_k} \right)^{\frac{m}{m-1}} s_{j,k}^{\frac{m}{m-1}} \tag{16}$$

with Lagrange multiplier γ_k. Summing over the cluster index $j \in \{1,\cdots,C\}$ and considering Eq. (14)$\Leftrightarrow$Eq. (2), we have

$$\left(\frac{1}{m\gamma_k} \right)^{\frac{m}{m-1}} \sum_{j=1}^{C} s_{j,k}^{\frac{m}{m-1}} = 1 \Leftrightarrow \left(\frac{1}{m\gamma_k} \right)^{\frac{m}{m-1}} = \frac{1}{\sum_{j=1}^{C} s_{j,k}^{\frac{m}{m-1}}}. \tag{17}$$

By inserting Eq. (16) into this equation, we can eliminate γ_k, yielding

$$u_{i,k} = \frac{s_{i,k}^{\frac{m}{m-1}}}{\sum_{j=1}^{C} s_{j,k}^{\frac{m}{m-1}}}. \tag{18}$$

The optimal cluster center is obtained using Eq. (13) as

$$W_i = \frac{1}{2\nu_i} \sum_{k=1}^{N} u_{i,k}^{\frac{1}{m}} \Phi(x_k) \tag{19}$$

with Lagrange multiplier ν_i. By considering the squared norm and taking Eq. (15)$\Leftrightarrow$ Eq. (10) into account, we have

$$\frac{1}{(2\nu_i)^2} \left\| \sum_{k=1}^{N} u_{i,k}^{\frac{1}{m}} \Phi(x_k) \right\|_{\mathbb{H}}^2 = 1 \Leftrightarrow \frac{1}{2\nu_i} = \frac{1}{\left\| \sum_{k=1}^{N} u_{i,k}^{\frac{1}{m}} \Phi(x_k) \right\|_{\mathbb{H}}}. \tag{20}$$

Inserting Eq. (20) into Eq. (19), eliminating ν_i, we have

$$W_i = \frac{\sum_{k=1}^{N} u_{i,k}^{\frac{1}{m}} \Phi(x_k)}{\left\| \sum_{k=1}^{N} u_{i,k}^{\frac{1}{m}} \Phi(x_k) \right\|_{\mathbb{H}}}. \tag{21}$$

Generally, Φ cannot be given explicitly, so the K-msbFCM algorithm assumes that a kernel function $\mathcal{K} : \mathbb{R}^p \times \mathbb{R}^p \to \mathbb{R}$ is given. This function describes the inner product value of pairs of the objects in the feature space as $\mathcal{K}(x_k, x_j) = \langle \Phi(x_k), \Phi(x_j) \rangle$. However, it can be interpreted that Φ is given explicitly by allowing $\mathbb{H} = \mathbb{R}^N$, $\Phi(x_k) = e_k$, where e_k is the N-dimensional unit vector whose ℓ-th element is the Kronecker delta $\delta_{k,\ell}$, and by introducing $K \in \mathbb{R}^{N \times N}$ such that

$$K_{k,j} = \langle \Phi(x_k), \Phi(x_j) \rangle. \tag{22}$$

Using this kernel matrix K, $s_{i,k}$ is described as

$$\begin{aligned} s_{i,k} &= \left\langle \Phi(x_k), \frac{\sum_{\ell=1}^{N} u_{i,\ell}^{\frac{1}{m}} \Phi(x_\ell)}{\left\| \sum_{\ell=1}^{N} u_{i,\ell}^{\frac{1}{m}} \Phi(x_\ell) \right\|_{\mathbb{H}}} \right\rangle = \frac{\sum_{\ell=1}^{N} u_{i,\ell}^{\frac{1}{m}} \langle \Phi(x_k), \Phi(x_\ell) \rangle}{\sqrt{\left\langle \sum_{\ell=1}^{N} u_{i,\ell}^{\frac{1}{m}} \Phi(x_\ell), \sum_{r=1}^{N} u_{i,r}^{\frac{1}{m}} \Phi(x_r) \right\rangle}} \\ &= \frac{\sum_{\ell=1}^{N} u_{i,\ell}^{\frac{1}{m}} K_{k,\ell}}{\sqrt{\sum_{\ell=1}^{N} \sum_{r=1}^{N} u_{i,\ell}^{\frac{1}{m}} u_{i,r}^{\frac{1}{m}} \langle \Phi(x_\ell), \Phi(x_r) \rangle}} = \frac{\sum_{\ell=1}^{N} u_{i,\ell}^{\frac{1}{m}} K_{k,\ell}}{\sqrt{\sum_{\ell=1}^{N} \sum_{r=1}^{N} u_{i,\ell}^{\frac{1}{m}} u_{i,r}^{\frac{1}{m}} K_{\ell,r}}}. \end{aligned} \tag{23}$$

Therefore, the K-msbFCM algorithm consists of updating (u, s) as follows:

Algorithm 2 (K-msbFCM)

STEP 1. Give the number of clusters C and the fuzzification parameter m, and set the initial partition matrix as u.

STEP 2. Calculate s by Eq. (23).

STEP 3. Calculate u by Eq. (18).

STEP 4. Check the stopping criterion for (u, s). If the criterion is not satisfied, go to Step. 2.   □

This algorithm can be applied to any kernel matrix, satisfying Eq. (8) such as the Gaussian kernel and its variants (e.g., [13]).

3.3   Spectral Clustering Approach to K-msbFCM

In this subsection, a spectral clustering approach to K-msbFCM is proposed. First, we obtain an equivalent objective function to Eq. (9) with $m = 2$ as

$$
\begin{aligned}
\sum_{i=1}^{C} \sum_{k=1}^{N} u_{i,k}^{1/2} s_{i,k} &= \sum_{i=1}^{C} \sum_{k=1}^{N} \frac{s_{i,k}}{(\sum_{j=1}^{C} s_{j,k}^2)^{1/2}} s_{i,k} \\
&= \sum_{i=1}^{C} \sum_{k=1}^{N} \frac{s_{i,k}^2}{(\sum_{j=1}^{C} s_{j,k}^2)^{1/2}} \\
&= \sum_{k=1}^{N} \frac{\sum_{i=1}^{C} s_{i,k}^2}{(\sum_{j=1}^{C} s_{j,k}^2)^{1/2}} \\
&= \sum_{k=1}^{N} \left(\sum_{i=1}^{C} s_{i,k}^2 \right)^{\frac{1}{2}}
\end{aligned}
\tag{24}
$$

by substituting the membership update equation (18) into the original problem (9). Next, assuming the extremely well-separated situation where

$$
\begin{aligned}
\langle W_i, W_j \rangle &= \delta_{i,j}, & (25) \\
\Phi(x_k) &\in \{W_i\}_{i=1}^{C}, & (26)
\end{aligned}
$$

$s_{i,k}$ is described as

$$
s_{i,k}^2 = \langle \Phi(x_k), W_i \rangle^2 =
\begin{cases}
1 & (\Phi(x_k) = W_i), \\
0 & (\text{otherwise}),
\end{cases}
\tag{27}
$$

from which we have

$$
\sum_{i=1}^{C} s_{i,k}^2 = 1,
\tag{28}
$$

and hence

$$\left(\sum_{i=1}^{C} s_{i,k}^2\right)^{\frac{1}{2}} = \sum_{i=1}^{C} s_{i,k}^2. \tag{29}$$

Thus, the objective function (24) is described as

$$\sum_{k=1}^{N}\sum_{i=1}^{C} s_{i,k}^2. \tag{30}$$

Furthermore, rewriting the cluster center W_i as a linear combination, similar to Eq. (21), i.e.,

$$W_i = \sum_{\ell=1}^{N} a_{i,\ell}\Phi(x_\ell) \tag{31}$$

with coefficients $a_{i,\ell}$, $s_{i,k}$ can be written as

$$s_{i,k} = \langle \Phi(x_k), W_i\rangle = \sum_{\ell=1}^{N} a_{i,\ell}\langle \Phi(x_k), \Phi(x_\ell)\rangle = \sum_{\ell=1}^{N} a_{i,\ell}K_{k,\ell}. \tag{32}$$

Therefore, the objective function (30) is given by

$$\sum_{k=1}^{N}\sum_{i=1}^{C} s_{i,k}^2 = \sum_{k=1}^{N}\sum_{i=1}^{C}\left(\sum_{\ell=1}^{N} a_{i,\ell}K_{k,\ell}\right)^2 = \sum_{k=1}^{N}\sum_{i=1}^{C}\sum_{\ell=1}^{N}\sum_{r=1}^{N} a_{i,\ell}a_{i,r}K_{k,\ell}K_{k,r}$$
$$= \text{trace}(A^\mathsf{T} K^2 A), \tag{33}$$

where the (ℓ, i)-th element of $A \in \mathbb{R}^{N \times C}$ is $a_{i,\ell}$. Additionally, Eqs. (25) and (31) imply that

$$\langle W_i, W_j\rangle = \left\langle \sum_{\ell=1}^{N} a_{i,\ell}\Phi(x_\ell), \sum_{r=1}^{N} a_{j,r}\Phi(x_r)\right\rangle = \sum_{\ell=1}^{N}\sum_{r=1}^{N} a_{i,\ell}a_{j,r}\langle \Phi(x_\ell), \Phi(x_r)\rangle$$
$$= \sum_{\ell=1}^{N}\sum_{r=1}^{N} a_{i,\ell}a_{j,r}K_{\ell,r} = \delta_{i,j}, \tag{34}$$

that is,

$$A^\mathsf{T} KA = E, \tag{35}$$

where E is the N-dimensional unit matrix. Therefore, the optimization problem of K-msbFCM under assumptions (25) and (26) is simply

$$\underset{A}{\text{maximize}}\,\text{trace}(A^{\mathsf{T}}K^2A) \tag{36}$$
$$\text{subject to } A^{\mathsf{T}}KA = E. \tag{37}$$

Using $B = K^{\frac{1}{2}}A$, the above problem can be described as

$$\underset{B}{\text{maximize}}\,\text{trace}(B^{\mathsf{T}}KB)$$
$$\text{subject to } B^{\mathsf{T}}B = E, \tag{38}$$

whose globally optimal solution can be obtained from the first C eigenvectors $\{b_i\}_{i=1}^{C}$ of K, written in descending order as $B = (b_1, \ldots, b_C)$, from which we have $A = K^{-\frac{1}{2}}B$. Then, $s_{i,k}$ is given by

$$s_{i,k} = \sum_{\ell=1}^{N} a_{i,\ell}K_{k,\ell} = e_k^{\mathsf{T}}Ka_i = e_k^{\mathsf{T}}KK^{-\frac{1}{2}}b_i = e_k^{\mathsf{T}}K^{\frac{1}{2}}b_i = \sqrt{\lambda_i}e_k^{\mathsf{T}}b_i = \sqrt{\lambda_i}b_{i,k}, \tag{39}$$

where λ_i is the eigenvalue corresponding to b_i and e_k is the k-th unit vector. With this expression for $s_{i,k}$, the membership $u_{i,k}$ is described as

$$u_{i,k} = \frac{s_{i,k}^2}{\sum_{j=1}^{C}s_{j,k}^2} = \frac{\lambda_i b_{i,k}^2}{\sum_{j=1}^{C}\lambda_j b_{j,k}^2}. \tag{40}$$

Because the actual situation is not well-separated, some iterations must be executed to update the memberships according to Eqs. (18) and (23). The above analysis suggests the following algorithm:

Algorithm 3 (Spectral Clustering Approach to K-msbFCM)

STEP 1. Give the number of clusters C, obtain the first C eigenpairs $\{(\lambda_i, b_i)\}_{i=1}^{C}$ of K in descending order, and set the initial partition according to Eq. (40).

STEP 2. Calculate s by Eq. (23).

STEP 3. Calculate u by Eq. (18).

STEP 4. Check the stopping criterion for (u, s). If the criterion is not satisfied, go to Step. 2. □

In this algorithm, a random initial value setting is not needed, solving the local convergence problem of the original algorithm. Similar to spectral clustering techniques, this algorithm consists of two stages: (1)solving the eigenproblem, and (2)updating the optimal solutions.

3.4   Modified K-msbFCM

We consider a slightly generalized objective function with additional parameter as,

$$\underset{u,W}{\text{maximize}} \sum_{k=1}^{N} \left(\sum_{i=1}^{C} u_{i,k}^{\frac{1}{m_1}} s_{i,k} \right)^{m_2} \tag{41}$$

subject to Eqs. (2) and (10), where m_1 is the fuzzification parameter m of K-msbFCM, and m_2 is another introduced parameter. In the present study, we consider $(m_1, m_2) = (2, 2)$.

In this case, the Lagrangian for $u_{i,k}$ is

$$\sum_{k=1}^{N} \left(\sum_{i=1}^{C} u_{i,k}^{\frac{1}{2}} s_{i,k} \right)^2 + \sum_{k=1}^{N} \gamma_k \left(1 - \sum_{i=1}^{C} u_{i,k} \right) \tag{42}$$

with Lagrangian multipliers $\{\gamma_k\}_{k=1}^{N}$. This is concave for $u_{i,k}$, and the necessary conditions of optimality are

$$\frac{\partial L(u)}{\partial u_{i,k}} = 0, \tag{43}$$

$$\frac{\partial L(u)}{\partial \gamma_k} = 0. \tag{44}$$

Optimal membership is obtained from Eq. (43) as

$$u_{i,k} = \left(\frac{J_k}{\gamma_k} \right)^2 s_{i,k}^2 \tag{45}$$

where

$$J_k = \sum_{i=1}^{C} u_{i,k}^{\frac{1}{2}} s_{i,k}. \tag{46}$$

Summing over the cluster index $j \in \{1, \cdots, C\}$ and considering Eq. (44)$\Leftrightarrow$Eq. (2), we have

$$\left(\frac{J_k}{\gamma_k} \right)^2 \sum_{j=1}^{C} s_{j,k}^2 = 1 \Leftrightarrow \left(\frac{J_k}{\gamma_k} \right)^2 = \frac{1}{\sum_{j=1}^{C} s_{j,k}^2}. \tag{47}$$

By inserting Eq. (45) into this equation, we can eliminate γ_k to yield

$$u_{i,k} = \frac{s_{i,k}^2}{\sum_{j=1}^{C} s_{j,k}^2}, \tag{48}$$

which is the same as the optimal membership expression of Eq. (18) in Algorithms 2 (K-msbFCM) and 3 (sK-msbFCM). Substituting this equation into the original objective function (41), we obtain an equivalent objective function

$$\sum_{k=1}^{N}\sum_{i=1}^{C} s_{i,k}^2 \tag{49}$$

leading the trace maximization problem in Eqs. (36) and (37) with the globally optimal solution obtained by solving the corresponding eigenproblem. Unlike the discussion in the previous subsection, without assuming the extremely well-separated situation (i.e., Eq. (25) holds, but Eq. (26) does not), the globally optimal membership can be obtained from the optimization problem

$$\underset{u,W}{\text{maximize}} \sum_{k=1}^{N} \left(\sum_{i=1}^{C} u_{i,k}^{\frac{1}{2}} s_{i,k} \right)^2 \tag{50}$$

$$\text{subject to Eqs. (2) and (25)}$$

as the following algorithm:

Algorithm 4 (Modified sK-msbFCM (msK-msbFCM))

STEP 1. Give the number of clusters C, obtain the first C eigenpairs $\{(\lambda_i, b_i)\}_{i=1}^{C}$ of K in descending order, and set the partition according to Eq. (40).   □

4   Numerical Example

This section shows the validify of the proposed methods (Algorithm 2–4) using the artificial datasets shown in Figs. 1(a) and 1(b), both of which consist of two nonlinearly bordered clusters.

In all the algorithms, the Gaussian kernel

$$K_{k,\ell} = \exp(-\sigma \|x_k - x_\ell\|_2^2) \tag{51}$$

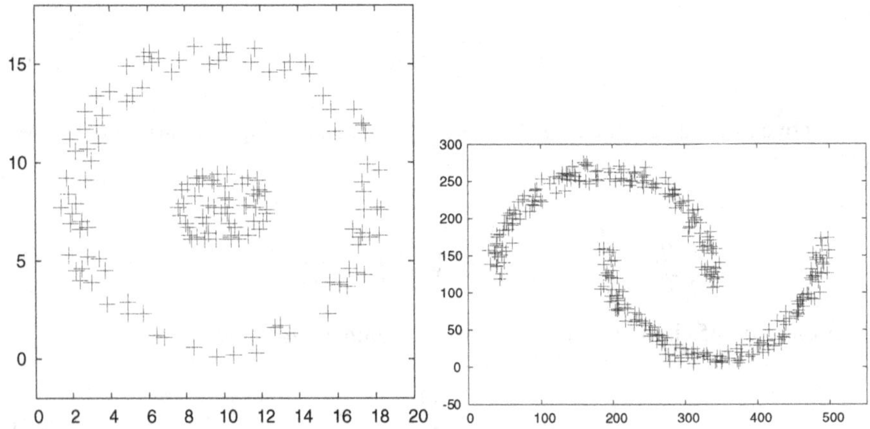

(a) Data#1   (b) Data#2

Fig. 1. Artificial Dataset #1 (left) and #2 (right)

is used, where the parameter $\sigma \in \{0.1, 0.2, 0.3, 0.4, 0.5\}$ for dataset #1 and $\sigma \in \{1.0 \times 10^{-4}, 2.0 \times 10^{-4}, 3.0 \times 10^{-4}, 4.0 \times 10^{-4}, 5.0 \times 10^{-4}\}$ for dataset #2. Algorithm 2 (K-msbFCM) is applied to the two datasets with $C = 2$, $m \in \{10, 3, 2, 1.5, 1.1\}$, and 100 different initial settings. Algorithms 3 (sK-msbFCM) and 4 (msk=msbFCM) are also applied to the same datasets with $C = 2$, $m = 2$, and no initial setting is needed.

We use normalized mutual information (NMI) [14] to evaluate the clustering results. NMI takes a value from zero to one, and higher values are preferred. The frequency ratio of NMI $= 1$ produced by all the algorithms is shown in Table 2 for dataset #1 and in Table 3 for dataset #2. The results for Algorithms 3 (sK-msbFCM) and 4 (msK-msbFCM) are 0.00 or 1.00, because these algorithms are not affected by the initial setting.

Table 2. Frequency ratio of NMI $= 1$ for dataset #1

		σ				
	m	0.1	0.2	0.3	0.4	0.5
Algorithm 2 (K-msbFCM)	1.1	0.60	0.28	0.03	0.00	0.00
	1.5	1.00	0.78	0.81	0.41	0.06
	2.0	0.00	0.00	0.01	0.04	0.02
	3.0	0.00	0.00	0.00	0.00	0.00
	10	0.00	0.00	0.00	0.00	0.00
Algorithm 3 (sK-msbFCM)	2.0	0.00	1.00	1.00	1.00	1.00
Algorithm 4 (msK-msbFCM)	2.0	0.00	1.00	1.00	1.00	1.00

Table 3. Frequency ratio of NMI $= 1$ for dataset #2

		σ				
	m	1×10^{-4}	2×10^{-4}	3×10^{-4}	4×10^{-4}	5×10^{-4}
Algorithm 2 (K-msbFCM)	1.1	0.00	0.04	0.01	0.00	0.01
	1.5	0.00	0.06	0.10	0.01	0.03
	2.0	0.00	0.00	0.03	0.00	0.04
	3.0	0.00	0.00	0.00	0.02	0.01
	10	0.00	0.00	0.01	0.00	0.00
Algorithm 3 (sK-msbFCM)	2.0	0.00	0.00	1.00	1.00	1.00
Algorithm 4 (msK-msbFCM)	2.0	0.00	0.00	0.00	1.00	1.00

These tables show that Algorithm 2 (K-msbFCM) achieves NMI $= 1$ at least once for several pairs (m, σ). However, only $(m, \sigma) = (1.5, 0.1)$ does the frequency ratio of NMI $= 1$ reach 1.00. This suggests that it is difficult to set both an adequate initial partition matrix and adequate parameter values (m, σ). On the other hand, we can see that Algorithm 3 (sK-msbFCM) achieves NMI $= 1$ for most values of σ, from which we deduce that the initial setting in the first stage of Algorithm 3 (sK-msbFCM) is effective. Because Algorithm 3 (sK-msbFCM) does not always achieve NMI $= 1$ (e.g., $\sigma \leq 0.1$ for dataset #1 and $\sigma \leq 2 \times 10^{-4}$ for dataset #2), it is not easy to set a value for σ even for Algorithm 3. These tables also show that Algorithm 4 (msK-msbFCM) achieves NMI = 1 for many values of σ, though not as often as Algorithm 3 (sK-msbFCM). Although we cannot say that the iterative refinement in the second stage of Algorithm 3 (sK-msbFCM) is unnecessary, we find that Algorithm 4 (msK-msbFCM), which does not employ such iteration, produces good clustering results in many cases.

5   Conclusions

This study proposed and implemented three modifications to msbFCM. Numerical examples demonstrated that the proposed methods produce good results for nonlinearly bordered clusters with adequate parameter values. The limitation in Algorithm 2 (K-msbFCM) is its initialization technique, which inherits the property of msbFCM originated from HCM. Algorithm 2 (K-msbFCM) follows randomly generated initial starting points which often result in poor clustering results. The better clustering results can be accomplished after several iterations. However, it is very complicated to decide the computation limit for obtaining better results. On the other hand, the framework of spectral clustering can free itself from initialization by incorporating the clustering problem into an eigenproblem. Algorithms 3 (sK-msbFCM) and 4 (msK-msbFCM), with the help of spectral clustering approach, overcome the restriction of local convergence suffered by Algorithm 2 (K-msbFCM).

In future research, the proposed method will be compared with conventional methods using many large and complex real datasets and various kernel parameter selection methods [15],[16]. The generalized objective function (41) with other parameter valuses than $(m_1, m_1) = (2, 2)$ will be also investigated. Next, the proposed method will be extended as follows. By introducing a variable to control the cluster sizes, the proposed method will capture clusters of various sizes, similar to the extensions of HCM and eFCM [7]. Using the technique proposed in this study, the linear membership weights of msHCM can be replaced with a less-than-one power of membership, and this will be applied to other methods, such as fuzzy nonmetric model [17], Windham's AP algorithm [18], possibilistic c-means [19], and co-clustering [20].

Acknowledgment. This work has partly been supported by the Grant-in-Aid for Scientific Research, Japan Society for the Promotion of Science, No. 00298176.

References

1. MacQueen, J.B.: Some Methods of Classification and Analysis of Multivariate Observations. In: Proc. 5th Berkeley Symposium on Math. Stat. and Prob., pp. 281–297 (1967)
2. Bezdek, J.: Pattern Recognition with Fuzzy Objective Function Algorithms. Plenum Press, New York (1981)
3. Miyamoto, S., Mukaidono, M.: Fuzzy c-Means as a Regularization and Maximum Entropy Approach. In: Proc. 7th Int. Fuzzy Systems Association World Congress (IFSA 1997), vol. 2, pp. 86–92 (1997)
4. Dhillon, I.S., Modha, D.S.: Concept Decompositions for Large Sparse Text Data Using Clustering. Machine Learning 42, 143–175 (2001)

5. Miyamoto, S., Mizutani, K.: Fuzzy Multiset Model and Methods of Nonlinear Document Clustering for Information Retrieval. In: Torra, V., Narukawa, Y. (eds.) MDAI 2004. LNCS (LNAI), vol. 3131, pp. 273–283. Springer, Heidelberg (2004)

6. Mizutani, K., Inokuchi, R., Miyamoto, S.: Algorithms of Nonlinear Document Clustering based on Fuzzy Set Model. International Journal of Intelligent Systems 23(2), 176–198 (2008)

7. Miyamoto, S., Ichihashi, H., Honda, K.: Algorithms for Fuzzy Clustering. Springer (2008)

8. Kanzawa, Y.: Maximizing Model of Bezdek-like Spherical Fuzzy c-Means Clustering. In: Proc. WCCI 2014 (to appear, 2014)

9. Miyamoto, S., Suizu, D.: Fuzzy c-Means Clustering Using Kernel Functions in Support Vector Machines. J. Advanced Computational Intelligence and Intelligent Informatics 7(1), 25–30 (2003)

10. Vapnik, V.N.: Statistical Learning Theory. Wiley, New York (1998)

11. Miyamoto, S., Kawasaki, Y., Sawazaki, K.: An Explicit Mapping for Kernel Data Analysis and Application to Text Analysis. In: Proc. IFSA-EUSFLAT 2009, pp. 618–623 (2009)

12. Miyamoto, S., Sawazaki, K.: An Explicit Mapping for Kernel Data Analysis and Application to c-Means Clustering. In: Proc. NOLTA 2009, pp. 556–559 (2009)

13. Zelnik-Manor, L., Perona, P.: Self-tuning Spectral Clustering. In: Advances in Neural Information Processing Systems, vol. 17, pp. 1601–1608 (2005)

14. Ghosh, G., Strehl, A., Merugu, S.: A Consensus Framework for Integrating Distributed Clusterings under Limited Knowledge Sharing. In: Proc. NSF Workshop on Next Generation Data Mining, pp. 99–108 (2002)

15. Wang, L., Chan, K.L.: Learning kernel parameters by using class separability measure. In: Proc. NIPS 2002 (2002)

16. Lu, C., Zhu, Z., Gu, X.: Kernel Parameter Optimization in Stretched Kernel-Based Fuzzy Clustering. In: Zhou, Z.-H., Schwenker, F. (eds.) PSL 2013. LNCS (LNAI), vol. 8183, pp. 49–57. Springer, Heidelberg (2013)

17. Roubens, M.: Pattern Classification Problems and Fuzzy Sets. Fuzzy Sets and Syst. 1, 239–253 (1978)

18. Windham, M.P.: Numerical Classification of Proximity Data with Assignment Measures. J. Classification 2, 157–172 (1985)

19. Krishnapuram, R., Keller, J.M.: A Possibilistic Approach to Clustering. IEEE Trans. on Fuzzy Systems 1, 98–110 (1993)

20. Oh, C., Honda, K., Ichihashi, H.: Fuzzy Clustering for Categorical Multivariate Data. In: Proc. IFSA World Congress and 20th NAFIPS International Conference, pp. 2154–2159 (2001)

A Note on Objective-Based Rough Clustering with Fuzzy-Set Representation

Ken Onishi[1], Naohiko Kinoshita[1], and Yasunori Endo[2]

[1] Graduate School of Sys. and Info. Eng., University of Tsukuba
Tennodai 1-1-1, Tsukuba, Ibaraki, 305-8573, Japan
{s1320619,s1220594}@u.tsukuba.ac.jp
[2] Faculty of Eng., Info. and Sys., University of Tsukuba
Tennodai 1-1-1, Tsukuba, Ibaraki, 305-8573, Japan
endo@risk.tsukuba.ac.jp

Abstract. Clustering is a method of data analysis. Rough k-means (RKM) by Lingras et al. is one of rough clustering algorithms[3]. The method does not have a clear indicator to determine the most appropriate result because it is not based on objective function. Therefore we proposed a rough clustering algorithm based on optimization of an objective function [7]. This paper will propose a new rough clustering algorithm based on optimization of an objective function with fuzzy-set representation to obtain better lower approximation, and estimate the effectiveness through some numerical examples.

Keywords: clustering, rough clustering, optimization, fuzzy set.

1 Introduction

Data have become large-scale and complex in recent years. We cannot get useful information without computers. The importance of data analysis techniques has been increasing accordingly and various data analysis methods have been proposed. Clustering is one of the major techniques in pattern recognition. Clustering is a technique automatically classifying data into some clusters. Many researchers have been interested in clustering as a significant data analysis method.

Types of clustering are divided broadly into hierarchical and non-hierarchical clustering. The standard techniques of non-hierarchical clustering are called objective-based clustering. The objective-based clustering is constructed to minimize a given objective function. Therefore, the objective function plays many important role in objective-based clustering.

From the viewpoint of the membership of an object to each cluster, called membership grade, types of clustering are divided into crisp and fuzzy. The value of membership grade is 0 or 1 in crisp clustering. The value is included into the unit interval [0,1] in fuzzy clustering. Fuzzy clustering allows an object to belong more than one cluster at the same time. That is why fuzzy clustering can be regarded as more flexible than crisp clustering. On the other hand, it is pointed out that the fuzzy degree of membership may be too descriptive for

V. Torra et al. (Eds.): MDAI 2014, LNAI 8825, pp. 122–134, 2014.

interpreting clustering results. In such cases, rough set representation is a more useful and powerful tool[1][2].

Recently, clustering based on rough set theory has attracted some attention[3]. Rough clustering represents a cluster by using two layers, upper and lower approximations. We can regard rough clustering as three-value clustering, that is, into the cluster, out of the cluster and unknown. The lower approximation means that an object surely belongs to the set and the upper one means that an object possibly belongs to the set. Clustering based on rough-set representation could provide a solution that is less restrictive than conventional clustering and less descriptive than fuzzy clustering, and therefore clustering based on rough set representation has attracted increasing interest of researchers[4][5][6].

However, traditional rough clustering does not have an objective function. For that reason the problem is pointed out that we cannot evaluate the result quantitatively. In order to solve this problem, a rough clustering algorithm based on optimization of an objective function was proposed[7]. But the algorithm has a problem that an object cannot belong to more than two upper approximations.

This paper proposes new rough clustering algorithms based on optimization of an objective function with fuzzy-set representation and estimate the effectiveness through some numerical examples.

2   Conventional Rough Clusterings

2.1   Rough Sets

Let U be the universe and $R \subseteq U \times U$ be an equivalence relation on U. R is also called equivalence relation. The pair $X = (U, R)$ is called approximation space. If $x, y \in U$ and $(x, y) \in R$, we say that x and y are indistinguishable in X.

Equivalence class of the relation R is called elementary set in X. The family of all elementary sets is denoted by U/R. The empty set is also elementary in every X.

Since it is impossible to distinguish each element in an equivalence class, we may not be able to get a precise representation for an arbitrary subset $A \subseteq U$. Instead, any A can be represented by its lower and upper bounds. The upper bound $\overline{A}$ is the least composed set in X containing A, called the best upper approximation or, in short, upper approximation. The lower bound $\underline{A}$ is the greatest composed set in X containing A, called the best lower approximation or, briefly, lower approximation. The set $\mathrm{Bnd}(A) = \overline{A} - \underline{A}$ is called the boundary of A in X.

The pair $(\underline{A}, \overline{A})$ is the representation of an ordinary set A in the approximation space X, or simply a rough set of A. The elements in the lower approximation of A definitely belong to A, while elements in the upper bound of A may or may not belong to A.

2.2   Rough k-Means

In this section, we explain rough k-means (RKM) by Lingras. From the above section of rough sets, we can define the following conditions for clustering.

(C1) An objet x can be part of at most one lower approximation.

(C2) If $x \in \underline{A} \Longrightarrow x \in \overline{A}$

(C3) An object x is not part of any lower approximation if and only if x belongs to two or more boundaries.

Cluster centers are updated by

$$
v_i = \begin{cases} \underline{\omega} \times \dfrac{\sum_{x_k \in \underline{A_i}} x_k}{|\underline{A_i}|} + \overline{\omega} \times \dfrac{\sum_{x_k \in \mathrm{Bnd}(A_i)} x_k}{|\mathrm{Bnd}(A_i)|}, & (\underline{A_i} \neq \emptyset \wedge \mathrm{Bnd}(A_i) \neq \emptyset) \\ \dfrac{\sum_{x_k \in \overline{A_i}} x_k}{|\overline{A_i}|}. & (\text{otherwise}) \end{cases}
$$

The coefficients $\underline{\omega}$ and $\overline{\omega}$ are weights of lower approximations and boundaries, respectively. $\underline{\omega}$ and $\overline{\omega}$ satisfy as follows:

$$
\underline{\omega} > 0, \quad \overline{\omega} > 0, \quad \underline{\omega} + \overline{\omega} = 1, \quad 1 \leq k \leq n, \quad 1 \leq i \leq c.
$$

Lower approximations and boundaries are calculated as follows:

$$
d_{ki} = \|x_k - v_i\|^2, \quad d_{km} = \min_{1 \leq i \leq c} d_{ki}
$$
$$
T = \{i \mid d_{ki} - d_{km} \leq \text{threshold}\} \qquad (i \neq m)
$$
$$
T \neq \emptyset \Rightarrow x_k \in \overline{A_m} \text{ and } x_k \in \overline{A_i} \qquad (\forall i \in T)
$$
$$
T = \emptyset \Rightarrow x_k \in \underline{A_m} .
$$

Algorithm 1. RKM

RKM0 Give initial cluster centers.

RKM1 Calculate lower approximations and boundaries.

RKM2 Calculate cluster centers.

RKM3 If the stop criterion satisfies, finish. Otherwise back to **RKM1**.

2.3   Rough c-Means

RKM has the following problems.

- Since RKM does not have objective function, there is no guidance to estimate the validity of the obtained results.
- There is no guidance to determine the threshold.

Rough c-means (RCM) which is based on optimization of an objective function was proposed to solve the above problems by Endo et al[7].

The objective function of RCM is defined as follows:

$$
J_{RCM} = \sum_{i=1}^{c} \sum_{k=1}^{n} \sum_{l=1}^{n} (\nu_{ki} u_{li} (\underline{\omega} d_{ki} + \overline{\omega} d_{li}) + (\nu_{ki} \nu_{li} + u_{ki} u_{li}) D_{kl}) .
$$

ν_{ki} represents a membership grade of x_k to a lower approximation of cluster of i. u_{li} represents a membership grade of x_l to a boundary of cluster of i. Here, $d_{si} = \|x_s - v_i\|^2$ and $D_{kl} = \|x_k - x_l\|^2$.

The constraints are as follows:

$$\underline{\omega} + \overline{\omega} = 1, \quad \nu_{ki} \in \{0,1\}, \quad u_{ki} \in \{0,1\}, \quad \sum_{i=1}^{c} \nu_{ki} \in \{0,1\}, \quad \sum_{i=1}^{c} u_{ki} \neq 1,$$

$$\sum_{i=1}^{c} \nu_{ki} = 1 \Longleftrightarrow \sum_{i=1}^{c} u_{ki} = 0.$$

Those constraints obviously satisfy the above conditions **C1**, **C2** and **C3**. Actually, those constraints are rewritten as:

$$\sum_{i=1}^{c} \nu_{ki} = 0 \Longleftrightarrow \sum_{i=1}^{c} u_{ki} = 2.$$

The cluster center v_i is calculated as follows:

$$v_i = \underline{\omega} \times \frac{\sum_{x_k \in \underline{A_i}} x_k}{|\underline{A_i}|} + \overline{\omega} \times \frac{\sum_{x_k \in \mathrm{Bnd}(A_i)} x_k}{|\mathrm{Bnd}(A_i)|}.$$

The optimal solutions to N and U are updated as follows:

$$\nu_{ki} = \begin{cases} 1, & (J_k^{\nu} < J_k^{u} \wedge i = p_k) \\ 0, & (\text{otherwise}) \end{cases}$$

$$u_{ki} = \begin{cases} 1, & (J_k^{\nu} > J_k^{u} \wedge (i = p_k \vee i = q_k)) \\ 0. & (\text{otherwise}) \end{cases}$$

Here p_k, q_k, J_k^{ν} and J_k^{u} are calculated as follows:

$$p_k = \arg\min_{i} d_{ki}, \quad q_k = \arg\min_{i \neq p_k} d_{ki},$$

$$J_k^{\nu} = \sum_{l=1,l\neq k}^{n} \nu_{kp_k}(u_{lp_k}(\underline{\omega} d_{kp_k} + \overline{\omega} d_{lp_k}) + 2\nu_{lp_k} D_{kl}),$$

$$J_k^{u} = \sum_{i=p_k,q_k} \sum_{l=1,l\neq k}^{n} u_{ki}(\nu_{li}(\underline{\omega} d_{li} + \overline{\omega} d_{ki}) + 2u_{li} D_{kl}).$$

Algorithm 2. RCM

RCM0 Give initial cluster centers.

RCM1 Calculate lower approximations and boundaries.

RCM2 Update cluster centers.

RCM3 Calculate $\min_{V} J_{\mathrm{RCM}}$ and update V.

RCM4 If the stop criterion satisfies, finish. Otherwise back to **RCM1**.

---

3   Proposed Method 1 — RCM-FU

We propose RCM-FU (RCM with fuzzy upper approximation) which is constructed by introducing fuzzy-set representation into membership of boundary.

3.1   Objective Function

The objective function of RCM-FU is defined as follows:

$$J_{\text{RCM-FU}} = \sum_{k=1}^{n} \sum_{l=1}^{n} \sum_{i=1}^{c} (u_{ki}^m \nu_{li} (\underline{\omega} d_{li} + \overline{\omega} d_{ki}) + (\nu_{ki} \nu_{li} + u_{ki}^m u_{li}^m) D_{kl}).$$

The constraints are as follows:

$$\underline{\omega} + \overline{\omega} = 1, \quad \nu_{ki} \in \{0,1\}, \quad u_{li} \in [0,1], \quad \sum_{i=1}^{c} \nu_{ki} \in \{0,1\},$$

$$\sum_{i=1}^{c} \nu_{ki} = 1 \Longleftrightarrow \sum_{i=1}^{c} u_{ki} = 0,$$

$$\sum_{i=1}^{c} \nu_{ki} = 0 \Longleftrightarrow \sum_{i=0}^{c} u_{ki} = 1.$$

3.2   Derivation of the Optimal Solution and Algorithm

A cluster center v_i is calculated as follows:

$$v_i = \begin{cases} \dfrac{\sum_{x_k \in \underline{A_i}} x_k}{|\underline{A_i}|}, & (\text{Bnd}(A_i) = \emptyset) \\[2ex] \dfrac{\sum_{k=1}^{n} u_{ki}^m x_k}{\sum_{k=1}^{n} u_{ki}^m}, & (\underline{A_i} = \emptyset) \\[2ex] \underline{\omega} \times \dfrac{\sum_{x_k \in \underline{A_i}} x_k}{|\underline{A_i}|} + \overline{\omega} \times \dfrac{\sum_{k=1}^{n} u_{ik}^m x_k}{\sum_{k=1}^{n} u_{ik}^m}. & (\text{otherwise}) \end{cases}$$

The optimal solutions to N and U are updated as follows:

In case that x_k belongs to the lower approximation of a cluster, the cluster is C_{p_k} ($p_k = \arg\min_i d_{ki}$) and

$$u_{ki} = 0, \quad (\forall i)$$

$$\nu_{ki} = \begin{cases} 1, & (i = p_k) \\ 0. & (\text{otherwise}) \end{cases}$$

Therefore, we calculate u_{ki} that minimizes J_k^ν as follows:

$$J_k^\nu = \sum_{l=1}^{n} (u_{lp_k}^m (\underline{\omega} d_{kp_k} + \overline{\omega} d_{lp_k}) + 2\nu_{lp_k} D_{kl}).$$

In case that x_k belongs to boundaries of some clusters, $\nu_{ki} = 0$. Thus, the objective function is represented by

$$J_k^u = \sum_{l=1}^{n} \sum_{i=1}^{c} (u_{ki}^m \nu_{li}(\underline{\omega} d_{li} + \overline{\omega} d_{ki}) + 2u_{ki}^m u_{li}^m D_{kl}).$$

We calculate the optimal solutions by Lagrange multiplier as follows:

$$u_{ki} = \frac{\left(\frac{1}{\sum_{l=1}^{n}(\nu_{li}(\underline{\omega} d_{li}+\overline{\omega} d_{ki})+4u_{li}^m D_{kl})} \right)^{\frac{1}{m-1}}}{\sum_{j=1}^{c} \left(\frac{1}{\sum_{l=1}^{n}(\nu_{lj}(\underline{\omega} d_{lj}+\overline{\omega} d_{kj})+4u_{lj}^m D_{kl})} \right)^{\frac{1}{m-1}}}.$$

In comparison with the above two cases, we obtain the optimal solutions on ν_{ki} and u_{ki} as follows:

$$\nu_{ki} = \begin{cases} 1, & (J_k^\nu < J_k^u \wedge i = p_k) \\ 0, & (\text{otherwise}) \end{cases}$$

$$u_{ki} = \begin{cases} 0, & (J_k^\nu < J_k^u \wedge i = p_k) \\ \dfrac{\left(\frac{1}{\sum_{l=1}^{n}(\nu_{li}(\underline{\omega} d_{li}+\overline{\omega} d_{ki})+4u_{li}^m D_{kl})} \right)^{\frac{1}{m-1}}}{\sum_{j=1}^{c} \left(\frac{1}{\sum_{l=1}^{n}(\nu_{lj}(\underline{\omega} d_{lj}+\overline{\omega} d_{kj})+4u_{lj}^m D_{kl})} \right)^{\frac{1}{m-1}}} & (\text{otherwise}) \end{cases}$$

Algorithm 3. RCM-FU

RCM-FU0 Give initial cluster centers.

RCM-FU1 Calculate lower approximations and boundaries.

RCM-FU2 Calculate cluster centers.

ERCM-FU3 Calculate $\min_{V} J_{\text{RCM-FU}}$ and update V.

ERCM-FU4 If the stop criterion satisfies, finish. Otherwise back to **RCM-FU1**.

---

4   Proposed Method 2 — Entropy RCM-FU

We propose Entropy RCM-FU(ERCM-FU) by introducing an entropy regularizer into RCM.

4.1   Objective Function

The objective function of ERCM-FU is defined as follows:

$$J_{\text{ERCM-FU}} = \sum_{k=1}^{n} \sum_{l=1}^{n} \sum_{i=1}^{c} (u_{ki}\nu_{li}(\underline{\omega} d_{li} + \overline{\omega} d_{ki}) + (\nu_{ki}\nu_{li} + u_{ki}u_{li})D_{kl})$$

$$+ \lambda \sum_{k=1}^{n} \sum_{i=1}^{c} u_{ki} \log u_{ki}.$$

The constraints are the same as ones of RCM-FU.

4.2 Derivation of the Optimal Solution and Algorithm

The cluster center v_i is calculated as follows:

$$
v_i = \begin{cases}
\dfrac{\sum_{x_k \in \underline{A}_i} x_k}{|\underline{A}_i|}, & (\mathrm{Bnd}(A_i) = \emptyset) \\[3ex]
\dfrac{\sum_{k=1}^{n} u_{ki}^m x_k}{\sum_{k=1}^{n} u_{ki}^m}, & (\underline{A}_i = \emptyset) \\[3ex]
\underline{\omega} \times \dfrac{\sum_{x_k \in \underline{A}_i} x_k}{|\underline{A}_i|} + \overline{\omega} \times \dfrac{\sum_{k=1}^{n} u_{ik}^m x_k}{\sum_{k=1}^{n} u_{ik}^m}. & (\text{otherwise})
\end{cases}
$$

The optimal solutions to N and U are updated as follows:

In case that x_k belongs to the lower approximation of a cluster, the cluster is C_{p_k} ($p_k = \arg \min_i d_{ki}$) and

$$
u_{ki} = 0, \quad (\forall i)
$$

$$
\nu_{ki} = \begin{cases}
1, & (i = p_k) \\
0. & (\text{otherwise})
\end{cases}
$$

Therefore, we calculate u_{ki} that minimizes J_k^{ν} as follows:

$$
J_k^{\nu} = \sum_{l=1}^{n} (u_{li^*} (\underline{\omega} d_{ki^*} + \overline{\omega} d_{li^*}) + 2\nu_{li^*} D_{kl} + \lambda u_{li^*} \log u_{li^*}).
$$

In case that x_k belongs to boundaries of some clusters, $\nu_{ki} = 0$. Thus, the objective function is represented by

$$
J_k^{u} = \sum_{l=1}^{n} \sum_{i=1}^{c} (u_{ki} \nu_{li} (\underline{\omega} d_{li} + \overline{\omega} d_{ki}) + 2 u_{ki} u_{li} D_{kl} + \lambda u_{ki} \log u_{ki}).
$$

We calculate the optimal solutions by Lagrange multiplier as follows:

$$
u_{ki} = \exp(\lambda^{-1}(\sum_{l=1}^{n} (-\nu_{li}(\underline{\omega} d_{li} + \overline{\omega} d_{ki}) - 4 u_{li} D_{kl}) - \lambda
$$
$$
- \lambda \log \sum_{j=1}^{c} (\exp(\lambda^{-1}(-\sum_{l=1}^{n} (\nu_{lj}(\underline{\omega} d_{lj} + \overline{\omega} d_{kj}) - 4 u_{lj} D_{kl}) - \lambda))))
$$

In comparison with the above two cases, we obtain the optimal solutions to ν_{ki} and u_{ki} as follows:

$$\nu_{ki} = \begin{cases} 1, & (J_k^\nu < J_k^u \wedge i = p_k) \\ 0, & \text{(otherwise)} \end{cases}$$

$$u_{ki} = \begin{cases} 0, (J_k^\nu < J_k^u \wedge i = p_k) \\ \exp(\lambda^{-1}(\sum_{l=1}^{n}(-\nu_{li}(\underline{\omega}d_{li} + \overline{\omega}d_{ki}) - 4u_{li}D_{kl}) - \lambda \\ -\lambda \log \sum_{j=1}^{c}(\exp(\lambda^{-1}(-\sum_{l=1}^{n}(\nu_{lj}(\underline{\omega}d_{lj} + \overline{\omega}d_{kj}) - 4u_{lj}D_{kl}) - \lambda)))))). \\ \text{(otherwise)} \end{cases}$$

The Algorithm of ERCM-FU is as same as the one of RCM-FU.

5   Numerical Examples

In this section, we use two artificial datasets (Fig. 1 and Fig. 4) and one real dataset to compare the proposed methods with the conventional ones. We examine the effectiveness of proposed methods (RCM-FU and ERCM-FU).RKM has no evaluation criteria so that we cannot evaluate the outputs of RKM. Therefore, we consider an objective function based on the objective function of HCM as the evaluation criterion as follows:

$$J = \underline{\omega} \times \sum_{i=1}^{c} \sum_{x_k \in \underline{A_i}} d_{ki} + \overline{\omega} \times \sum_{i=1}^{c} \sum_{x_k \in \text{Bnd}(A_i)} d_{ki}.$$

Table 1. Algorithms used for comparison

Algorithm	Parameters
RCM-FU	proposed method: $\underline{\omega} = 0.55$, $m = 2.0$
ERCM-FU	proposed method: $\underline{\omega} = 0.35$, $\lambda = 0.7$
RCM	$\underline{\omega} = 0.55$
RKM	$\underline{\omega} = 0.55$, threshold=0.01
HCM	−
FCM[8]	fuzzy parameter : 2.0

5.1   Artificial Dataset

We show result of artificial dataset in Fig. 1. The membership of each objects to boundaries in Table 2. The membership of boundary is fuzzy.

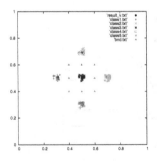

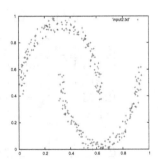

Fig. 1. Original data   **Fig. 2.** Artificial data(RCM-FU, $\underline{\omega} = 0.55$)

Table 2. The membership of each objects to upper approximations (RCM-FU)

(x,y)	class1	class2	class3	class4	class5
(0.4,0.6)	0.258989	0.327004	0.237458	0.078667	0.097882
(0.4,0.5)	0.310618	0.150450	0.304497	0.136976	0.097460
(0.4,0.4)	0.254435	0.090088	0.241740	0.315333	0.098405
(0.5,0.6)	0.155314	0.405250	0.241697	0.064693	0.133046
(0.5,0.4)	0.156579	0.075916	0.248490	0.386302	0.132713
(0.6,0.6)	0.123476	0.296644	0.213485	0.073845	0.292549
(0.6,0.5)	0.121365	0.135507	0.255305	0.120102	0.367722
(0.6,0.4)	0.127512	0.087803	0.222790	0.277909	0.283987

We show the result by ERCM-FU in Fig. 3.

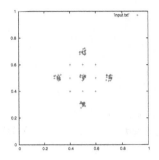

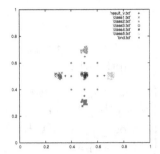

Fig. 3. Artificial data (ERCM-FU, $\underline{\omega} = 0.35$, $\lambda = 0.7$)

Fig. 4. Original data

The membership of each objects to boundaries in Table 3. The membership of boundary is fuzzy.

Table 3. The membership of each objects to boundaries(ERCM-FU)

(x,y)	class1	class2	class3	class4	class5
(0.4,0.6)	0.169420	0.625175	0.202058	0.000057	0.000067
(0.4,0.5)	0.631929	0.129328	0.216806	0.019321	0.000191
(0.4,0.4)	0.174968	0.006356	0.064091	0.754525	0.000059
(0.5,0.6)	0.004080	0.657372	0.329219	0.000075	0.006546
(0.5,0.4)	0.003527	0.005084	0.080845	0.905825	0.004677
(0.6,0.6)	0.000020	0.427396	0.384281	0.000026	0.188276
(0.6,0.5)	0.000072	0.082251	0.392387	0.010462	0.514828
(0.6,0.4)	0.000029	0.005078	0.146973	0.620485	0.223996

We show the results of crescents data by proposed methods and RCM in Fig 5, 6 and 7.

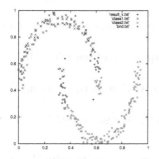
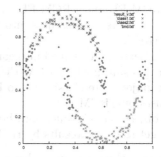

Fig. 5. Crescents data(RCM, $\underline{\omega} = 0.55$)   **Fig. 6.** Crescents data(RCM-FU, $\underline{\omega} = 0.55$)

From the above results, we can find as follows:

- Some objects are classified into incorrect lower approximation by RCM and ERCM-FU.
- There are more objects which are classified into correct lower approximation by RCM-FU than RCM.
- There are more objects which we classified into boundary by RCM-FU than RCM.

5.2 Comparison of Proposed Methods with Conventional Ones

We compare the proposed methods to conventional ones through Iris dataset (150 objects, 4 dimensions, 3 clusters). We define the ratio of correct answers as (the number of correct answers)/(the number of objects). We assign the objects which were classified into boundaries to the cluster to which the membership is maximum. Table 4 shows as follows:

Table 4. The ratio of correct answers

Algorithm	lower approximation			boundary			total	
	numbers	correct	ratio	numbers	correct	ratio	numbers of correct	ratio
RCM-FU (m=1.5)	116	115	0.991	34	20	0.588	135	0.9
RCM-FU (m=2.0)	68	67	0.985	82	62	0.756	129	0.86
ERCM-FU ($\lambda = 0.5$)	145	137	0.944	5	5	1.0	142	0.947
ERCM-FU ($\lambda = 2.0$)	109	109	1.0	41	32	0.780	141	0.94
RCM ($\underline{\omega} = 0.55$)	139	128	0.921	11	5	0.455	133	0.887
RCM ($\underline{\omega} = 0.75$)	135	126	0.933	15	11	0.733	137	0.913
RKM (threshold=0.01)	150	134	0.893	0	0	—	134	0.893
RKM (threshold=3.0)	76	75	0.987	74	40	0.541	115	0.767
FCM	150	134	0.893	0	0	—	134	0.893
HCM	150	134	0.893	0	0	—	134	0.893

- There are more objects which are classified into correct lower approximations by RCM-FU and ERCM-FU than RCM.
- More objects are classified into boundaries as the parameter m increases by RCM-FU.
- All objects are classified into correct lower approximations when the parameter λ is suitable by ERCM-FU.
- Less objects are classified into lower approximation as the parameter λ increases by ERCM-FU.
- No objects are classified into boundaries as the threshold=0.01 by RKM.
- We get the same results by HCM and FCM.

The optimal parameter for lower approximation is different from the optimal one for the whole in both RCM-FU and ERCM-FU.

5.3   Consideration of Parameters

We consider the relation between parameters and the number of objects which are classified into boundaries.

Fig. 8 shows the relation between $\underline{w}$ and the number of objects which are classified into boundaries by RCM. Horizontal- and vertical-axes mean $\underline{w}$ and the number of objects which are classified into boundaries, respectively. Fig. 8 shows that $\underline{w}$ is ineffective at the number of objects which are classified into boundaries. This means that it is difficult to adjust the ratio of the number of objects into boundaries to the number of all objects by the parameter $\underline{w}$.

Fig. 9 shows the relation between m and the number of objects which are classified into boundary by RCM-FU. Horizontal- and vertical-axes mean m and the number of objects which are classified into boundaries, respectively. Fig. 9 shows that m is effective at the number of objects which are classified into boundaries. This means that it is easy to adjust the ratio of the number of objects into boundaries to the number of all objects by the parameter m.

Fig. 10 shows the relation between λ and the number of objects which are classified into boundaries by RCM-FU. Horizontal- and vertical-axes mean λ

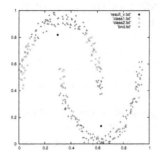 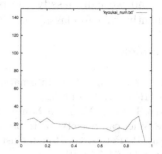

Fig. 7. Crescents data(ERCM-FU, $\underline{\omega} = 0.35, \ \lambda = 1.0$)

Fig. 8. The relation between $\underline{w}$ and the number of objects which are classified into boundaries by RCM

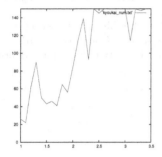 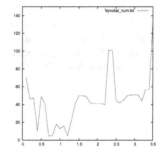

Fig. 9. The relation between m and the number of objects which are classified into boundaries by RCM-FU

Fig. 10. The relation between λ and the number of objects which are classified into boundaries by ERCM-FU

and the number of objects which are classified into boundaries, respectively. We fixed $\underline{w} = 0.35$.

6 Conclusion

This paper proposed new rough clustering algorithms based on optimization of objective functions. The proposed methods based on optimization of objective functions with fuzzy-set representation can classify more objects into correct lower approximations than conventional ones proposed by Endo et al [7], and we can adjust the ratio of the number of objects into boundaries to the number of all objects by parameters. The conventional rough clustering algorithm [7] has a problem that an object cannot belong to more than two upper approximations. We introduced fuzzy-set representation into membership of boundary to solve such a problem. Thus, each object into boundaries has a membership grade in [0,1], and we can classify the object according to the value of the grade. Consequently, we can classify all objects into clusters like FCM [8].

References

1. Pawlak, Z.: Rough Sets. International Journal of Computer and Information Sciences 11(5), 341–356 (1982)
2. Lingras, P., Yao, Y.: Data Mining Using Extensions of Rough Set Model. Journal of the American Society for Information Science, 415–422 (1998)
3. Lingras, P., Peters, G.: Rough clustering. WIREs Data Mining and Knowledge Discovery 1(1), 64–72 (2011)
4. Hirano, S., Tsumoto, S.: An Indiscernibility-Based Clustering Method with Iterative Refinement of Equivalence Relations. Journal of Advanced Computational Intelligence and Intelligent Informatics 7(2), 169–177 (2003)
5. Lingras, P., West, C.: Interval Set Clustering of Web Users with Rough K-Means. Journal of Information Systems, 5–16 (2004)
6. Peters, G.: Some refinements of rough k-means clustering. Pattern Recognition 39, 1481–1491 (2006)
7. Endo, Y., Kinoshita, N.: On Objective-Based Rough c-Means Clustering. In: Proc. of the 2012 IEEE International Conference on Granular Computing (2012)
8. Bezdek, J.C.: Pattern Recognition with Fuzzy Objective Function Algorithms. Plenum Press, New York (1981)

Semi-Supervised Hard and Fuzzy c-Means with Assignment Prototype Term

Yukihiro Hamasuna[1] and Yasunori Endo[2]

[1] Department of Informatics, School of Science and Engineering,
Kinki University,
Kowakae 3-4-1, Higashi-osaka, Osaka, 577-8502, Japan
yhama@info.kindai.ac.jp
[2] Faculty of Engineering, Information and Systems,
University of Tsukuba, Tennodai 1-1-1, Tsukuba, Ibaraki, 305-8573, Japan
endo@risk.tsukuba.ac.jp

Abstract. Semi-supervised learning is an important task in the field of data mining. Pairwise constraints such as must-link and cannot-link are used in order to improve clustering properties. This paper proposes a new type of semi-supervised hard and fuzzy c-means clustering with assignment prototype term. The assignment prototype term is based on the Windham's assignment prototype algorithm which handles pairwise constraints between objects in the proposed method. First, an optimization problem of the proposed method is formulated. Next, a new clustering algorithm is constructed based on the above discussions. Moreover, the effectiveness of the proposed method is shown through numerical experiments.

Keywords: pairwise constraint, hard c-means, fuzzy c-means, assignment prototype algorithm, semi-supervised learning.

1 Introduction

The aim of cluster analysis which is well known as clustering is to discover important structures and features from massive and complex databases. Clustering is one of the data analysis methods which divides a set of objects into some groups called clusters. Objects classified in the same cluster are considered similar, while the ones classified in a different cluster are considered dissimilar. Hard c-means which is known as k-means [6,9] and fuzzy c-means clustering (FCM) is one of the most well-known clustering methods [2]. Entropy based fuzzy c-means clustering (eFCM) and other variants of FCM are also famous and important techniques in the field of clustering [10,11]. Semi-supervised learning has been remarked and studied in many research fields such as clustering, support vector machine and so on [3]. Pairwise constraints, that is, must-link and cannot-link are frequently used in order to improve clustering performance [14]. Semi-supervised clustering which is also referred to as constrained clustering handles a prior knowledge as pairwise constraints in the clustering framework [1].

V. Torra et al. (Eds.): MDAI 2014, LNAI 8825, pp. 135–144, 2014.
© Springer International Publishing Switzerland 2014

Semi-supervised clustering which are based on k-means [14], fuzzy c-means clustering [5], kernel methods [8], and hierarchical clustering [7] have been widely discussed and proposed. Pairwise constraints referred to must-link and cannot-link are used as a prior knowledge about which data should be in the same or a different cluster in semi-supervised clustering. These constraints are given between objects and handled in clustering procedures. Significant techniques such as a regularization, kernel method, and probabilistic model are used in order to handle pairwise constraints in the clustering algorithms. It is widely known that regularization technique is useful and effective for not only clustering but also other data analysis methods.

Fuzzy non-metric model (FNM) [4,13] is one of the clustering methods in which the membership degrees of each datum to each cluster is calculated directly from dissimilarities between data. Assignment prototype (AP) algorithm [15] is proposed as a improved version of FNM in order to overcome the initial value dependence of FNM. The cluster center which is referred to as representative of cluster and used in hard and fuzzy c-means is not considered in these algorithms. FNM and AP can handle relational data such as a table of distance between objects. The way of handling pairwise constraints as regularization term by using assignment prototype algorithm is described from that sense. A new semi-supervised hard and fuzzy c-means with assignment prototype based regularization term is proposed in this paper.

The rest of this paper is organized as follows: In section 2, we introduce some symbols and hard conventional clustering methods. In section 3, we propose a new semi-supervised hard and fuzzy $c-$means with assignment prototype term. In section 4, we show the effectiveness of proposed method. In section 5, we conclude this paper.

2   Preparation

A set of objects to be clustered is given and denoted by $X = \{x_1, \ldots, x_n\}$ in which x_k ($k = 1, \ldots, n$) is an object. In most cases, $x_1, \ldots, x_n$ are p-dimensional vectors $\Re^p$, that is, an object $x_k \in \Re^p$. A cluster, its cluster center, and a set of cluster center are denoted as $C_i(i = 1, \ldots, c)$, $v_i \in \Re^p$, $V = \{v_i, \ldots, v_c\}$. A membership degrees of x_k belonging to C_i and a partition matrix is also denoted as u_{ki} and $U = (u_{ki})_{1 \leq k \leq n, \ 1 \leq i \leq c}$.

2.1   Hard and Entropy-Based Fuzzy c-Means

Hard c-means (HCM) [6,9] and entropy based fuzzy c-means (eFCM) [10,11] are based on optimizing an objective function under the constraint for membership degrees.

We consider the following two objective functions J_h and J_e.

$$J_h(U, V) = \sum_{i=1}^{c} \sum_{k=1}^{n} u_{ki} \|x_k - v_i\|^2,$$

$$J_e(U, V) = \sum_{i=1}^{c} \sum_{k=1}^{n} u_{ki} \|x_k - v_i\|^2 + \lambda_u \sum_{i=1}^{c} \sum_{k=1}^{n} u_{ki} \log u_{ki}.$$

Here, $\lambda_u > 0.0$ is fuzzification parameter and $\|x_k - v_i\|^2$ means a dissimilarity measure between an object x_k and a cluster center v_i. J_h is the objective function of HCM and J_e is the one of eFCM, respectively.

Constraints for membership degrees u_{ki} considered in HCM and eFCM are described $\mathcal{U}_h$ and $\mathcal{U}_f$ as follow:

$$\mathcal{U}_h = \left\{ (u_{ki}) : u_{ki} \in \{0, 1\}, \ \sum_{i=1}^{c} u_{ki} = 1, \ ^{\forall}k \right\}, \tag{1}$$

$$\mathcal{U}_f = \left\{ (u_{ki}) : u_{ki} \in [0, 1], \ \sum_{i=1}^{c} u_{ki} = 1, \ ^{\forall}k \right\}. \tag{2}$$

Clustering algorithms of HCM and eFCM are constructed based on alternating optimization with u_{ki} and v_i.

2.2 Assignment Prototype Algorithm

The objective function of assignment prototype algorithm which handles relational data is similar to and a bit different from the one of fuzzy non-metric model [13]. Two variables, that is, membership degrees u_{ki} and prototype weight w_{ti} are used in assignment prototype algorithm. w_{ti} stands for the prototype weight of object t to cluster i. Windham considers the following objective function:

$$J_{ap}(U, W) = \sum_{i=1}^{c} \sum_{k=1}^{n} \sum_{t=1}^{n} (u_{ki})^2 (w_{ti})^2 r_{kt}.$$

Here, r_{kt} is a distance between objects and $W = (w_{ti})_{1 \leq t \leq n, \ 1 \leq i \leq c}$ is prototype weight matrix and satisfies the following constraint:

$$\mathcal{W}_f = \left\{ (w_{ti}) : w_{ti} \in [0, 1], \ \sum_{t=1}^{n} w_{ti} = 1, \ ^{\forall}i \right\}. \tag{3}$$

Hard assignment prototype algorithm (HAP) and entropy based assignment prototype one are also considered in the same manner as HCM and eFCM. The objective functions of HAP and eFAP are as follows:

$$J_{hap}(U, W) = \sum_{i=1}^{c} \sum_{k=1}^{n} u_{ki} w_{ti} r_{kt}, \tag{4}$$

$$J_{efap}(U, W) = \sum_{i=1}^{c} \sum_{k=1}^{n} u_{ki} w_{ti} r_{kt}$$
$$+ \lambda_u \sum_{i=1}^{c} \sum_{k=1}^{n} u_{ki} \log u_{ki} + \lambda_w \sum_{i=1}^{c} \sum_{t=1}^{n} w_{ti} \log w_{ti}. \tag{5}$$

Here, λ_u and λ_w are fuzzification parameters. Constraints for membership degrees u_{ki} are the same as (1) and (2). Also, constraint for prototype weight w_{ti} for eFAP is the same as (3) and one for HAP is as follows:

$$\mathcal{W}_h = \left\{ (w_{ti}) : w_{ti} \in \{0, 1\}, \ \sum_{t=1}^{n} w_{ti} = 1, \ ^\forall i \right\}. \tag{6}$$

Clustering algorithms of HAP and eFAP are also constructed based on alternating optimization with u_{ki} and w_{ti}.

2.3   Pairwise Constraint

Typical examples of pairwise constraints are must-link and cannot-link [14]. These constraints are considered as a prior knowledge about which data should be in the same or a different cluster. A set $ML = \{(x_k, x_l)\} \subset X \times X$ consists of must-link pairs so that x_k and x_l should be in the same cluster, while another set $CL = \{(x_q, x_r)\} \subset X \times X$ consists of cannot-link pairs so that x_q and x_r should be in different clusters. ML and CL are assumed to be symmetric, that is, if $(x_k, x_l) \in ML$ then $(x_l, x_k) \in ML$, and if $(x_q, x_r) \in CL$ then $(x_r, x_q) \in CL$. Obviously, ML and CL are supposed to be disjoint.

In semi-supervised clustering, these pairwise constraints are considered as hard or soft constraints. In hard constraints approach, pairwise constraints ML and CL are always satisfied in clustering procedures and results, while they are not always satisfied in soft constraints approach. Many semi-supervised clustering methods have been proposed and discussed in order to improve clustering properties and features by using pairwise constraints in many forms [1,5,7,8,14].

3   Proposed Method

3.1   Semi-Supervised Hard c-Means with Hard Assignment Prototype Term

The objective function of semi-supervised hard c-means with hard assignment prototype term (HCM-HAP) is based on the HCM and HAP. We consider the following objective function for HCM-HAP:

$$J_{hhap}(U, V, W) = \sum_{i=1}^{c} \sum_{k=1}^{n} u_{ki} \|x_k - v_i\|^2 - \sum_{i=1}^{c} \sum_{k=1}^{n} \sum_{t=1}^{n} u_{ki} w_{ti} \alpha_{kt} + \sum_{i=1}^{c} \sum_{k=1}^{n} \sum_{t=1}^{n} u_{ki} w_{ti} \beta_{kt}.$$

Here, the costraints for u_{ki} and w_{ti} are the same as (1) and (6). α_{kt} and β_{kt} means must-link and cannot-link described as follows:

$$\alpha_{kt} = \begin{cases} \alpha & (\ (x_k, x_t) \in ML\ ) \\ 0 & (\ \text{otherwise}\ ) \end{cases} \quad \left(\because \alpha = \gamma_{ml} \max_{p,q} \|x_p - x_q\|^2 \right),$$

$$\beta_{kt} = \begin{cases} \beta & (\ (x_k, x_t) \in CL\ ) \\ 0 & (\ \text{otherwise}\ ) \end{cases} \quad \left(\because \beta = \gamma_{cl} \max_{p,q} \|x_p - x_q\|^2 \right).$$

γ_{ml} and γ_{cl} are parameters for adjusting the degree of pairwise constraints. The larger the value of γ_{ml} and γ_{cl} are, the stronger the effect of ML and CL is. In the case with large γ_{ml} and γ_{cl}, ML and CL are considered as hard constraints that means pairwise constraints are always satisfied in the clustering procedures. The optimal solution of v_i is derived from partial derivative $\frac{\partial J}{\partial v_i} = 0$. The ones of u_{ki} and w_{ti} are derived from objective function and constraints by considering linear programming problem as follows:

$$u_{ki} = \begin{cases} 1 & (\ i = \arg\min_j \left\{ \|x_k - v_j\|^2 - \sum_{t=1}^n w_{tj}\alpha_{kt} + \sum_{t=1}^n w_{tj}\beta_{kt} \right\}\ ) \\ 0 & (\ \text{otherwise}\ ) \end{cases}. \quad (7)$$

$$v_i = \frac{\sum_{k=1}^n u_{ki}x_k}{\sum_{k=1}^n u_{ki}}. \quad (8)$$

$$w_{ti} = \begin{cases} 1 & (\ t = \arg\min_q \left\{ -\sum_{k=1}^n u_{ki}\alpha_{kq} + \sum_{k=1}^n u_{ki}\beta_{kq} \right\}\ ) \\ 0 & (\ \text{otherwise}\ ) \end{cases}. \quad (9)$$

3.2 Semi-Supervised Entropy Based Fuzzy c-Means with Fuzzy Assignment Prototype Term

The objective function of semi-supervised entropy based fuzzy c-means with fuzzy assignment prototype term (eFCM-eFAP) is based on the eFCM and eFAP. We consider following objective function for eFCM-eFAP:

$$J_{eefap}(U, V, W) = \sum_{i=1}^c \sum_{k=1}^n u_{ki}\|x_k - v_i\|^2 - \sum_{i=1}^c \sum_{k=1}^n \sum_{k=1}^n u_{ki}w_{ti}\alpha_{kt} + \sum_{i=1}^c \sum_{k=1}^n \sum_{k=1}^n u_{ki}w_{ti}\beta_{kt}$$
$$+ \lambda_u \sum_{i=1}^c \sum_{k=1}^n u_{ki}\log u_{ki} + \lambda_w \sum_{i=1}^c \sum_{t=1}^n w_{ti}\log w_{ti}$$

λ_u, λ_w, α_{kt}, and β_{kt} are the same as above discussions. Constraints for u_{ki} and w_{ti} are also the same as (2) and (3). The optimal solution of v_i is the same as (8). The ones for u_{ki} and w_{ti} are derived from Lagrangian as follows:

$$u_{ki} = \frac{\exp\left(-\lambda_u^{-1}d_{ki}\right)}{\sum_{l=1}^c \exp\left(-\lambda_u^{-1}d_{kl}\right)}, \quad (10)$$

$$w_{ti} = \frac{\exp\left(-\lambda_w^{-1}g_{ti}\right)}{\sum_{q=1}^n \exp\left(-\lambda_w^{-1}g_{qi}\right)}. \quad (11)$$

Here, d_{ki} and g_{ti} are as follows:

$$d_{ki} = \|x_k - v_i\|^2 - \sum_{t=1}^{n} w_{ti}\alpha_{kt} + \sum_{t=1}^{n} w_{ti}\beta_{kt},$$

$$g_{ti} = -\sum_{k=1}^{n} u_{ki}\alpha_{kt} + \sum_{k=1}^{n} u_{ki}\beta_{kt}.$$

3.3   Algorithm Based on the Proposed Method

The clustering algorithm of HCM-HAP and eFCM-eFAP are constructed based on above discussions. HCM-eFAP and eFCM-HAP are also considered as the same procedure. Objective functions of HCM-eFAP J_{hfap} and eFCM-HAP J_{ehap} are as follows:

$$J_{hefap}(U, V, W) = J_{hhap} + \lambda_w \sum_{i=1}^{c} \sum_{t=1}^{n} w_{ti} \log w_{ti},$$

$$J_{ehap}(U, V, W) = J_{hhap} + \lambda_u \sum_{i=1}^{c} \sum_{k=1}^{n} u_{ki} \log u_{ki}.$$

The optimal solutions of J_{hefap} and J_{ehap} are also derived from those objective functions and constraints as the same procedures. The clustering algorithm of proposed method is described as Algorithm 1. Eqs. **A**, **B**, and **C** used in each algorithm follow Table 1.

Algorithm 1. Algorithm of the proposed method.

Step1 Set initial values and fuzzification parameters.
Step2 Calculate $u_{ki} \in U$ by using Equation **A**.
Step3 Calculate $v_i \in V$ by using Equation **B**.
Step4 Calculate $w_{ti} \in W$ by using Equation **C**.
Step5 If convergence criterion is satisfied, stop. Otherwise go back to **Step2**.

Table 1. The equations of u_{ki}, v_i, and w_{ti} used in the algorithms

Algorithm	Eq. A	Eq. B	Eq. C
HCM-HAP	(7)	(8)	(9)
HCM-eFAP	(7)	(8)	(11)
eFCM-HAP	(10)	(8)	(9)
eFCM-eFAP	(10)	(8)	(11)

4   Numerical Experiments

We show the results of proposed methods with Iris data set published in UCI machine learning repository (http://archive.ics.uci.edu/ml/). Iris data set consists of 150 objects with 4 attributes and should be classified into three clusters. Each attribute is standardized, that is, mean and standard derivation are equal to 0 and 1, respectively. We show the effectiveness of proposed method by evaluating the results of average and standard derivation of Rand Index (RI) [12] and number of violated constraints out of 100 trials. Pairwise constraints are generated by class information in benchmark data set. If class label of two objects are the same, then must-link is generated between objects. Also, cannot-link is generated between objects if class label of two objects are different. We fix $\lambda_u = 1.0$ and $\lambda_w = 100.0$ and set $\gamma_{ml}, \gamma_{cl} = 0.50$ or 1.00 in these experiments.

Tables 2, 3, 4, 5 are the results of RI by HCM-HAP, HCM-eFAP, eFCM-HAP, and eFCM-eFAP, respectively. Tables 6, 7, 8, 9 are the results of number of violated constraints by HCM-HAP, HCM-eFAP, eFCM-HAP, and eFCM-eFAP, respectively. The value described in these tables denotes average $\pm$ standard derivation of RI and the number of violated constraints out of 100 trials. Results of RI by conventional HCM is 0.818 ± 0.044 and eFCM is 0.832 ± 0.000.

These results show the effectiveness and important properties of proposed methods as follows:

- eFAP regularization term obtains better results than HAP term by comparing with these results. HAP is strongly depended on initial value and data sequence, that is, only one object takes the value of $w_{ti} = 1$. Many pairwise constraints are ignored in clustering procedure because of this property.
- Must-link much more affective than cannot-link in the proposed method by considering the parameter of γ_{ml} and γ_{cl} and number of pairwise constraints. Especially, HCM-eFAP and eFCM-eFAP takes better results with large value of γ_{ml} and large number of must-link.
- The larger the number of must-link is, the larger the standard derivation is in the proposed method. These results show that the choice of objects which are given pairwise constraints is important for clustering results from the viewpoint of robustness of algorithm.

Table 2. Results of RI by HCM-HAP out of 100 trials

Number of pairwise constraints	$\gamma_{ml}, \gamma_{cl} = 0.50$		$\gamma_{ml}, \gamma_{cl} = 1.00$	
	ML	CL	ML	CL
100	0.802 ± 0.045	0.819 ± 0.038	0.802 ± 0.045	0.819 ± 0.038
300	0.806 ± 0.050	0.819 ± 0.032	0.806 ± 0.050	0.820 ± 0.030
500	0.814 ± 0.057	0.824 ± 0.017	0.814 ± 0.057	0.824 ± 0.017

Table 3. Results of RI by HCM-eFAP out of 100 trials

Number of pairwise constraints	$\gamma_{ml}, \gamma_{cl} = 0.50$		$\gamma_{ml}, \gamma_{cl} = 1.00$	
	ML	CL	ML	CL
100	0.821 ± 0.041	0.825 ± 0.036	0.830 ± 0.041	0.825 ± 0.036
300	0.835 ± 0.048	0.827 ± 0.036	0.933 ± 0.057	0.834 ± 0.034
500	0.916 ± 0.049	0.834 ± 0.034	0.969 ± 0.073	0.858 ± 0.040

Table 4. Results of RI by eFCM-HAP out of 100 trials

Number of pairwise constraints	$\gamma_{ml}, \gamma_{cl} = 0.50$		$\gamma_{ml}, \gamma_{cl} = 1.00$	
	ML	CL	ML	CL
100	0.825 ± 0.020	0.832 ± 0.001	0.824 ± 0.020	0.832 ± 0.001
300	0.816 ± 0.038	0.832 ± 0.001	0.815 ± 0.038	0.832 ± 0.002
500	0.830 ± 0.043	0.831 ± 0.006	0.830 ± 0.043	0.829 ± 0.006

Table 5. Results of RI by eFCM-eFAP out of 100 trials

Number of pairwise constraints	$\gamma_{ml}, \gamma_{cl} = 0.50$		$\gamma_{ml}, \gamma_{cl} = 1.00$	
	ML	CL	ML	CL
100	0.835 ± 0.005	0.831 ± 0.003	0.847 ± 0.008	0.834 ± 0.004
300	0.859 ± 0.007	0.838 ± 0.004	0.955 ± 0.028	0.845 ± 0.010
500	0.933 ± 0.015	0.845 ± 0.004	0.988 ± 0.031	0.878 ± 0.021

Table 6. Results of number of violated constraints by HCM-HAP out of 100 trials

Number of pairwise constraints	$\gamma_{ml}, \gamma_{cl} = 0.50$		$\gamma_{ml}, \gamma_{cl} = 1.00$	
	ML	CL	ML	CL
100	27.32 ±6.75	14.72 ± 7.05	27.32 ± 6.75	14.72 ± 7.05
300	79.71 ± 20.73	43.75 ± 16.83	79.71 ± 20.73	43.20 ± 15.96
500	124.51 ± 35.94	67.55 ± 12.26	124.51 ± 35.94	67.62 ± 12.18

Table 7. Results of number of violated constraints by HCM-eFAP out of 100 trials

Number of pairwise constraints	$\gamma_{ml}, \gamma_{cl} = 0.50$		$\gamma_{ml}, \gamma_{cl} = 1.00$	
	ML	CL	ML	CL
100	23.12 ±4.77	13.92 ± 6.44	21.31 ± 4.67	13.82 ± 6.48
300	62.48 ± 9.67	42.66 ± 18.89	20.21 ± 16.85	40.24 ± 16.38
500	50.87 ± 20.32	67.86 ± 29.97	13.11 ± 33.39	56.69 ± 31.15

Table 8. Results of number of violated constraints by eFCM-HAP out of 100 trials

Number of pairwise constraints	$\gamma_{ml}, \gamma_{cl} = 0.50$		$\gamma_{ml}, \gamma_{cl} = 1.00$	
	ML	CL	ML	CL
100	28.00 ± 6.49	12.46 ± 3.34	28.16 ± 6.46	12.46 ± 3.34
300	80.49 ± 18.88	38.54 ± 5.89	80.92 ± 18.95	38.60 ± 5.88
500	120.90 ± 28.03	63.70 ± 7.36	120.90 ± 28.03	64.58 ± 7.50

Table 9. Results of number of violated constraints by eFCM-eFAP out of 100 trials

Number of pairwise constraints	$\gamma_{ml}, \gamma_{cl} = 0.50$		$\gamma_{ml}, \gamma_{cl} = 1.00$	
	ML	CL	ML	CL
100	24.96 ± 4.22	12.47 ± 3.45	22.21 ± 4.63	12.15 ± 3.44
300	62.35 ± 8.19	36.74 ± 5.73	12.83 ± 7.96	33.91 ± 6.79
500	42.83 ± 11.82	57.38 ± 6.71	5.20 ± 12.72	41.32 ± 14.48

5 Conclusions

In this paper, we have proposed semi-supervised clustering with assignment prototype term. We have shown the effectiveness of proposed method through numerical experiments. We have also shown that entropy based fuzzy assignment prototype term is suitable for regularization term which handle pairwise constraints.

In future works, we will consider the way to obtain better results with small number of pairwise constraints. We can obtain better results in case of large number of pairwise constraints is handled. It however takes much costs to collect large number of pairwise constraints in real world problems. That is why we have to consider the above problem. We will, moreover, verify the effectiveness of proposed method with various kinds of benchmark data sets and real ones.

References

1. Basu, S., Davidson, I., Wagstaff, K. (eds.): Constrained Clustering: Advances in Algorithms, Theory and Applications. Data Mining and Knowledge Discovery, vol. 3. Chapman & Hall/CRC (2008)
2. Bezdek, J.C.: Pattern Recognition with Fuzzy Objective Function Algorithms. Plenum Press, New York (1981)
3. Chapelle, O., Schoölkopf, B., Zien, A. (eds.): Semi-Supervised Learning. MIT Press (2006)
4. Endo, Y.: On Entropy Based Fuzzy Non Metric Model – Proposal, Kernelization and Pairwise Constraints. Journal of Advanced Computational Intelligence and Intelligent Informatics (JACIII) 16(1), 169–173 (2012)
5. Hamasuna, Y., Endo, Y.: On semi-supervised fuzzy c-means clustering for data with clusterwise tolerance by opposite criteria. Soft Computing 17(1), 71–81 (2013)
6. Jain, A.K.: Data clustering: 50 years beyond K-means. Pattern Recognition Letters 31(8), 651–666 (2010)
7. Klein, D., Kamvar, S., Manning, C.: From instance-level constraints to space-level constraints: making the most of prior knowledge in data clustering. In: Proc. of the 19th International Conference on Machine Learning (ICML 2002), pp. 307–314 (2002)
8. Kulis, B., Basu, S., Dhillon, I., Mooney, R.: Semi-supervised graph clustering: a kernel approach. Machine Learning 74(1), 1–22 (2009)
9. MacQueen, J.B.: Some methods for classification and analysis of multivariate observations. In: Proc. of Fifth Berkeley Symp. on Math. Statist. and Prob., pp. 281–297 (1967)

10. Miyamoto, S., Mukaidono, M.: Fuzzy c-means as a regularization and maximum entropy approach. In: Proc. of the 7th International Fuzzy Systems Association World Congress (IFSA 1997), vol. 2, pp. 86–92 (1997)
11. Miyamoto, S., Ichihashi, H., Honda, K.: Algorithms for Fuzzy Clustering. Springer, Heidelberg (2008)
12. Rand, W.M.: Objective criteria for the evaluation of clustering methods. Journal of the American Statistical Association 66(336), 846–850 (1971)
13. Roubens, M.: Pattern classification problems and fuzzy sets. Fuzzy Sets and Systems 1, 239–253 (1978)
14. Wagstaff, K., Cardie, C., Rogers, S., Schroedl, S.: Constrained k-means clustering with background knowledge. In: Proc. of the 18th International Conference on Machine Learning (ICML 2001), pp. 577–584 (2001)
15. Windham, M.P.: Numerical classification of proximity data with assignment measures. J. of Classification 2, 157–172 (1985)

Hard and Fuzzy c-means Algorithms
with Pairwise Constraints by Non-metric Terms

Yasunori Endo[1], Naohiko Kinoshita[2],
Kuniaki Iwakura[2], and Yukihiro Hamasuna[3]

[1] Faculty of Engineering, Information and Systems, University of Tsukuba Tennodai 1-1-1,
Tsukuba, Ibaraki, 305-8573, Japan
endo@risk.tsukuba.ac.jp
[2] Graduate School of Systems and Information Engineering,
University of Tsukuba Tennodai 1-1-1, Tsukuba, Ibaraki, 305-8573, Japan
{s1220594,s1320626}@u.tsukuba.ac.jp
[3] Department of Informatics, School of Science and Engineering, Kinki University 3-4-1,
Kowakae, Higashiosaka, Osaka 577-8502, Japan
yhama@info.kindai.ac.jp

Abstract. Recently, semi-supervised clustering has been focused, e.g.,
Refs. [2–5]. The semi-supervised clustering algorithms improve clustering results by incorporating prior information with the unlabeled data. This paper proposes three new clustering algorithms with pairwise constraints by introducing
non-metric term to objective functions of the well-known clustering algorithms.
Moreover, its effectiveness is verified through some numerical examples.

1 Introduction

As computer technologies develop, huge number of data has existed around us. Such
increases in the number of data have focused attention on machine learning and data
mining. We can find useful knowledge from such data with data mining techniques and
clustering is one of most effective technique of such methods (Ref. [1] etc.).

Recently, semi-supervised clustering has occupied an important place in the fields
of machine learning and data mining. While only unlabeled data are used to generate
clusters in conventional clustering algorithms, semi-supervised clustering algorithms
incorporate prior information with the unlabeled data to improve clustering results
(Refs. [2–5] etc.).

Research of semi-supervised clustering approaches this problem from two viewpoints: constraint-based and distance-based. In constraint-based, objective functions are
modified so as to satisfy constraints, and enforced constraints during the clustering process. In distance-based, a distance function is trained on the supervised dataset to satisfy
constraints or labels and applied to the complete dataset.

In both approaches, the concept of constraint plays very important role. The most
common constraints in recent research of semi-supervised clustering are pairwise must-
link constraint and cannot-link constraint. The former means that the pairs of objects
that should belong to the same cluster, and the latter means that the pairs of objects that

V. Torra et al. (Eds.): MDAI 2014, LNAI 8825, pp. 145–157, 2014.

should belong to different clusters. The pairwise constraints occur naturally in many fields.

In this paper, we consider terms of must-link constraint and cannot-link constraint for prior information to use non-metric model (NMM) [6], and then unify the terms objective functions of the well-known clustering algorithms: hard c-means (HCM) [7] and two types of fuzzy c-means (FCM) [1, 8]. Therefore, these proposed clustering algorithms are constraint-based. Moreover, we verify the algorithms through some numerical examples.

2 Preparation

We will construct three new clustering algorithms based on the following well-known methods: hard c-means (HCM), standard fuzzy c-means (sFCM), and entropy-based fuzzy c-means (eFCM) in a later section. Thus, we mention these methods.

First, let each object and the set of all objects be $x_k = (x_k^1, \ldots, x_k^p)^T \in \mathfrak{R}^p$ ($k = 1, \ldots, n$), and $X = \{x_1, \ldots, x_n\}$, respectively. C_i, $v_i = (v_i^1, \ldots, v_i^p)^T \in \mathfrak{R}^p$ and $V = \{v_1, \ldots, v_c\}$ mean the i-th cluster ($i = 1, \ldots, c$), the cluster center of C_i, and the set of all cluster centers, respectively. Moreover u_{ki} means the belongingness of x_k to the i-th clutter. $U = (u_{ki})_{k=1,\ldots,n, \ i=1,\ldots,c}$ is called partition matrix.

2.1 Hard, Standard Fuzzy, and Entropy-Based Fuzzy c-means

The objective functions of HCM, sFCM, and eFCM are defined as follows:

$$J_{\mathrm{HCM}}(U, V) = \sum_{k=1}^{n} \sum_{i=1}^{c} u_{ki} d_{ki}, \quad (d_{ki} = \|x_k - v_i\|^2)$$

$$J_{\mathrm{sFCM}}(U, V) = \sum_{k=1}^{n} \sum_{i=1}^{c} (u_{ki})^m d_{ki},$$

$$J_{\mathrm{eFCM}}(U, V) = \sum_{k=1}^{n} \sum_{i=1}^{c} u_{ki}(d_{ki} + \lambda \log u_{ki}).$$

$m \geq 1$ and $\lambda > 0$ mean fuzzification parameters. The constraints $\mathcal{U}_{\mathrm{HCM}}$ for HCM, and $\mathcal{U}_{\mathrm{FCM}}$ for sFCM and eFCM are as follows:

$$\mathcal{U}_{\mathrm{HCM}} = \left\{ (u_{ki}) \mid u_{ki} \in \{0, 1\}, \ \sum_{i=1}^{c} u_{ki} = 1, \ \forall k \right\},$$

$$\mathcal{U}_{\mathrm{FCM}} = \left\{ (u_{ki}) \mid u_{ki} \in [0, 1], \ \sum_{i=1}^{c} u_{ki} = 1, \ \forall k \right\}.$$

All of HCM, sFCM, and eFCM algorithms give us clustering results by minimizing J_{HCM}, J_{sFCM}, and J_{eFCM} through iterative optimization, respectively.

2.2 Non-Metric Model

We mention generalized Non-metric Model (NMM) proposed by Roubens [6]. NMM classifies datasets by using distance between objects instead of one between cluster centers and data.

The objective function of NMM is defined as follows:

$$J_{\text{NMM}}(U) = \sum_{k=1}^{n} \sum_{l=1}^{n} \sum_{i=1}^{c} (u_{ki})^m (u_{li})^m D_{kl}. \quad (D_{kl} = \|x_k - x_l\|^2)$$

$m = 2$ in Ref. [6]. In particular, when $m = 1$ and the constraint is $\mathcal{U}_{\text{HCM}}$, the optimal solution to u_{ki} is derived as follows:

$$u_{ki} = \begin{cases} 1, & (i = \arg\min_j \sum_{l=1}^{n} u_{lj} D_{kl}) \\ 0. & (\text{otherwise}) \end{cases}$$

Similar to HCM and FCM, NMM algorithm are also constructed based on iterative optimization.

2.3 Pairwise Constraint

Pairwise constraint is one of semi-supervised clustering method [2, 4] and pairwise relations naturally occur in various fields and applications. Some information of relationship between two objects is given in advance in the method. In this paper, we consider two typical pairwise constraints: must-link constraint and cannot-link constraint.

Let ML be the following set:

$$ML = \{(x_k, x_l) \in X \times X \mid x_k \text{ and } x_l \text{ should be assigned into one cluster}, k \neq l.\}.$$

The relation of pairs in ML is called must-link.

Let CL be the following set:

$$CL = \{(x_k, x_l) \in X \times X \mid x_k \text{ and } x_l \text{ should be assigned into different clusters}, k \neq l.\}.$$

The relation of pairs in CL is called cannot-link.

3 Proposed Algorithms

We propose three new semi-supervised clustering algorithms: hard c-means with pairwise constraints by non-metric term (HCM-NM), standard fuzzy c-means with pairwise constraints by non-metric term (sFCM-NM), and entropy-based fuzzy c-means with pairwise constraints by non-metric term (eFCM-NM).

3.1   Hard c-means with Pairwise Constraints by Non-metric Term

We define the objective function of the proposed algorithm as follows:

$$J_{\text{HCM-NM}}(U, V) = \sum_{k=1}^{n} \sum_{i=1}^{c} u_{ki} d_{ki} - \sum_{k=1}^{n} \sum_{l=1}^{n} \sum_{i=1}^{c} u_{ki} u_{li} (\alpha_{kl} - \beta_{kl}).$$

The constraint is $\mathcal{U}_{\text{HCM}}$. The first, second, and third terms mean the objective function of HCM, the non-metric term for must-link, and the non-metric term for cannot-link, respectively. Here α_{kl} and β_{kl} are as follows:

$$\alpha_{kl} = \begin{cases} \lambda_{ML} D_{\max}, & ((x_k, x_l) \in ML) \quad (D_{\max} = \max_{p,q} D_{pq}) \\ 0, & (\text{otherwise}) \end{cases}$$

$$\beta_{kl} = \begin{cases} \lambda_{CL} D_{\max}, & ((x_k, x_l) \in CL) \\ \varepsilon \lambda_{ML} D_{\max}. & (\text{otherwise}) \end{cases}$$

λ_{ML}, λ_{CL}, and ε ($\lambda_{CL} > \varepsilon \lambda_{ML} > \lambda_{ML} > 0$) are parameters by which we can control effectiveness of pairwise constraints. We can obtain optimal solutions which minimize the objective function $J_{\text{HCM-NM}}$ by the method of Lagrange multiplier.

First, we derive the optimal solution to v_i. It is sufficient to obtain v_i when the partial derivative of the objective function $J_{\text{HCM-NM}}$ with respect to v_i is equal to zero because the objective function $J_{\text{HCM-NM}}$ is convex for v_i. The partial derivative of the objective function $J_{\text{HCM-NM}}$ with respect to v_i is as follows:

$$\frac{\partial J}{\partial v_i} = \sum_{k=1}^{n} (-2) u_{ki} (x_k - v_i).$$

Therefore, the optimal solution to v_i is obtained as follows:

$$v_i = \frac{\sum_{k=1}^{n} u_{ki} x_k}{\sum_{k=1}^{n} u_{ki}}.$$

Second, we derive the optimal solution to u_{ki}. When we put $\zeta_{ki} = d_{ki} - \sum_{l=1}^{n} u_{li} \alpha_{kl} + \sum_{l=1}^{n} u_{li} \beta_{kl}$, the objective function $J_{\text{HCM-NM}}$ can be rewritten as follows:

$$J_{\text{HCM-NM}}(U, V) = \sum_{k=1}^{n} \sum_{i=1}^{c} u_{ki} \zeta_{ki}.$$

Because of $\mathcal{U}_{\text{HCM}}$, the optimal solution to u_{ki} is derived as follows:

$$u_{ki} = \begin{cases} 1, & (i = \arg\min_j \zeta_{ki}) \\ 0. & (\text{otherwise}) \end{cases}$$

3.2   Standard Fuzzy c-means with Pairwise Constraints by Non-metric Term

We define the objective function of the proposed algorithm as follows:

$$J_{\text{sFCM-NM}}(U, V) = \sum_{k=1}^{n} \sum_{i=1}^{c} (u_{ki})^m d_{ki} - \sum_{k=1}^{n} \sum_{l=1}^{n} \sum_{i=1}^{c} (u_{ki})^m (u_{li})^m (\alpha_{kl} - \beta_{kl}). \tag{1}$$

The constraint is $\mathcal{U}_{FCM}$. The first, second, and the third terms mean the objective function of sFCM, the non-metric term for must-link, and the non-metric term for cannot-link, respectively. We can obtain optimal solutions which minimize the objective function $J_{eFCM-NM}$ by the method of Lagrange multiplier.

First, we mention the optimal solution to v_i. We obtain the same optimal solution to v_i as HCM-NM in the same way of HCM-NM.

Second, we derive the optimal solution to u_{ki}. When we put $\eta_{ki} = d_{ki} - \sum_{l=1}^{n}(u_{li})^m \alpha_{kl} + \sum_{l=1}^{n}(u_{li})^m \beta_{kl}$, the objective function $J_{sFCM-NM}$ can be rewritten as follows:

$$J_{sFCM-NM}(U, V) = \sum_{k=1}^{n} \sum_{i=1}^{c}(u_{ki})^m \eta_{ki}.$$

Because of $\mathcal{U}_{FCM}$, the optimal solution to u_{ki} is derived as follows:

$$u_{ki} = \frac{(1/\eta_{ki})^{\frac{1}{m-1}}}{\sum_{j=1}^{c}(1/\eta_{kj})^{\frac{1}{m-1}}}.$$

3.3 Entropy-Based Fuzzy *c*-means with Pairwise Constraints by Non-metric Term

We define the objective function of the proposed algorithm as follows:

$$J_{eFCM-NM}(U, V) = \sum_{k=1}^{n} \sum_{i=1}^{c} u_{ki}(d_{ki} + \lambda \log u_{ki}) - \sum_{k=1}^{n} \sum_{l=1}^{n} \sum_{i=1}^{c} u_{ki}u_{li}(\alpha_{kl} - \beta_{kl}).$$

The constraint is $\mathcal{U}_{FCM}$. The first and second terms mean the objective function of eFCM. The third and forth terms mean the non-metric term for must-link, and the non-metric term for cannot-link, respectively. We can obtain optimal solutions which minimize the objective function $J_{eFCM-NM}$ by the method of Lagrange multiplier.

First, we derive the optimal solution to v_i. We obtain the same optimal solution to v_i as HCM-NM in the same way of HCM-NM.

Second, we derive the optimal solution to v_i. The objective function $J_{eFCM-NM}$ can be rewritten as follows:

$$J_{eFCM-NM}(U, V) = \sum_{k=1}^{n} \sum_{i=1}^{c} u_{ki}(\zeta_{ki} + \lambda \log u_{ki}).$$

Because of $\mathcal{U}_{FCM}$, the optimal solution to u_{ki} is derived as follows:

$$u_{ki} = \frac{\exp(-\lambda^{-1}\zeta_{ki})}{\sum_{j=1}^{c}\exp(-\lambda^{-1}\zeta_{kj})}.$$

3.4 Proposed Algorithms: HCM-NM and FCM-NM

We propose three new semi-supervised clustering algorithms HCM-NM and FCM-NM (sFCM-NM and eFCM-NM) by using the optimal solutions obtained above in Algorithm 1. These algorithms give clustering results through iterative optimization.

Algorithm 1. HCM-NM and FCM-NM Algorithms

Step1. Give ML and CL. Set the initial values of U and V to $U^{(0)}$ $V^{(0)}$. Let the iteration count be $L = 0$.

Step2. Calculate $V^{(L+1)}$ with fixing $U^{(L)}$ as follows:

$$v_i^{(L+1)} = \frac{\sum_{k=1}^n u_{ki}^{(L)} x_k}{\sum_{k=1}^n u_{ki}^{(L)}}.$$

Step3. Calculate $U^{(L+1)}$ with fixing $V^{(L+1)}$ as follows:

In case of HCM-NM:

$$u_{ki}^{(L+1)} = \begin{cases} 1, & (i = \arg\min_j \zeta_{ki}^{(L+1)}) \\ 0. & \text{(otherwise)} \end{cases}$$

Here $\zeta_{ki}^{(L+1)} = d_{kj}^{(L+1)} - \sum_{l=1}^n u_{lj}^{(L)} \alpha_{kl} + \sum_{l=1}^n u_{lj}^{(L)} \beta_{kl}$.

In case of sFCM-NM:

$$u_{ki}^{(L+1)} = \frac{(1/\eta_{ki}^{(L+1)})^{\frac{1}{m-1}}}{\sum_{j=1}^c (1/\eta_{kj}^{(L+1)})^{\frac{1}{m-1}}}.$$

Here $\eta_{ki}^{(L+1)} = d_{ki}^{(L+1)} - \sum_{l=1}^n (u_{li}^{(L)})^m \alpha_{kl} + \sum_{l=1}^n (u_{li}^{(L)})^m \beta_{kl}$.

In case of eFCM-NM:

$$u_{ki}^{(L+1)} = \frac{\exp(-\lambda^{-1}\zeta_{ki}^{(L+1)})}{\sum_{j=1}^c \exp(-\lambda^{-1}\zeta_{kj}^{(L+1)})}.$$

Step4. If the convergence criterion is satisfied, the algorithm is finished. Otherwise, $L = L + 1$ and back to **Step2.**

End of Algorithm.

---

4 Numerical Examples

4.1 Datasets

We use two artificial datasets in Fig. 1 and Fig. 2. The former has 120 objects and it consists of two clusters: upper and lower. The latter has 150 objects and it consists of two clusters: outside and inside. The parameters are shown in Table 1 and we give 1000 types of initial values to each clustering algorithm. We verify clustering results by four

Table 1. Parameters

m	λ	λ_{ML}	λ_{CL}	ε
2.0	2.0	1.0	2.0	1.1

typical validity indexes: normalized mutual information (NMI), Purity, Entropy, and Rand index as follows:

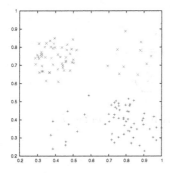

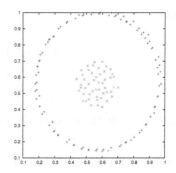

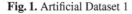

Fig. 1. Artificial Dataset 1   **Fig. 2.** Artificial Dataset 2

NMI Let C and $\mathcal{T}$ be a set of generated clusters $\{C_i \mid i = 1, \ldots, c\}$ and a set of correct clusters $\{T_i \mid i = 1, \ldots, c\}$. The validity function v_{NMI} of NMI is defined as follows:

$$v_{\text{NMI}}(C, \mathcal{T}) = \frac{H(C) + H(\mathcal{T}) - H(C, \mathcal{T})}{\max\{H(C), H(\mathcal{T})\}}.$$

Here

$$H(C) = \sum_{i=1}^{c} -P(C_i) \log P(C_i),$$

$$H(C, \mathcal{T}) = \sum_{i=1}^{c} \sum_{j=1}^{c} -P(C_i, T_j) \log P(C_i, T_j),$$

$$P(C_i) = \frac{|C_i|}{\text{Total number of all objects}},$$

$$P(C_i, T_j) = P(C_i)P(T_j|C_i).$$

$v_{\text{NMI}}(C, \mathcal{T}) \in [0, 1]$ and $v_{\text{NMI}}(C, \mathcal{T}) = 1$ if and only if $C_i = T_i$ for any i.

Purity The validity function v_{PT} of Purity is defined as follows:

$$v_{\text{PT}}(C, \mathcal{T}) = \sum_{j=1}^{c} P(T_j) \max_i P(C_i|T_j).$$

$v_{\text{PT}}(C, \mathcal{T}) \in [0, 1]$ and $v_{\text{PT}}(C, \mathcal{T}) = 1$ if and only if $C_i = T_i$ for any i.

Entropy The validity function v_{ET} of Entropy is defined as follows:

$$v_{\text{ET}}(C, \mathcal{T}) = -\frac{1}{\log c} \sum_{j=1}^{c} P(T_j) \sum_{i=1}^{c} P(C_i \mid T_j) \log P(C_i \mid T_j).$$

$v_{\text{PT}}(C, \mathcal{T}) \in [0, 1]$ and the smaller the value of vPT, the better the clustering result is.

Rand index The validity function v_{RI} of Rand index is defined as follows:

$$v_{\text{RI}}(C, \mathcal{T}) = \frac{1}{|N|} \left(\binom{|N|}{2} - \sum_{i=1}^{c} \binom{n_{i.}^2}{2} - \sum_{j=1}^{c} \binom{n_{.j}^2}{2} + 2 \sum_{i=1}^{c} \sum_{j=1}^{c} \binom{n_{ij}^2}{2} \right).$$

Here,

$$n_{ij} = |C_i \cap T_j|,$$
$$N = \{n_{ij} \mid i = 1, \ldots, c, \ j = 1, \ldots, c\},$$
$$n_{i.} = \sum_{j=1}^{c} n_{ij}, \quad n_{.j} = \sum_{i=1}^{c} n_{ij}.$$

$v_{\mathrm{RI}}(C, \mathcal{T}) \in [0, 1]$ and $v_{\mathrm{RI}}(C, \mathcal{T}) = 1$ if and only if $C_i = T_i$ for any i.

4.2   Results of Artificial Dataset 1

We show clustering results. Table 2 Table 3, and Table 4 show the number of objects which is truly classified into correct clusters, the ratio of the number to total number of data, and values of validity indexes.

Table 2. Results for Artificial Dataset 1 by HCM-NM

$(|CL|, |ML|)$	Correct Clustering		NMI	Purity	Entropy	Rand index
	Number	Ratio (%)				
HCM	109	0.90	0.57	0.91	0.43	0.80
$(0,0)$	100	0.83	0.35	0.83	0.65	0.72
$(10,0)$	100	0.83	0.35	0.83	0.65	0.72
$(20,0)$	120	1.00	1.00	1.00	0.00	1.00
$(50,0)$	120	1.00	1.00	1.00	0.00	1.00
$(0,10)$	120	1.00	1.00	1.00	0.00	1.00
$(0,20)$	120	1.00	1.00	1.00	0.00	1.00
$(0,50)$	120	1.00	1.00	1.00	0.00	1.00
$(10,10)$	120	1.00	1.00	1.00	0.00	1.00
$(20,20)$	120	1.00	1.00	1.00	0.00	1.00
$(50,50)$	120	1.00	1.00	1.00	0.00	1.00

Table 3. Results for Artificial Dataset 1 by sFCM-NM

$(|CL|, |ML|)$	Correct Clustering		NMI	Purity	Entropy	Rand index
	Number	Ratio (%)				
sFCM	110	0.92	0.59	0.92	0.40	0.85
$(0,0)$	78	0.65	0.07	0.65	0.93	0.54
$(10,0)$	70	0.58	0.09	0.58	0.33	0.51
$(20,0)$	80	0.67	0.19	0.67	0.46	0.55
$(50,0)$	71	0.59	0.10	0.59	0.34	0.51
$(0,10)$	110	0.92	0.65	0.92	0.33	0.85
$(0,20)$	68	0.57	0.07	0.57	0.28	0.50
$(0,50)$	107	0.89	0.59	0.89	0.38	0.81
$(10,10)$	63	0.53	0.03	0.53	0.15	0.50
$(20,20)$	94	0.78	0.37	0.78	0.49	0.66
$(50,50)$	86	0.72	0.26	0.72	0.49	0.59

Table 4. Results for Artificial Dataset 1 by eFCM-NM

$(|CL|, |ML|)$	Correct Clustering		NMI	Purity	Entropy	Rand index
	Number	Ratio (%)				
eFCM	120	1.00	1.00	1.00	0.00	1.00
$(0, 0)$	82	0.68	0.10	0.68	0.90	0.56
$(10, 0)$	67	0.56	0.01	0.56	0.99	0.50
$(20, 0)$	120	1.00	1.00	1.00	0.00	1.00
$(50, 0)$	120	1.00	1.00	1.00	0.00	1.00
$(0, 10)$	105	0.88	0.55	0.88	0.42	0.78
$(0, 20)$	120	1.00	1.00	1.00	0.00	1.00
$(0, 50)$	120	1.00	1.00	1.00	0.00	1.00
$(10, 10)$	118	0.98	0.89	0.98	0.11	0.97
$(20, 20)$	120	1.00	1.00	1.00	0.00	1.00
$(50, 50)$	120	1.00	1.00	1.00	0.00	1.00

Fig. 3, Fig. 4, and Fig. 5 show ratios of the number to total number of data. The values of the ratios when $|ML| = 0, 10, 20$ and 50 are connect with a solid line. The values of the ratios when $|CL| = 0, 10, 20$ and 50 are connect with a dashed line. The values of the ratios when $|ML| + |CL| = 0, 20, 40$ and 100 (plotted at $0, 10, 20$ and 50 on the horizontal axis) are connect with a dotted line.

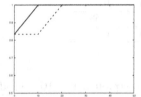
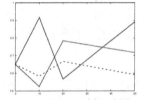
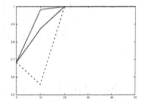

Fig. 3. Ratio for Artificial Dataset 1 by HCM-NM

Fig. 4. Ratio for Artificial Dataset 1 by sFCM-NM

Fig. 5. Ratio for Artificial Dataset 1 by eFCM-NM

We show the best result by each proposed algorithm in Fig. 6, Fig. 7, and Fig. 8. Solid and dashed lines mean must-link and cannot-link relations, respectively.

4.3   Results of Artificial Dataset 2

We show clustering results. Table 5, Table 6, and Table 7 show the number of objects which is truly classified into correct clusters, the ratio of the number to total number of data, and values of validity indexes. In Table 6, we find the symbol "nan". It means "division by zero" and the reason is that two cluster centers are calculated as the same.

Fig. 9, Fig. 10, and Fig. 11 show ratios of the number to total number of data. The values of the ratios when $|ML| = 0, 10, 20$ and 50 are connect with a solid line. The values of the ratios when $|CL| = 0, 10, 20$ and 50 are connect with a dashed line.

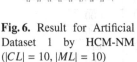 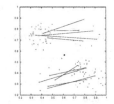 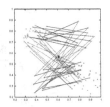

Fig. 6. Result for Artificial Dataset 1 by HCM-NM ($|CL| = 10, |ML| = 10$)   **Fig. 7.** Result for Artificial Dataset 1 by sFCM-NM ($|CL| = 0, |ML| = 10$)   **Fig. 8.** Result for Artificial Dataset 1 by eFCM-NM ($|CL| = 20, |ML| = 20$)

Table 5. Results for Artificial Dataset 2 by HCM-NM

$(|CL|, |ML|)$	Correct Clustering Number	Ratio (%)	NMI	Purity	Entropy	Rand index
HCM	85	0.57	0.03	0.57	0.96	0.51
$(0, 0)$	81	0.54	0.01	0.54	0.99	0.50
$(10, 0)$	77	0.51	0.00	0.51	1.00	0.50
$(20, 0)$	91	0.61	0.04	0.61	0.96	0.52
$(50, 0)$	114	0.76	0.36	0.76	0.63	0.63
$(0, 10)$	91	0.61	0.04	0.61	0.96	0.52
$(0, 20)$	94	0.63	0.06	0.63	0.94	0.53
$(0, 50)$	125	0.83	0.46	0.83	0.54	0.72
$(10, 10)$	94	0.63	0.06	0.63	0.94	0.53
$(20, 20)$	105	0.70	0.14	0.70	0.86	0.58
$(50, 50)$	125	0.83	0.46	0.83	0.54	0.72

The values of the ratios when $|ML| + |CL| = 0, 20, 40$ and 100 (plotted at 0, 10, 20 and 50 on the horizontal axis) are connect with a dotted line.

We show the best result by each proposed algorithm in Fig. 12, Fig. 13, and Fig. 14. Solid and dashed lines mean must-link and cannot-link relations, respectively.

4.4 Consideration

For Artificial Dataset 1, The best result is by HCM-NM in Table 2 and the result by eFCM-NM is similar to HCM-NM in Table 4. Besides, the result by sFCM-NM does not look good in any cases from in Fig. 4.

On the other hand, for Artificial Dataset 2, the best result is by eFCM-NM in Table 14. Moreover, the results by eFCM-NM are better than the other algorithms from clustering results in Table 5, Table 6 and Table 7, and the ratios of the number of objects which is truly classified into correct clusters to total number of objects in Fig. 9, Fig. 10 and Fig. 11.

Thus, we think the most useful algorithm is eFCM-NM. However all the proposed algorithms strongly depend on initial values. Therefore, the other algorithms may output better results.

Table 6. Results for Artificial Dataset 2 by sFCM-NM

$(|CL|, |ML|)$	Correct Clustering		NMI	Purity	Entropy	Rand index
	Number	Ratio (%)				
sFCM	83	0.55	0.017	0.55	0.97	0.50
$(0,0)$	84	0.56	0.17	0.56	0.62	0.50
$(10,0)$	100	0.67	nan	0.67	0.00	0.55
$(20,0)$	100	0.67	nan	0.67	0.00	0.55
$(50,0)$	100	0.67	nan	0.67	0.00	0.55
$(0,10)$	100	0.67	nan	0.67	0.00	0.55
$(0,20)$	100	0.67	nan	0.67	0.00	0.55
$(0,50)$	100	0.67	nan	0.67	0.00	0.55
$(10,10)$	84	0.56	0.00	0.56	0.84	0.50
$(20,20)$	100	0.67	nan	0.67	0.00	0.55
$(50,50)$	100	0.67	nan	0.67	0.00	0.55

Table 7. Results for Artificial Dataset 2 by eFCM-NM

$(|CL|, |ML|)$	Correct Clustering		NMI	Purity	Entropy	Rand index
	Number	Ratio (%)				
eFCM	79	0.53	0.00	0.53	0.99	0.50
$(0,0)$	78	0.52	0.00	0.52	0.99	0.50
$(10,0)$	93	0.62	0.08	0.62	0.91	0.53
$(20,0)$	97	0.65	0.14	0.65	0.82	0.54
$(50,0)$	116	0.77	0.38	0.77	0.62	0.65
$(0,10)$	81	0.54	0.01	0.54	0.97	0.50
$(0,20)$	95	0.63	0.07	0.63	0.92	0.53
$(0,50)$	150	1.00	1.00	1.00	0.00	1.00
$(10,10)$	90	0.60	0.09	0.60	0.86	0.52
$(20,20)$	125	0.83	0.34	0.83	0.33	0.72
$(50,50)$	125	0.83	0.46	0.83	0.54	0.72

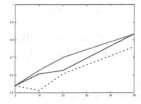

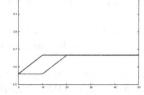

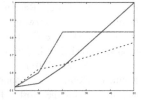

Fig. 9. Ratio for Artificial Dataset 2 by HCM-NM

Fig. 10. Ratio for Artificial Dataset 2 by sFCM-NM

Fig. 11. Ratio by for Artificial Dataset 2 eFCM-NM

sFCM-NM is a problem that all cluster centers are extremely closed and consequently, the belongingness of all objects are about the same value. The reason is that the values of the third term for the cannot-link relation in the objective function of sFCM-NM (1) are positive for all pairs. In case of not only pairs which are included in CL, but also pairs which are not included in CL, the closer to $1/c$ the belongingness of

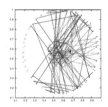   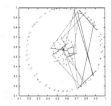   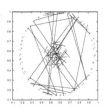

Fig. 12. Result for Artificial Dataset 2 by HCM-NM
($|CL| = 50, |ML| = 50$)

Fig. 13. Result for Artificial Dataset 2 by sFCM-NM
($|CL| = 10, |ML| = 10$)

Fig. 14. Result for Artificial Dataset 2 by eFCM-NM
($|CL| = 0, |ML| = 50$)

the objects in the pairs is, the smaller the value of the third term of (1). Therefore, all cluster centers are extremely closed to the centroid of all objects, and consequently, the belongingness of all objects are about the same value.

5   Conclusion

In this paper, we considered terms of must-link constraint and cannot-link constraint for prior information to use non-metric model, and then unified the terms objective functions of the well-known clustering algorithms: hard c-means, standard fuzzy c-means, and entropy-based fuzzy c-means. Moreover, we verified the algorithms through some numerical examples. We think the most useful proposed algorithm is sFCM-CM.

Acknowledgment. We would like to gratefully thank Professor Sadaaki Miyamoto of the University of Tsukuba, and Associate Professor Yuchi Kanzawa of Shibaura Institute of Technology for their advice.

References

1. Bezdek, J.C.: Pattern Recognition with Fuzzy Objective Function Algorithms. Kluwer Academic Publishers, Norwell (1981)
2. Wagstaff, K., Cardie, C., Rogers, S., Schroedl, S.: Constrained K-means Clustering with Background Knowledge. In: Proceedings of the 18th International Conference on Machine Learning, pp. 577–584 (2001)
3. Klein, D., Kamvar, D., Manning, C.: From Instance-level Constraints to Space-level Constraints: Making the Most of Prior Knowledge in Data Clustering. In: Proceedings of the 19th International Conference on Machine Learning, pp. 307–314 (2002)
4. Basu, S., Banerjee, A., Mooney, R.J.: Active Semi-Supervision for Pairwise Constrained Clustering. In: Proceedings of the SIAM International Conference on Data Mining, pp. 333–344 (2004)
5. Kulis, B., Basu, S., Dhillon, I., Mooney, R.: Semi-Supervised Graph Clustering: A Kernel Approach. In: Proceedings of the 22nd International Conference on Machine Learning, pp. 457–464 (2005)
6. Roubens, M.: Pattern Classification Problems and Fuzzy Sets. Fuzzy Sets and Systems 1, 239–253 (1978)

7. MacQueen, I.: Some Methods for Classification and Analysis of Multivariate Observations. In: Proceedings of the Fifth Berkeley Symposium on Mathematical Statistics and Probability, pp. 281–297 (1967)
8. Miyamoto, S., Umayahara, K., Mukaidono, M.: Fuzzy Classification Functions in the Methods of Fuzzy c-means and Regularization by Entropy. Journal of Japan Society for Fuzzy Theory and Systems 10(3), 548–557 (1998) (in Japanese)

Application of the Fuzzy-Possibilistic Product Partition in Elliptic Shell Clustering[*]

László Szilágyi[1,2], Zsuzsa Réka Varga[2], and Sándor Miklós Szilágyi[3]

[1] Budapest University of Technology and Economics, Hungary
Dept. of Control Engineering and Information Technology
lalo@ms.sapientia.ro
[2] Sapientia - Hungarian Science University of Transylvania, Romania
Faculty of Technical and Human Science of Tîrgu-Mureş
[3] Petru Maior University of Tîrgu-Mureş, Romania
Dept. of Mathematics - Informatics

Abstract. Creating accurate and robust clustering models is utmost important in pattern recognition. This paper introduces an elliptic shell clustering model aiming at accurate detection of ellipsoids in the presence of outlier data. The proposed fuzzy-possibilistic product partition c-elliptical shell algorithm (FP^{3}CES) combines the probabilistic and possibilistic partitions in a qualitatively different way from previous, similar algorithms. The novel mixture partition is able to suppress the influence of extreme outlier data, which gives it net superiority in terms of robustness and accuracy compared to previous algorithms, fact supported by cluster validity indices.

Keywords: fuzzy c-elliptical shell clustering, probabilistic partition, possibilistic partition, robust clustering.

1 Introduction

Robustness in clustering refers to the stability or reproducibility of the achieved partition, and insensitivity to several kinds of noise including severely outlier data. The fuzzy c-means (FCM) algorithm introduced by Bezdek [2] is a very popular clustering model due to the fine partitions it makes and its easily comprehensible alternating optimization (AO) scheme. However, the probabilistic constraints involved in FCM make it sensitive to outlier data. To combat this problem, several solutions have been proposed, which found their way to relax the probabilistic constraint.

An early solution was given by Davé [4], who introduced an extra, specially treated noisy class to attract feature vectors situated far from all normal cluster prototypes. This theory was later extended by Menard et al [14]. Alternately, Krishnapuram and Keller proposed the possibilistic c-means algorithm (PCM) [12], which distributes the partition matrix elements based on statistical rules.

---
[*] Research supported by the Hungarian National Research Funds (OTKA), Project no. PD103921 and the MTA János Bolyai Fellowship Program.

V. Torra et al. (Eds.): MDAI 2014, LNAI 8825, pp. 158–169, 2014.

This approach seemed to have solved the sensitivity to outliers, but it cannot be called a robust algorithm due to the coincident clusters it frequently produces [1]. Timm et al [19] set up a repulsive force between all couples of cluster prototypes of PCM, the strength of which decrease with distance. Their method succeeded in avoiding coincident clusters, but failed to accurately treat cases when two clusters are really close to each other. Two versions of fuzzy-possibilistic partition mixtures were proposed by Pal et al [15,16], out of which the second one (possibilistic fuzzy c-means - PFCM) appears to be a reliable clustering model.

In a recently published paper we introduced the novel fuzzy-possibilitistic product partition c-means clustering model (FP^{3}CM) [17], in which the degrees of membership are given as the product of a probabilistic and a possibilistic factor. This new approach proved to eliminate all adverse effects of distant outliers, while producing high quality partitions.

All algorithms discussed above work with point-type cluster prototypes that are computed as weighted means of the input data. However, frequently emerge situations when the shape of the clusters differ from the default. To cope with such scenarios, several solutions have been proposed. Linear manifolds are usually modeled via adaptive fuzzy c-varieties [9], while spherical ones via the fuzzy c-spherical shell (FCSS) algorithm model by Krishnapuram et al [11]. Elliptical prototypes were introduced by the adaptive fuzzy c-shells algorithm, and the fuzzy c-ellipsoidal shells clustering model [7]. Generalized versions of shell clusters to the quadric case were given by Krishnapuram et al [10]. Further quadric prototypes models include the fuzzy c-quadrics [6] and the fuzzy c-quadric shells [13]. The norm-induced shell prototypes introduced by Bezdek et al [3] can be adapted to detect ellipses, quadrics, and rectangles as well. Later, Höppner extended the palette of detectable shapes with his fuzzy c-rectangular shell models [8]. Out of these shaped cluster models, only FCSS has explicit expressions to compute the cluster prototypes in each iteration of the main AO loop. All others use nonlinear implicit expressions that need to be solved in an iterative way. The latter, besides being more complicated to implement, also represents a higher computational load.

In an earlier paper [18] we introduced the so-called fuzzy-possibilistic product partition c-spherical shell clustering model, which combines the robust and accurate FP^{3}CM algorithm [17] with the considerations of the fuzzy c-spherical shell clustering approach [11], and investigated its capabilities in identifying spheroidal shapes in two-, three-, and multidimensional environments. In this paper we introduce the fuzzy-possibilistic product partition c-elliptical shell (FP^{3}CES) clustering algorithm, as the combination of the fuzzy-possibilistic product partition with the elliptical shell clustering mechanism proposed by Dave [5], and will investigate its capabilities in identifying ellipsoidal clusters in noisy environment.

The rest of this paper is structured as follows. Section 2 summarizes the c-means clustering framework applying either a pure probabilistic or possibilistic, or a mixed partition. Section 3 introduces the novel FP^{3}CES clustering model.

Section 4 produces a numerical analysis and comparison of the proposed and earlier methods. Conclusions are given in the last section.

2 Fuzzy c-Means Clustering Models with Probabilistic, Possibilistic, and Mixed Partition

The generally formulated c-means clustering model partitions a set of n object data $\boldsymbol{x}_k \in \mathbb{R}^z$, $k = 1 \ldots n$, into a previously defined number of c clusters, based on the minimization of a quadratic objective function, formulated as:

$$J_c = \sum_{i=1}^{c} \sum_{k=1}^{n} f(u_{ik}, t_{ik}) ||\boldsymbol{x}_k - \boldsymbol{v}_i||_{\boldsymbol{A}}^2 + \sum_{i=1}^{c} \eta_i \sum_{k=1}^{n} g(u_{ik}, t_{ik}) , \tag{1}$$

where $\boldsymbol{v}_i$ represents the prototype or centroid value or representative element of cluster i ($i = 1 \ldots c$), $d_{ik\boldsymbol{A}} = ||\boldsymbol{x}_k - \boldsymbol{v}_i||_{\boldsymbol{A}} = \sqrt{(\boldsymbol{x}_k - \boldsymbol{v}_i)^T \boldsymbol{A}(\boldsymbol{x}_k - \boldsymbol{v}_i)}$ represents the generalized distance between input vector $\boldsymbol{x}_k$ and the cluster prototype $\boldsymbol{v}_i$, and η_i are the penalty terms that control the variance of the clusters [12]. Further on, f and g are two functions which depend on the employed algorithm, whose definitions are described in Table 1, where u_{ik} is the probabilistic fuzzy membership function showing the degree to which input vector $\boldsymbol{x}_k$ belongs to cluster i, t_{ik} is the degree of compatibility of vector $\boldsymbol{x}_k$ with cluster i, $m > 1$ and $p > 1$ are the fuzzy and possibilistic exponents, respectively. The probabilistic partition consists of u_{ik} fuzzy membership values, which are constrained by:

$$\begin{cases} 0 \le u_{ik} \le 1 & \forall i = 1 \ldots c \quad \forall k = 1 \ldots n \\ \sum_{i=1}^{c} u_{ik} = 1 & \forall k = 1 \ldots n \end{cases} . \tag{2}$$

The possibilistic partition consists of t_{ik} fuzzy membership values, which are constrained by:

$$\begin{cases} 0 \le t_{ik} \le 1 & \forall i = 1 \ldots c \quad \forall k = 1 \ldots n \\ 0 < \sum_{i=1}^{c} t_{ik} < c & \forall k = 1 \ldots n \end{cases} . \tag{3}$$

The optimization of J_c is performed within the framework of grouped coordinate minimization. In case of pure probabilistic (possibilistic) partition, this consists of alternately applying the optimization of J_c over $\{u_{ik}\}$ ($\{t_{ik}\}$) with $\boldsymbol{v}_i$ fixed, and the optimization of J_c over $\{\boldsymbol{v}_i\}$ with u_{ik} (t_{ik}) fixed [2,12]. In case of mixed partitions, each iteration of the optimization scheme consists of three steps: (1) optimization of J_c over $\{t_{ik}\}$ with $\{u_{ik}\}$ and $\boldsymbol{v}_i$ fixed, (2) optimization of J_c over $\{u_{ik}\}$ with $\{t_{ik}\}$ and $\boldsymbol{v}_i$ fixed, and (3) optimization of J_c over $\boldsymbol{v}_i$ with $\{u_{ik}\}$ and $\{t_{ik}\}$ fixed. PFCM allows the execution of the first two steps in any order. During each cycle, the optimal partitions and cluster prototypes are computed from the zero gradient conditions, and obtained as presented in the bottom three rows of Table 1.

We need to remark that the original PFCM algorithm [16] used $g(u_{ik}, t_{ik}) = (1 - t_{ik})^p$. We decided to use $g(u_{ik}, t_{ik}) = b(1 - t_{ik})^p$ because this way the tradeoff parameter b does not interfere with the penalty term values η_i.

Table 1. Various c-means algorithms defined with f_{ik} and g_{ik} functions, and the formulas of the alternating optimization framework

Algorithm	FCM [2]	PFCM [16]	FP^{3}CM [17]
Parameters	exponent $m > 1$	exponents $m, p > 1$ $\eta_i,\ i = 1 \dots c$ a, b for tradeoff	exponents $m, p > 1$ $\eta_i,\ i = 1 \dots c$
$f(u_{ik}, t_{ik})$	u_{ik}^m	$au_{ik}^m + bt_{ik}^p$	$u_{ik}^m t_{ik}^p$
$g(u_{ik}, t_{ik})$	0	$b(1 - t_{ik})^p$	$u_{ik}^m (1 - t_{ik})^p$
$t_{ik}^\star$	not applicable	$\left[1 + \left(\frac{d_{ik}^2}{\eta_i}\right)^{\frac{1}{p-1}}\right]^{-1}$	$\left[1 + \left(\frac{d_{ik}^2}{\eta_i}\right)^{\frac{1}{p-1}}\right]^{-1}$
$u_{ik}^\star$	$\dfrac{d_{ik}^{-2/(m-1)}}{\sum\limits_{j=1}^{c} d_{jk}^{-2/(m-1)}}$	$\dfrac{d_{ik}^{-2/(m-1)}}{\sum\limits_{j=1}^{c} d_{jk}^{-2/(m-1)}}$	$\dfrac{[d_{ik}^2 t_{ik}^p + \eta_i(1-t_{ik})^p]^{-1/(m-1)}}{\sum\limits_{j=1}^{c} [d_{jk}^2 t_{jk}^p + \eta_j(1-t_{jk})^p]^{-1/(m-1)}}$
$v_i^\star$	$\dfrac{\sum\limits_{k=1}^{n} u_{ik}^m \boldsymbol{x}_k}{\sum\limits_{k=1}^{n} u_{ik}^m}$	$\dfrac{\sum\limits_{k=1}^{n} [au_{ik}^m + bt_{ik}^p]\boldsymbol{x}_k}{\sum\limits_{k=1}^{n} [au_{ik}^m + bt_{ik}^p]}$	$\dfrac{\sum\limits_{k=1}^{n} u_{ik}^m t_{ik}^p \boldsymbol{x}_k}{\sum\limits_{k=1}^{n} u_{ik}^m t_{ik}^p}$

3 Methods

The fuzzy-possibilistic product partition proved to be a successful tool in c-means clustering of noisy data [17] and in the detection of hyperspheres in noisy environment [18]. In the following, we will extend the use of the product partition to identify ellipsoidal shell clusters in the presence of outliers.

Ellipsoids in computational geometry are usually described by a center point, one radius value in each dimension, and a rotation vector which defines the orientation of the ellipsoid. Here we will define an ellipsoid as a collection of points x that satisfy the equation:

$$(\boldsymbol{x} - \boldsymbol{v})\mathbf{A}(\boldsymbol{x} - \boldsymbol{v})^T = r^2 \ ,$$

where $\boldsymbol{v}$ is the center of the ellipsoid, r is a variable that controls the size of the ellipsoid (radii vary proportionally with r), and matrix $\mathbf{A}$ is a positive definite matrix that describes the shape and orientation of the ellipsoid. Under such circumstances, the distance of any data point $\boldsymbol{x}_k$ ($k = 1 \dots n$) from cluster prototype number i ($i = 1 \dots c$), defined by $\boldsymbol{v}_i$, r_i, and $\mathbf{A}_i$ is computed as:

$$D_{ik}^2 = (d_{ik} - r_i)^2 \ , \tag{4}$$

where

$$d_{ik} = ||\boldsymbol{x}_k - \boldsymbol{v}_i||_\mathbf{A} = \sqrt{(\boldsymbol{x}_k - \boldsymbol{v}_i)\mathbf{A}(\boldsymbol{x}_k - \boldsymbol{v}_i)^T} \ . \tag{5}$$

The objective function of the proposed elliptic shell clustering algorithm is:

$$J_{\text{FP3CES}} = \sum_{i=1}^{c} \sum_{k=1}^{n} u_{ik}^m [t_{ik}^p D_{ik}^2 + (1 - t_{ik})^p \eta_i] \ ,$$

which will be optimized under the probabilistic and possibilistic constraints given in Eqs. (2) and (3), and a further constraint $\det(\mathbf{A}_i) = \rho_i$ fixed, which assures that each possible ellipsoid has a unique description using its own v_i, r_i, and $\mathbf{A}_i$ values.

Using the same notations and the ones given in Table 1, the objective functions of counter candidate algorithms FCES and PFCES will be:

$$J_{\text{FCES}} = \sum_{i=1}^{c} \sum_{k=1}^{n} u_{ik}^m D_{ik}^2 \ ,$$

$$J_{\text{PFCES}} = \sum_{i=1}^{c} \sum_{k=1}^{n} [(au_{ik}^m + bt_{ik}^p)D_{ik}^2 + b(1 - t_{ik})^p \eta_i] \ .$$

In order to optimize any of the above objective functions, we need to find the optimal cluster prototypes (v_i, r_i, $\mathbf{A}_i$ for any $i = 1 \ldots c$) and optimal partition (u_{ik} and t_{ik} where applicable, for any $i = 1 \ldots c$ and $k = 1 \ldots n$). The optimum is reached via grouped coordinate minimization by alternately optimizing the partition with fixed cluster prototypes, and then optimizing the cluster prototypes keeping the partition fixed. The alternating optimization is stopped when the norm of variation of cluster prototypes during an iteration stays below a predefined constant ε. The alternately applied optimization formulas are obtained from zero crossing of the objective function's partial derivatives, using Lagrange multipliers where necessary. The optimization formulas thus obtained are necessary conditions of finding the objective function's optimum.

The optimization formulas are obtained as:

– Partition update formulas (for $t_{ik}^\star$ and $u_{ik}^\star$, where applicable) are those indicated in Table 1, using distances D_{ik} defined in Eq. (4) instead of d_{ik}. In the case of FP^{3}CES algorithm, these formulas become:

$$\begin{cases} t_{ik}^\star = \left[1 + \left(\frac{D_{ik}^2}{\eta_i}\right)^{\frac{1}{p-1}}\right]^{-1} & \forall i = 1 \ldots c \\ u_{ik}^\star = \dfrac{[D_{ik}^2 t_{ik}^p + \eta_i(1 - t_{ik})^p]^{-1/(m-1)}}{\sum\limits_{j=1}^{c} [D_{jk}^2 t_{jk}^p + \eta_j(1 - t_{jk})^p]^{-1/(m-1)}} & \forall k = 1 \ldots n \end{cases} \tag{6}$$

– Cluster prototype centers $v_i^\star$ and radii $r_i^\star$ ($\forall i = 1 \ldots c$) need to be extracted via Newton's method from implicit equations:

$$\begin{cases} \sum\limits_{k=1}^{n} f(u_{ik}, t_{ik}) \frac{D_{ik}}{d_{ik}}(\boldsymbol{x}_k - \boldsymbol{v}_i) = \mathbf{0} \\ \sum\limits_{k=1}^{n} f(u_{ik}, t_{ik}) D_{ik} = 0 \end{cases} \tag{7}$$

– Ellipse orientation matrices $\mathbf{A}_i$ ($i = 1 \ldots c$) are obtained as:

$$A_i^\star = \sqrt[z]{\rho_i \det(\mathbf{S}_i)} \mathbf{S}_i^{-1} \ , \tag{8}$$

Table 2. The alternating optimization algorithm of FP³CES clustering algorithm

---

1. Fix the number of clusters c, $2 \leq c \leq n/3$.
2. Set fuzzy exponent m and possibilistic exponent p, both greater than 1.
3. Set possibilistic penalty terms η_i, $i = 1 \ldots c$, as recommended by Krishnapuram and Keller in [12].
4. Initialize partition using the partition update formula of the FP³CM algorithm, as shown in Table 1.
5. Compute cluster centers v_i and radii r_i ($i = 1 \ldots c$) using Newton's method, with D_{ik} and d_{ik} defined in Eqs. (4) and (5), respectively.
6. Update D_{ik} values d_{ik} ($i = 1 \ldots c$, $k = 1 \ldots n$) according to Eqs. (4) and (5), respectively.
7. Update matrices $\mathbf{S}_i$ ($i = 1 \ldots c$) using Eq. (9), and then matrices $\mathbf{A}_i$ ($i = 1 \ldots c$) using Eq. (8).
8. Update partition using Eq. (6).
9. Repeat steps 5-8 until cluster prototypes converge.
10. The degree of membership of input vector x_k with respect to elliptic cluster i is given by $\sqrt[m+p]{u_{ik}^m t_{ik}^p}$.
11. Vector x_k is an identified outlier, if $\sum_{i=1}^{c} \sqrt[m+p]{u_{ik}^m t_{ik}^p} < \varepsilon$, where ε is a predefined small constant (e.g. $\varepsilon = 0.01$).

---

where

$$\mathbf{S}_i = \sum_{k=1}^{n} f(u_{ik}, t_{ik}) \frac{D_{ik}}{d_{ik}} (x_k - v_i)(x_k - v_i)^T . \tag{9}$$

and z is the number of dimensions.

The proposed elliptic shell clustering algorithm is summarized in Table 2.

4 Results and Discussion

In the following, we will perform some numerical tests to evaluate the robustness and accuracy of the proposed algorithm. We will compare its performance with counter candidates like the FCES [5], and the unpublished possibilistic-fuzzy c-elliptical shell (PFCES) clustering we derived from the ultimate robust and accurate PFCM [16]. Elliptic shell clustering based on pure possibilistic partition is excluded from these tests due to its frequently coincident cluster prototypes.

It is necessary to remark that most shell clustering algorithms were published without being tested in noisy environment. This study investigates the accuracy of algorithms in the presence of noise, in order to emphasize the advantages of the proposed method.

4.1 Three Ellipses and One Outlier

Let us consider three sets of $\nu = 18$ data points each, uniformly distributed in the proximity of three ellipses centered at $\overline{v}_1 = (-5, 0)^T$, $\overline{v}_2 = (0, 5)^T$, and

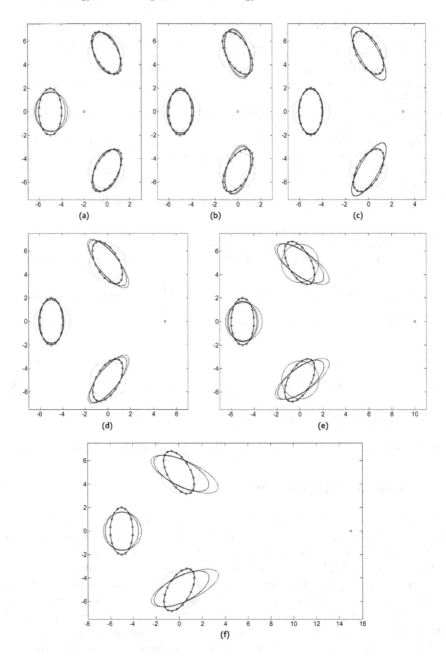

Fig. 1. Identified ellipses in case of $3\nu + 1 = 55$ input vectors, $c = 3$ clusters, and outlier position (a) $\delta = -2$; (b) $\delta = 0$; (c) $\delta = 3$; (d) $\delta = 5$; (e) $\delta = 10$; (f) $\delta = 15$. Cluster prototypes were initialized as circles in the proximity of the ideal solution. Ellipses of FCES are drawn with dotted lines, the result of PFCES with dashed lines, while the identified ellipses of the proposed FP^{3}CES are represented with continuous lines. Dotted lined circles indicate the initial cluster prototypes.

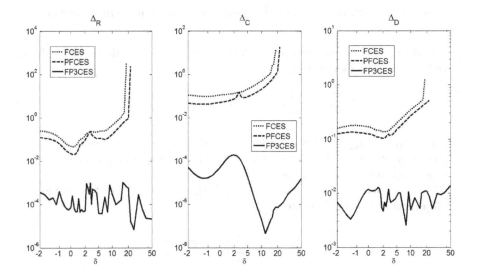

Fig. 2. Cluster validity indices Δ_R, Δ_C, and Δ_D, plotted against the position δ of the outlier data vector. The fuzzy-possibilistic product partition performs better than others by orders of magnitudes.

$\overline{v}_3 = (0, -5)^T$, respectively. All three ellipses have longer and shorter radii of 2 and 1 units, respectively. The ellipses are rotated with respect to each other as shown in Fig. 1. The input data set also includes an outlier, situated at $x_{3\nu+1} = (\delta, 0)^T$, where δ will vary during the experiment. We will attempt to find $c = 3$ two-dimensional elliptic shell clusters among these $n = 3\nu + 1$ vectors. We have fed these input vectors to three algorithms: FCES, PFCES, and the proposed FP³CES. Initial clusters prototypes were set as unit radius circles centered at v_1, v_2, and v_3. The question is, how these algorithms will identify the three ellipses within the data set, as parameter δ varies.

Figure 1 exhibits the obtained ellipses for various positions δ of the outlier data point. The ellipses identified by FCES are drawn with dotted lines, the ones of PFCES with dashed lines, while the ellipses of the proposed method are represented with continuous lines. This convention will persist throughout all graphical representations in this paper. The probabilistic fuzzy partition of FCES gives the outlier approximately $1/c = 0.333$ probability to belong to all classes, and that single outlier visibly distorts the found ellipses in case of every tested outlier position. The possibilistic fuzzy partition gives lower fuzzy membership values to the outlier with respect to all classes due to its possibilistic component (t_{ik}), enabling the PFCES algorithm to better approximate the three ellipses. PFCES visibly performs better than FCES, but the found ellipses are still far from the ideal ones. On the other hand, the proposed algorithm finds the three ellipses correctly in all visualized cases.

In order to numerically evaluate the results, we need to introduce some cluster validation indices (CVI):

- The defect of radii, denoted by Δ_R, will characterize the differences of ellipse radii from their ideal value:

$$\Delta_R = \frac{1}{c}\sum_{i=1}^{c}\left(\frac{r_{i1} - \overline{r}_{i1}}{\overline{r}_{i1}}\right)^2 + \left(\frac{r_{i2} - \overline{r}_{i2}}{\overline{r}_{i2}}\right)^2 ,$$

where r_{i1} and r_{i2} stand for the longest and shortest radius, respectively, of cluster prototype (ellipse) number i, and the corresponding overlined variables indicate the ideal values.
- The defect of centers, denoted by Δ_C, will characterize the distance of found ellipse centers from their ideal value:

$$\Delta_C = \frac{1}{c}\sum_{i=1}^{c}||\boldsymbol{v}_i - \overline{\boldsymbol{v}}_i||^2 ,$$

where $\overline{\boldsymbol{v}}_i$ is the ideal center position of cluster prototype (ellipse) number i, and the norm is computed as Euclidean distance.
- Δ_D will characterize the distance of data points (excluding the outlier) from the closest found ellipse:

$$\Delta_D = \frac{1}{c\nu}\sum_{k=1}^{c\nu}\min_{i}\{\mathrm{dist}(\boldsymbol{x}_k, \mathrm{Ellipse}_i)|i = 1\ldots c\}^2 .$$

All three indices introduced above give low values for valid clusters. Zero value of these indices is possible for the ideal solution only. These three CVI values are represented in Fig. 2, for all three tested algorithms and variable outlier coordinate δ. The curves for FCES and PFCES end at $\delta = 18$ and $\delta = 23$, respectively, because the algorithms crash at these values of the outlier coordinate. On the other hand, the algorithm using fuzzy-possibilistic product partition leads to fine solution for much further outliers as well. CVI values for FP³CES are lower by orders of magnitude, indicating the superiority of the proposed algorithm in accuracy and robustness.

4.2   Twelve Ellipses Around the Clock and Five Outliers

Let us now define another numerical example with $c = 12$ ellipses to be identified, of random radius and orientation, each represented by $\nu = 18$ data points situated along the boundary, as shown in the middle panel of Fig. 3. The ellipse centers are situated on a circle of 10-unit radius, thus the center of ellipse number i ($i = 1\ldots c$) is given by $\overline{\boldsymbol{v}}_i = (10\cos(2\pi i/c), 10\sin(2\pi i/c))^T$. Five further data points (not too distant outliers) are added to the input data set, namely $\{(0,0)^T, (0,3)^T, (0,-3)^T, (3,0)^T, (-3,0)^T\}$. The resulting data set is clustered to $c = 12$ elliptic shell clusters via FCES, PFCES and FP³CES algorithms. Initial cluster prototypes are set as unit radius circles placed at $\overline{\boldsymbol{v}}_i$ ($i = 1\ldots c$),

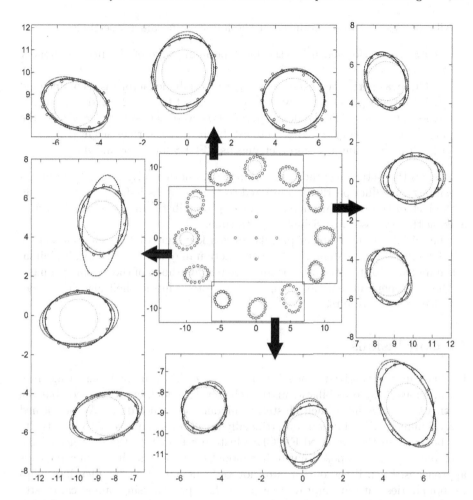

Fig. 3. Magnified view of identified ellipses in case of $12\nu + 5 = 221$ input vectors and $c = 12$ elliptic shell clusters. Cluster prototypes were initialized as unit radius circles centered in ideal position. Ellipses identified by FCES are drawn with dotted lines, the result of PFCES with dashed lines, while the identified ellipses of the proposed FP^3CES are represented with continuous lines.

drawn in Fig. 3 with light and narrow dotted lines. The four outer panels of Fig. 3 exhibits the outcome of all three algorithms.

Again, we can easily notice the superiority of the proposed algorithm as it identifies all ellipses with hardly visible mistakes, while the outliers bring considerable obstacles for the counter candidate algorithms. In general, FCES produces stronger distortions than PFCES, but none of them can be called accurate in the presence of outliers.

All results so far were produced at the following parameter settings:

- FCES was performed using the most popular value of the fuzzy exponent $m = 2$;
- PFCES was performed using a stronger possibilistic factor (as recommended by its authors [16]) caused by fuzzy exponent set to $m = 2.5$ and possibilistic exponent to $p = 1.5$. Trade-off parameters were set equal: $a = b = 1$; the possibilistic penalty terms η_i were established by the rules set in [12].
- FP³CES was using the same settings as PFCES for m, p, and η_i.

Besides the better accuracy and robustness, another advantage of the fuzzy-possibilistic product partition compared to the possibilistic-fuzzy partition [16], is the lack of trade-off parameters between probabilistic and possibilistic terms, making the proposed algorithm easier to tune.

The initialization of cluster prototypes is a key issue that strongly influences the outcome of the clusters. As the cost function may have several local minima, it is important to start the algorithm in the neighborhood of the global optimum. Without proper initialization, none of the tested elliptic shell algorithms can produce accurate results.

5   Conclusions

In this paper we derived a novel approach to c-elliptical shell clustering from our previous fuzzy-possibilistic mixture clustering model, in order to combat the sensitivity of existing c-shell clustering models to outlier data. We performed several numerical evaluations on artificially created test data sets, to investigate the behavior of the proposed FP³CES clustering model. In the presence of distant outliers, the proposed clustering model outperforms all existing c-means approaches. Even if the outliers are not very distant, FP³CES identifies elliptic boundaries with an improved accuracy, fact proved using numerical cluster validity indices. Further works will focus on providing cluster prototype initialization techniques to support the proposed accurate and robust shell clustering algorithm.

References

1. Barni, M., Capellini, V., Mecocci, A.: Comments on a possibilistic approach to clustering. IEEE Trans. Fuzzy Syst. 4, 393–396 (1996)
2. Bezdek, J.C.: Pattern recognition with fuzzy objective function algorithms. Plenum, New York (1981)
3. Bezdek, J.C., Hathaway, R.J., Pal, N.R.: Norm induced shell prototype (NISP) clustering. Neur. Parall. Sci. Comput. 3, 431–450 (1995)
4. Davé, R.N.: Characterization and detection of noise in clustering. Patt. Recogn. Lett. 12, 657–664 (1991)
5. Davé, R.N.: Generalized fuzzy c-shells clustering and detection of circular and elliptical boundaries. Pattern Recogn. 25(7), 713–721 (1992)

6. Davé, R.N., Bhaswan, K.: Adaptive fuzzy c-shells clustering and detection of ellipses. IEEE Trans. Neural Netw. 3(5), 643–662 (1992)
7. Frigui, H., Krishnapuram, R.: A comparison of fuzzy shell clustering methods for the detection of ellipses. IEEE Trans. Fuzzy Syst. 4, 193–199 (1996)
8. Höppner, F.: Fuzzy shell clustering algorithms in image processing: fuzzy c-rectangular and 2-rectangular shells. IEEE Trans. Fuzzy Syst. 5, 599–613 (1997)
9. Gunderson, R.: An adaptive FCV clustering algorithm. Int. J. Man-Mach. Stud. 19, 97–104 (1983)
10. Krishnapuram, R., Frigui, H., Nasraoui, O.: New fuzzy shell clustering algorithms for boundary detection and pattern recognition. In: SPIE Proc. Robot. Comp. Vis., vol. 1607, pp. 1460–1465 (1991)
11. Krishnapuram, R., Nasraoui, O., Frigui, H.: A fuzzy c spherical shells algorithm: a new approach. IEEE Trans. Neur. Netw. 3, 663–671 (1992)
12. Krishnapuram, R., Keller, J.M.: A possibilistic approach to clustering. IEEE Trans. Fuzzy Syst. 1, 98–110 (1993)
13. Krishnapuram, R., Frigui, H., Nasraoui, O.: Fuzzy and possibilistic shell clustering algorithms and their application to boundary detection and surface approximation - Part I. IEEE Trans. Fuzzy Syst. 3, 29–43 (1995)
14. Menard, M., Damko, C., Loonis, P.: The fuzzy $c + 2$ means: solving the ambiguity rejection in clustering. Patt. Recogn. 33, 1219–1237 (2000)
15. Pal, N.R., Pal, K., Bezdek, J.C.: A mixed c-means clustering model. In: Proc. IEEE Int'l Conf. Fuzzy Systems (FUZZ-IEEE), pp. 11–21 (1997)
16. Pal, N.R., Pal, K., Keller, J.M., Bezdek, J.C.: A possibilistic fuzzy c-means clustering algorithm. IEEE Trans. Fuzzy Syst. 13, 517–530 (2005)
17. Szilágyi, L.: Fuzzy-Possibilistic Product Partition: A Novel robust approach to c-means clustering. In: Torra, V., Narakawa, Y., Yin, J., Long, J. (eds.) MDAI 2011. LNCS, vol. 6820, pp. 150–161. Springer, Heidelberg (2011)
18. Szilágyi, L.: Robust spherical shell clustering using fuzzy-possibilistic product partition. Int. J. Intell. Syst. 28, 524–539 (2013)
19. Timm, H., Borgelt, C., Döring, C., Kruse, R.: An extension to possibilistic fuzzy cluster analysis. Fuzzy Sets and Systems 147, 3–16 (2004)

A Multiobjective Evolutionary Optimized Recurrent Neural Network for Defects Detection on Flat Panel Displays

H.A. Abeysundara[1], Hiroshi Hamori[1], Takeshi Matsui[2], and Masatoshi Sakawa[2]

[1] Research & Development Department, OHT Incorporation, Fukuyama, Japan
[2] Department of Electrical Systems and Mathematical Engineering, Faculty of Engineering, Hiroshima University, Japan
ha_abey@ohtinc.jp

Abstract. Thin film transistor (TFT) lines on glass substrates of flat panel displays (FPD) often contain many electrical defects such as open circuits and short circuits that have to be inspected and detected in early manufacturing stages in order to repair and restore them. This paper proposes a multiobjective evolutionary optimized recurrent neural network for inspection of such electrical defects. The inspection is performed on digitized waveform data of voltage signals that are captured by a capacitor based non-contact sensor through scanning over TFT lines on the surface of mother glass of FPD. Waveform data that were captured over TFT lines, which contain open or short circuits, show irregular patterns and the proposed RNN is capable of classifying and detecting them. A multiobjective evolutionary optimization process is employed to determine the parameters of the best suited topology of the RNN. This method is an extension to address the drawbacks in our previous work, which utilizes a feed-forward neural network. Experimental results show that this method is capable of detecting defects on more realistic and noisy data than both of the previous method and the conventional threshold based method.

Keywords: Recurrent neural networks, Evolutionary optimization, Non-contact Defects Inspection, Open short detection.

1 Introduction

Amid growing demand for flat panel displays (FPD) in recent years, having a wide range of applications in almost every available consumer electronics such as TVs, computers, cameras, mobile phones, medical equipment, toys, etc., the manufacturers face a stiff competition for high throughput product lines and low cost manufacturing. The demand for larger sizes of mother glass (ex: Generation 10,11,12) as well as the demand for high density thin film transistor (TFT) pitch patterning of FPDs with the emergence of ultra-high definition 4K and 8K TVs have also been increasing. With increasing pitch pattern density there is a tendency for having defects, such as inter -layer short circuits between TFT lines, to increase. Under these circumstances,

V. Torra et al. (Eds.): MDAI 2014, LNAI 8825, pp. 170–181, 2014.

detection and repair of defects in early manufacturing stages have become significantly important. As a result, the speed and the precision of defects detection have been major issues for the manufactures of FPDs and researchers alike.

Currently there are several methods used by manufacturers for detection of defetcs on FPDs. Automatic optical inspection (AOI) methods, which are based on still or video images, have been mainly used in the past for detection of defects during intermediate processes of fabrication lines of FPDs [1-5]. Though the AOI methods are fast, totally non-contact and devoid of any damage to glass substrates as in pin probe methods, non-electrical defects such as particles on panel surface (micro dusts) and even slight color changes on TFT wirings can also be falsely detected as defects. Another major drawback in AOI method is that it is extremely difficult to distinguish non-electrical defects from certain electrical defects (open NG and short NG) that need to be repaired and restored.

Another commonly used method is the pin probe method where electrode pins make direct contacts with selected points of the entire circuitry on panel surface and measure the current flown after applying a known voltage. Though this method has the advantage of detecting nothing but electrical defects, it also has disadvantages such as very low-speed of inspection, poor adjustability for changes of TFT circuit design and line pitch and the necessity of frequent replacing of pin probing fixtures, which is an expensive process.

The non-contact FPD inspection method proposed by Hamori et al. [6-9] is the most promising technique to-date, which is totally non-contact, utilizing a capacitor based sensor that scans over the TFT lines of mother glass panels of FPDs. The detection of defects is based on analyzing peaks and troughs on a waveform of a voltage signal captured by a sensor using a threshold method after noise cleaning. However determining a proper threshold to correctly indentify such peaks and troughs on waveforms is still not easy as the measured voltage signal is mixed with various noises such as random noises, external vibrations and noises due to environmental effects such as fluctuations of machine temperature.

In our previous work [10,11], using a feed-forward neural network (FNN), we proposed an alternative detection method to the non-contact inspection method [6-9]. Since the problem is highly data driven and non-linear on a set of non-stationary waveform data, an intelligent approach, which can learn and adapt to varying patterns of data must be more appropriate. In that a 4 layered FNN to classify candidate points, which were selected from waveform data and presented to the network, as defective (NG) or non-defective (OK) was employed. There was no threshold selection ambiguity involved in that method instead some local and neighborhood characteristics of candidate points were entered as inputs to the network and the network itself classified if they were defective or not based on prior learning. The system produced promising results and looked more feasible than the existing thresholding method. The drawbacks of this method include network's inability to response correctly for more noisy data and irregular wave fluctuations.

In this paper, as an alternative, we propose a recurrent neural network (RNN) based defect inspection method to address and overcome the above mentioned drawbacks in our previous method based on an FNN. Still determining the best structure or the topology of a RNN for a particular problem was a major obstacle since the trial and error approach can never be productive. In order to overcome this, a multi-objective evolutionary optimization process [12,13] to optimize the topology of the RNN is adopted. Experimental results show that this method is superior to the previous FNN based method in the context of detecting defects on more noisy data. The three local features at candidate input points on waveforms, namely, signal to noise ratio (SNR), residual difference and change of wave length within a neighborhood are used as inputs to the network.

2   Non-Contact FPD Inspection

In the non-contact inspection method for FPD proposed by Hamori et al. [6-9] the capacitor based non-contact sensor utilizes two electrodes, a feeding electrode and a receiving electrode, that scan parallel to each other across TFT lines over the mother glass of FPD panel (Fig. 1 (a)). During scanning, a known voltage is applied to TFT lines on the panel surface through the feeding electrode and is received through the receiving electrode capturing the voltage signal through an AD converter, which is sent to the host computer as a digitized waveform for analysis.

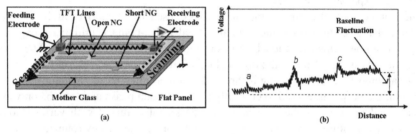

Fig. 1. (a) Non-contact FPD inspection system; (b) Typical pattern of a waveform captured by a non-contact sensor

Fig. 1 (b) shows a typical waveform pattern of a captured voltage waveform through a non-contact sensor. The detection of defects on TFT lines will manifest themselves as peaks and toughs on the waveform, detection of which in effect produce the basis for detection of defects. Generally such waveforms are mixed with lot of random noises, external vibrations and other artifacts as shown in the figure. The large deviation at point *a* may be a random electrical noise, at point *b* may be a deviation caused due to a real electrical defects and at points *c* may be a vibration caused by an external force. Due to practical reasons in real production environments the gap between the surfaces of the scanning electrodes and the flat panel are not uniformly even. This unevenness causes low frequency swinging or baseline fluctuations on the captured voltage waveform as shown in the above figure since the input is a small voltage signal.

2.1   Threshold Based Inspection Method and Its Drawbacks

Since the original waveform data captured by a non-contact sensor are full of noises (Fig. 2a), a moving average filter is applied initially to reduce high frequency random noises (Fig. 2b). Then the low frequency swinging and baseline fluctuations of the waveform due to the unevenness of the gap between the panel surface and the sensor surface are neutralized by applying a derivative operator (Fig. 2c). The resulting waveform has undergone again a moving average operator to remove remaining spike noises. Finally magnitude values of the waveform are compared with a pre-determined threshold value and the points that exceed the threshold level are considered as defect points (Fig. 2c).

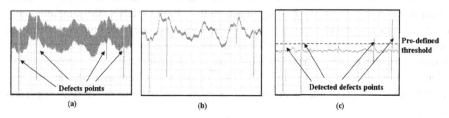

Fig. 2. Defects detection by thresholding method; (a) Original waveform; (b) After noise suppression; (c) Thresholding on differential waveform

However, as shown in Fig. 2c, determining a threshold line for discriminating defects points on the waveform is difficult due to the fact that original data are full of noises and consist of varying levels of baseline fluctuations. In the context of these circumstances the operators always have to resort to looking at data manually and determining proper threshold levels.

2.2   Feed-Forward Neural Network Based Method and Its Drawbacks

In our previous work [10,11], we proposed an FNN to classify and detect defect points on waveform data. The structure of the FFN was a 4-layer feed-forward network containing an input layer with two units, two hidden layers with 2 units and 3 units and an output layer with one unit. The inputs to the network were signal to noise ratio (SNR), residual difference, and change of wave length at candidate points on the input waveform. There were 27 weights associated with each input of units, which were optimized by commonly used back propagation training.

Though this method proved to be much more superior to the existing thresholding method in terms of rate of missed detections and rate of false detections, it still has its own drawbacks, such as difficulty of picking candidates to the network from noisy data environment and, therefore, the obvious difficulty of detection. Determining a proper topology of the FNN was also a difficult task as it was trial and error based.

3   Proposed Method Using an Evolutionary Optimized Recurrent Neural Network

As described above, the existing criteria of using a threshold for determining defect points on waveforms in non-contact defect inspection method is lacking the appropriateness since the feature variations on and around defect points are largely local features while the deciding threshold is a global value. Also as described in section 2.2, our previous method based an FNN has its own drawbacks such as difficulties in topology determination process and poor performance in noisy data environment.

After careful observation of the variation of patterns of waveform data from various environments with varying levels of noise levels, it was clear that a recurrent neural network would be the best approach to the problem. RNNs are fundamentally different from feed-forward architecture in the sense that they not only operate on an input space but also on an internal state space, a trace of which has already been processed by the network. In other words, an internal feedback can be processed together with external inputs in an RNN. One of the major reasons, why an RNN is brought into this problem, is because it is more resilient on noisy and imperfect inputs.

3.1   Inputs to the Recurrent Neural Network

Since a neural network can learn from experience, i.e. a neural network can be trained by feeding a known set of data, any feature around an input point on the input waveform that can be considered as influential to the output must be considered as an input to the network. By looking at neighborhood characteristics around defective points on waveforms and comparing with normal area, following three features have been identified as inputs (input vector $\underline{x}$) to the network, namely Signal to Noise Ratio (SNR), Residual difference, and Change of wave length. All of these input parameters $x(x_1, x_2, x_3)$ are picked within a pre-determined length of neighborhoods of possible candidate points on the waveform, where candidate points are selected by a simple low level threshold in order to avoid missing any defect points though it may include false detections.

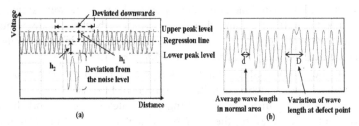

Fig. 3. Picking input parameters from a neighborhood of a point; (a) Measurement of SNR and Residual difference; (b) Measurement of change of wave length on differential waveform

1. Signal to Noise Ratio (SNR)

As shown in Fig. 3a level of the signal at a particular point shows a considerable deviation against the level of background noise, i.e. the SNR is noticeably high. SNR is considered as the first component (x_1) of the input vector $\underline{x}$ and is taken as:

$$x_1 = SNR = \frac{\mu}{\sigma} \tag{1}$$

where μ is the mean value and σ is the standard deviation of the waveform within the selected neighborhood.

2. Residual Difference

Besides the sharp deviation at a defect point, the neighboring area consisting of a few wave lengths that can also be seen deviated towards the same direction as main deviated point (Fig. 3a). This particular feature of the waveform within the neighborhood is measured as the difference of average upper peak level with the regression line (h_1) and the difference of average lower peak level with the regression line (h_2). In other words the residual difference of upper and lower peek levels in the neighborhood is taken as the second component (x_2) of the input vector $\underline{x}$ and is taken as:

$$x_2 = h_1 - h_2 \tag{2}$$

3. Change of Wave Length

In the original waveform it shows a significant change of wave length at a defect point (Fig. 3a) and is taken as the next input to the network. Though change of wave length appears in the normal waveform, it is easier to measure on the differentiated waveform as shown in Fig. 3b. The rate of change of wave length of a defect point from average wave length in the normal area is taken as the third component (x_3) of the input vector $\underline{x}$ and is taken as:

$$x_3 = \frac{D-d}{d} \tag{3}$$

where D is the wave length at the input point and the d is the average wave length in the neighborhood (Fig. 3b)

3.2   Multiobjective Topology Optimization of RNN

Chosing an appropriate topology of an RNN was again a major hurdle as it was even difficult than the FNN method by using the trial and error approach. Therefore a genetic algorithm based evolutionary optimization approach to determine the topology of the RNN is employed. It is clear that topologies generated by a genetic algorithm may contain many superfluous [20,21] components such as single nodes or even whole clusters of nodes isolated from the network input. Such units, called passive nodes, do not process the signals applied to the network sensors and produce only constant responses that are caused by the relevant biases. A simplification procedure or a list of constraints can be used to remove passive nodes and links so that the input/output mapping of the network remains unchanged.

Among various methods of topology optimization in the literature [12-19], Delgado et al. [12] proposed a technique for simultaneous training and topology optimization of RNNs using a multiobjective hybrid process based on SPEA2 and NSGA2 [14]. Katagiri et al. [13] introduced some improvements to the Delgado et al. method by introducing an elite preservation strategy, a self-adaptive mutation probability and preservation of local optimal solutions and their efficiency have been verified with benchmark time series data. For this reason, multiobjective evolutionary optimization of training and topology of RNN [13] are adopted for optimizing an appropriate topology of an RNN with the capability of addressing our problem.

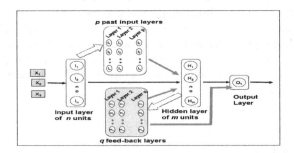

Fig. 4. Initial topology of the RNN

The initial topology of the RNN to be optimized will be constructed with one input layer with *n* units, one hidden layer with *m* units and a single unit output layer. And a *p* number of consecutive previous input layers, called past input layers, are copied and kept and are considered as inputs to the hidden layer during each epoch. Similarly a *q* number of consecutive hidden layers, called feedback layers, are copied and kept and are considered as inputs to both the hidden layer itself and the output layer (Fig. 4). In the figure black arrow lines indicate the inputs to respective layers whereas dashed arrow lines indicate copying of layers to past input layers and feedback layers.

3.3   Optimization Algorithm

The multiobjective optimization algorithm is described in the following 8 steps.

Step 1- Initialization of population: Create a population of chromosomes of size *N* such that each chromosome contains an RNN with the topology as shown in Fig. 4. All of topology parameters in all RNNs of chromosomes are set randomly within their respective ranges. A set of training data is also assigned to the system initially.

Step 2 – Evaluation of Solutions: Each RNN of chromosomes in the population is trained with back propagation through time (BPTT) algorithm using the given training data set. Since this is the most time consuming step, the patterns of error graphs of each RNN is checked frequently during training. If the error graph of any RNN goes out of shape from the expected convergence pattern, the training process of that particular RNN will be immediately terminated without continuing for the rest of the pre-set number of iterations.

Step 3 – Measurement of fitness: A fitness value is assigned to each chromosome according to a pre-set marking scheme. The marking scheme assigns percentage of marks to each chromosome with the following criteria.

      a)   Converged error value
      b)   Convergence pattern of error graph of its RNN (the better the convergence the higher the marks it earns)
      c)   Number of neurons in input layer (the fewer the better)
      d)   Number of neurons in hidden layer (the fewer the better)
      e)   Number of past input layers (the fewer the better)
      f)   Number of feedback layers (the fewer the better)

Step 4 – Checking for the pass mark limit: If the total fitness level of a chromosome exceeds the pass mark (a pre set value, i.e. 90%), the evolution process is terminated.

Step 5 – Discarding week chromosomes: If the fitness level of a chromosome is lower than a pre-set level, for example 20%, it is discarded from the populations for not to allow it to mutate or cross over with others members.

Step 6 – Preservation: If the fitness level of a chromosome is bigger than a pre-set level it is considered to be good enough to carry forward to the next generation without mutation or cross over, and the chromosome is flagged and kept.

Step 7 – Mutation and Cross Over: Remaining chromosomes are allocated a mutation probability and a cross over probability based on a roulette wheel based selection procedure. Accordingly the chromosomes that mutates will select mutation points randomly and carries on, and similarly the cross over pairs selects their cross over points randomly and performs cross over.

Step 8 – Cycle of evolution: Create new chromosomes for discarded chromosomes in step 5 and return to step 2.

4   Experimental Results

4.1   Optimization Results

The evolutionary optimization process, mentioned in section 3.4, was performed with a population of size 40. We observed that the average fitness level of generations was gradually increasing during the evolution and after about 100 generations we found a chromosome with a fitness level over 90%.

The topology of the RNN of optimized chromosome was consisting of one input layer with 2 units and one hidden layer with 6 units. The number of past input layers were optimized to 3, and the number of previous hidden layers were optimized to 1 (Fig. 5). The black arrows indicate the inputs to a certain layer from a certain layer whereas dotted arrows indicate the layer copying.

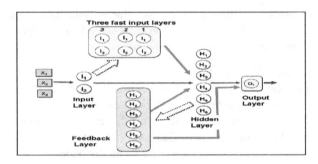

Fig. 5. Optimized topology of the RNN

With this RNN, if the output of the network is $Y(t)$ for a given set of inputs X at any given time t, then $Y(t)$ can be explicitly expressed as:

$$Y(t) = f(\sum_{i=1}^{6} w_{ol}^{H}(t).H_{l}(t) + \sum_{m=1}^{6} w_{om}^{F}(t).H_{m}(t-1)), \tag{4}$$

where

$$H_{l}(t) = f(\sum_{i=1}^{2} w_{li}^{IH}(t).I_{i}(t) + \sum_{j=1}^{6} w_{lj}^{FH} H_{j}(t-1) + \sum_{r=1}^{3}\sum_{k=1}^{2} w_{lkr}^{PH}(t).I_{k}(t-r)), \tag{5}$$

$$\text{where } I_{i}(t) = f(\sum_{s=}^{3} w_{is}^{XI}(t).X_{s}) \tag{6}$$

$$\text{and } f(x) = \frac{1}{1+e^{-x}} \tag{7}$$

In equations (4) and (5), $H_{l}(t)$, $l=1,\ldots.6$, are outputs from the hidden layer and $I_{i}(t)$, $i=1,2$, are outputs from the input layer at time t. w_{ol}^{H} and w_{om}^{F} in (4) are weights associated with connections between l^{th} unit of the hidden layer and the output unit and between m^{th} unit of the feedback layer and the output layer respectively. Similarly, w_{li}^{IH}, w_{lj}^{FH} and w_{lkr}^{PH} in (5) are the weights associated with the connections between the ith unit in input layer and the lth unit in hidden layer, between the jth unit in feedback layer and the lth unit in hidden layer, and between k^{th} unit in the rth past input layer and the lth unit in the hidden layer respectively. In equation (6), w_{is}^{XI} is the weight associated with the connection between s^{th} input and the ith unit in the input layer.

4.2   Detection Results

The above optimized RNN was used and Fig. 6 shows some detection results on 2 different waveforms captured from 2 different machines in different locations. Both of those voltage waveforms consist of different levels of random noises, external vibrations and baseline fluctuations on them but were able to detect using our method correctly.

The red circles in Fig.6 show real defect points while blue dotted lines are candidates and red dotted lines are detected defect points by the RNN among candidates. The left side shows detection in FNN method while the right side shows detection in RNN method in both upper and lower graphs. It shows in both cases that there are some defects points that cannot be detected in FNN method but were possible in RNN method. We have tested several sets of data and compared results with both thresholding method and our previous FNN based method.

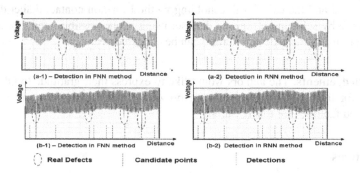

Fig. 6. Detection results on two different waveforms

Table 1. Comparison of Results

Total defects		Missed detection			False detections		
		Threshold method	FNN method	RNN method	Threshold method	FNN method	RNN method
Data set 1	48	14(29.1%)	6(12.5%)	3(6.25%)	13(27%)	8(16.6%)	5(10.4%)
Data set 2	60	16(26.5%)	5(8.3%)	3(5.0%)	18(30.0%)	9(15.0%)	4(6.6%)
Data set 2	50	11(22.0%)	4(8.0%)	2(4.0%)	14(28.0%)	8(16.0%)	3(6.0%)
Total	158	41(25.9%)	8(5.0%)	8(5.0%)	45(28.4%)	25(15.8%)	12(7.5%)

5   Conclusion

In this paper, we proposed a multiobjective evolutionary optimized recurrent neural network for detection of defects on TFT lines on flat panel displays. This method was able to address a major shortcoming of topology determination in our previously proposed FNN based method.

Further this method, being based on a more noise resilient recurrent neural network, reacts to the problem much better than the feed-forward network based method and the threshold based method.

By comparing the performance of our method with the existing threshold based method and the FNN based method following results were confirmed (Table 1).

- The existing ratio of missed detection (20%∼30%) in thresholding method and 10% in FNN method was able to decrease below 5%. This is because the evolutionary optimized RNN is more resilient to noise than an FNN.
- The existing ratio of false detection (20%∼30%) in thresholding method and 16% in FNN method was able to decrease below 8%. This is also due to the usage of a trained recurrent neural network for the purpose.

This proves that this method is more feasible and superior than our previous FNN based method and the existing thresholding method for non-contact defect detection. Furthermore by selecting a training data set using every possible scenario, the miss-detection ratio and false detection ratio can be expected near zero.

Acknowledgements. The authors would like to express their gratitude to OHT Incorporation for their generous support and providing their actual data and all the resources and facilities for this research.

References

1. Liu, Y.H., Lin, S.H., Hsueh, Y.L., Lee, M.J.: Automatic Target Defects Identification for TFT-LCD array process using FCM based fuzzy SVDD ensemble. International Journal of Expert Systems with Applications 36(2), 1978–1998 (2009)
2. Liu, Y.H., Liu, Y.C., Chen, Y.Z.: High-Speed inline defect detection for TFT-LCD array process using a novel support vector data description. Expert Systems with Applications 38(5), 6222–6231
3. Liu, Y.H., Chen, Y.J.: Automatic Defect Detection for TFT-LED Array Process using Quasi-conformal Kernel Support Vector Data Description. International Journal Neural of Molecular Science, 5762–5781 (December 2011)
4. Lu, C.-J., Tsai, D.-M.: Automatic Defects Inspections for LCD using Singular Value Decomposition. International Journal of Advanced Manufacturing Technology 25(1-2), 53–61 (2005)
5. Lu, C.-J., Tsai, D.-M.: Defects Inspections of Patterned TFT-LCD Panels Using a Fast Sub-Image base SVD. In: Proceedings of Fifth Asia Pacific Industrial Engineering and Management Systems Conference, pp. 3.3.1–3.3.16 (2004)

6. Hamori, H., Sakawa, M., Katagiri, H., Matsui, T.: A fast non-contact inspection system based on a dual channel measurement system. Journal of Japan Institute of Electronic Packaging 13(7), 562–568 (2010) (in Japanese)
7. Hamori, H., Sakawa, M., Katagiri, H., Matsui, T.: Fast non-contact flat panel inspection through a dual channel measurement system. In: Proceedings of International Conference on Computers and Industrial Engineers, CD-ROM (July 2010)
8. Hamori, H., Sakawa, M., Katagiri, H., Matsui, T.: A Defect position Identification System based on a dual channel measurement system. Journal of Japan Institute of Electronics, Information and Communication Engineers J94-C(10), 323–333 (2011) (in Japanese)
9. Hamori, H., Sakawa, M., Katagiri, H., Matsui, T.: A Dual channel defect position identification method for touch panel manufacturing process. In: Proceedings of International Conference on Electronics Packaging, pp. 732–736 (2011)
10. Abeysundara, H.A., Hamori, H., Matsui, T., Sakawa, M.: Defects Detection on TFT lines of Flat Panels using a Feed Forward Neural Network. International Journal of Artificial Intelligence Research 2(4), 1–12 (2013)
11. Abeysundara, H.A., Hamori, H., Matsui, T., Sakawa, M.: A Neural Network Approach for Non-Contact Defects Inspection of Flat Panel Displays. In: 17th International Conference in Knowledge Based and Intelligent Information and Engineering Systems, Kita Kyushu, September 9-11, pp. 28–38 (2013)
12. Delgado, M., Cuellar, M.P., Pegalajar, M.C.: Multiobjective Hybrid Optimization and Training of Recurrent Neural Networks. IEEE Transactions on Systems, Man and Cybernetics – Part B Cybernetics 38(2), 381–403 (2008)
13. Katagiri, H., Nishizaki, I., Hayashida, T., Kadoma, T.: Multiobjective Evolutionary Optimization of Training and Topology of Recurrent Neural Networks for Time Series Prediction. The Computer Journal 55(3), 325–336 (2011)
14. Deb, K., Pratap, A., Agarwal, S., Meyarivan, T.: A Fast and Elitist Multiobjective Genetic Algorithm: NSGA-II. IEEE Transactions on Evolutionary Computation 6(2), 182–197 (2002)
15. Huang, B.Q., Rashid, T., Kechadi, M.T.: Multi-Context Recurrent Neural Networks for Time Series Applications. World Academy of Science, Engineering and Technology 10, 4448–4457 (2007)
16. Ang, J.H., Goh, C.K., Teoh, E.J., Mamum, A.A.: Multi-Objective Evolutionary Recurrent Neural Network for System Identification. In: Proceedings of IEEE Congress on Evolutionary Computations, pp. 1586–1592 (2007)
17. Husken, M., Stagge, P.: Recurrent Neural Networks for Time series Classification. Neurocomputing 50(C), 223–235
18. Dolinsky, J., Takagi, H.: RNN with Recurrent Output Layer for Learning of Naturalness. Neural Information Processing – Letters and Reviews 12(1-3), 31–42 (2008)
19. Zang, G.P.: Neural Networks for Classification: A Survey. IEEE Transactions on System, Man and Cybernetics 30(4), 451–462 (2000)
20. Rumelhart, D.E., Mcclelland, D.E.: Parallel distributed processing: Explorations in the microstructure of cognition Foundations, vol. 1. The MIT Press
21. Rojas, P.: Neural Networks - A Systematic Introduction. Springer (1996)

Applying Representative Uninorms
for Phonetic Classifier Combination

Gábor Gosztolya[1,*] and József Dombi[2]

[1] MTA-SZTE Research Group on Artificial Intelligence
of the Hungarian Academy of Sciences and University of Szeged
Szeged, Hungary
ggabor@inf.u-szeged.hu
[2] University of Szeged, Hungary
dombi@inf.u-szeged.hu

Abstract. When combining classifiers, we aggregate the output of different machine learning methods, and base our decision on the aggregated probability values instead of the individual ones. In the phoneme classification task of speech recognition, small excerpts of speech need to be identified as one of the pre-defined phonemes; but the probability value assigned to each possible phoneme also hold valuable information. This is why, when combining classifier output in this task, we must use a combination scheme which can aggregate the output probability values of the basic classifiers in a robust way. We tested the representative uninorms for this task, and were able to significantly outperform all the basic classifiers tested.

Keywords: uninorms, aggregation function, additive generator function, speech recognition, classifier combination, phoneme classification.

1 Introduction

In Artificial Intelligence, perhaps the most intensively investigated area is that of machine learning in general, and classification in particular. In it the goal is to assign one of the pre-defined class labels to a given example. One of the many areas where classification techniques are applied is that of speech recognition, where small acoustic portions of speech have to be identified as phonemes. Historically, several methods were applied in this given task. First Gaussian Mixture Models (GMMs [9]) were employed, especially due to their good probability estimation capabilities; but their low computational complexity also played a role in their popularity. Later Artificial Neural Networks (ANNs [3]) were used, and with the discovery of Deep Neural Networks (DNNs [12,20]) they have become

* This publication is supported by the European Union and co-funded by the European Social Fund. Project title: Telemedicine-oriented research activities in the fields of mathematics, informatics and medical sciences. Project number: TÁMOP-4.2.2.A-11/1/KONV-2012-0073. This study was also partially supported by TÁMOP-4.2.1/B-09/1/KONV-2010-0005.

V. Torra et al. (Eds.): MDAI 2014, LNAI 8825, pp. 182–191, 2014.

the dominant method. In spite of this, other methods such as Support-Vector Machines (SVM [19]) and AdaBoost [18] were also tested.

For a given task the accuracy of a classification method can usually be characterized by one numerical value. However, experiments show that the sets of patterns misclassified by the different classifiers do not necessarily overlap, so a good combination of the original classifiers may reinforce their strong points. This may account for the interest in classifier combination in several areas of Artificial Intelligence (e.g. [17,22,2]), and also in speech recognition [10].

Speech recognition and phoneme classification in particular, however, have another requirement for classifier combination. Here, besides finding the correct class label (phoneme), the estimated probability value of each possible class label is also important, so a method is required that can aggregate the output likelihood values of the basic classifiers into one likelihood value. It is logical to look for operators like this in the field of fuzzy logic; yet fuzzy logic has quite a big range of possible operators. It would also be logical to expect this operator to output high values if most of its input values are high, and from mostly low scores it should produce a score that is close to zero. Uninorms are a set of operators that have these kind of properties, so next we will examine them.

2 Representable Uninorms

The aggregative operators were first introduced in [5] by selecting a set of minimal concepts that must be fulfilled by an evaluation-like operator. In 1982, Dombi [5] defined the aggregative operator in the following way:

Definition 1. *An aggregative operator is a function* $a : [0,1]^2 \to [0,1]$ *with the properties:*

1. *Continuous on* $[0,1]^2 \backslash \{(0,1),(1,0)\}$
2. $a(x,y) < a(x,y')$ *if* $y < y'$, $x \neq 0$, $x \neq 1$
   $a(x,y) < a(x',y)$ *if* $x < x'$, $y \neq 0$, $y \neq 1$
3. $a(0,0) = 0$ *and* $a(1,1) = 1$ *(boundary conditions)*
4. $a(x,a(y,z)) = a(a(x,y),z)$ *(associativity)*
5. *There exists a strong negation operator* η *such that* $a(x,y) = \eta(a(\eta(x),\eta(y)))$ *(self-DeMorgan identity) if* $\{x,y\} \neq \{0,1\}$ *or* $\{x,y\} \neq \{1,0\}$
6. $a(1,0) = a(0,1) = 0$ *or* $a(1,0) = a(0,1) = 1$

The definition of uninorms, originally given by Yager and Rybalov [21] in 1996, is the following:

Definition 2. *A uninorm* U *is a mapping* $U : [0,1]^2 \to [0,1]$ *that has the following properties:*

1. $U(x,y) = U(y,x)$ *(commutativity)*
2. $U(x_1,y_1) \geq U(x_2,y_2)$ *if* $x_1 \geq x_2$ *and* $y_1 \geq y_2$ *(monotonicity)*
3. $U(x,U(y,z)) = U(U(x,y),z)$ *(associativity)*
4. $\exists \nu_* \in [0,1] \ \forall x \in [0,1] \ U(x,\nu_*) = x$ *(neutral element)*

Uninorms are a generalization of t-norms and t-conorms. By adjusting the value of its neutral element ν_*, a uninorm is a t-norm if $\nu_* = 1$ and a t-conorm if $\nu_* = 0$. The main difference between the definition of aggregative operators and uninorms is that the self-DeMorgan identity requirement does not appear in uninorms, and the neutral element property is not in the definition for the aggregative operators. Fodor [11] showed that uninorms which are strict and continuous on $[0,1] \times [0,1] \setminus (\{0,1\},\{1,0\})$ (also called *representative uninorms*) are equivalent to aggregative operators, and they can be represented by the *additive generator* function g as $U(x,y) = g^{-1}(g(x)+g(y))$. Next, we will briefly explain the application of representative uninorms in expert opinion aggregation, following the work of Dombi [8].

Theorem 1. *Let g be an additive generator of an aggregative operator (i.e. representative uninorm) and consider $\nu_* \in (0,1)$; then $a_{\nu_*} : [0,1]^2 \to [0,1]$ defined by*

$$a_{\nu_*}(x,y) = g^{-1}\big(g(x) + g(y) - g(\nu_*)\big) \tag{1}$$

is an aggregation operator (i.e. representative operator) with neutral element ν_*. The extension to n arguments is given by the formula

$$a_{\nu_*}(\mathbf{x}) = g^{-1}\left(g(\nu_*) + \sum_{i=1}^{n} \big(g(x_i) - g(\nu_*)\big) \right). \tag{2}$$

With this, we can construct an aggregative operator from any given generator function that has the desired neutral value. For example, for the Dombi operator case we get

$$a_{\nu_*}(\mathbf{x}) = \cfrac{1}{1 + \frac{1-\nu_*}{\nu_*} \prod\limits_{i=1}^{n} \left(\frac{1-x_i}{x_i} \frac{\nu_*}{1-\nu_*} \right)}. \tag{3}$$

By weighting each parameter with a w_i factor ($0 \le w_i \le 1$), we get the general form in the additive case:

$$a_{\nu_*}(\mathbf{w},\mathbf{x}) = g^{-1}\left(\sum_{i=1}^{n} w_i g(x_i) + \big(1 - \sum_{i=1}^{n} w_i\big) g(\nu_*) \right). \tag{4}$$

In general, a weighted aggregative operator lacks associativity and commutativity and it is not a representable uninorm. Note that ν_* can be treated as the $n+1$th input, in which approach the $n+1$ weights sum up to one. Therefore, if the original n weights sum up to one, we get the following, simplified form:

$$a_{\nu_*}(\mathbf{w},\mathbf{x}) = g^{-1}\left(\sum_{i=1}^{n} w_i g(x_i) \right). \tag{5}$$

In the Dombi operator case, we get

$$a_{\nu_*}(\mathbf{w},\mathbf{x}) = \cfrac{1}{1 + \frac{1-\nu_*}{\nu_*} \prod\limits_{i=1}^{n} \left(\frac{1-x_i}{x_i} \frac{\nu_*}{1-\nu_*} \right)^{w_i}}, \tag{6}$$

or

$$a_{\nu_*}(\mathbf{w}, \mathbf{x}) = \frac{\nu_*(1-\nu_*)^{\sum_{i=1}^{n} w_i} \prod_{i=1}^{n} x_i^{w_i}}{\nu_*(1-\nu_*)^{\sum_{i=1}^{n} w_i} \prod_{i=1}^{n} x_i^{w_i} + (1-\nu_*)\nu_*^{\sum_{i=1}^{n} w_i} \prod_{i=1}^{n} (1-x_i)^{w_i}}. \tag{7}$$

If $\sum w_i = 1$, then we get

$$a_{\nu_*}(\mathbf{w}, \mathbf{x}) = \frac{\prod_{i=1}^{n} x_i^{w_i}}{\prod_{i=1}^{n} x_i^{w_i} + \prod_{i=1}^{n} (1-x_i)^{w_i}}. \tag{8}$$

Using these formulas we can readily aggregate expert probability values [8]. That is, given the probability vector $\mathbf{x}$ (the opinions of experts) as input and their weight vector $\mathbf{w}$, we can readily calculate the resulting likelihood $a_{\nu_*}(\mathbf{w}, \mathbf{x})$. Performing this for all options (classes), we get one likelihood score for each of them, and we can base our decision on the values of this vector.

3   The Phoneme Classification Task

Speech recognition seeks to transcribe audio data; that is, given an audio recording (*utterance*), we would like to find its correct textual representation. This is not a pure classification task, as both the input and the output are of variable length. To overcome this problem, the input utterance A is usually divided into small, equal-sized parts (*frames*) typically 10ms in length; that is, $A = a_1, a_2, \ldots, a_n$. The a_i frames can be classified into one of the previously defined (and language-dependent) phonemes, using specific features extracted from the frame and its near neighbourhood. For the phonetic labels $ph_1, ph_2, \ldots ph_m$, classification produces the likelihood values $P(ph_i|a_j) \; \forall \; 1 \leq i \leq m, 1 \leq j \leq n$, where

$$\sum_{i=1}^{m} P(ph_i|a_j) = 1 \tag{9}$$

holds for all $1 \leq j \leq n$. In the next step, a search is performed based on these posterior probability values, where the intention is to find the most probable *phoneme sequence*. Assuming that the neighbouring frames are independent, we basically look for a phoneme sequence such that the product of the appropriate frame-level posterior probability values is maximal. (In the actual implementation there are a number of restrictions on the allowed phoneme sequences, which were omitted for sake of clarity. We also did not discuss the incorporation of other aspects like phonetic and language models.) The output of this process is the optimal phoneme sequence, from which we remove duplicate neighbouring phonemes.

Although standard classification algorithms are used for the classification of these small speech excerpts, it is not a pure classification task, as we are not only interested in the resulting class label, but also the posterior probability values contain valuable information. (For example, a few incorrectly classified frames can be corrected if the probability value of the surrounding frames for the given phoneme is sufficiently high.) Despite there being a clear connection between the posterior scores and the class labels (e.g. we normally choose the class which has the highest posterior score), it is common to have a classification method that produces accurate class labels, but supplies only inaccurate posterior estimates (e.g. AdaBoost.MH).

3.1   Aggregating Phoneme Classifiers

Classifier combination can be easily incorporated into this scheme as well: instead of relying on the output of a single classification method, we will treat the output of the aggregated probability value as the $P(ph_i|a_j)$ posterior scores. For this, practically any aggregation method can be used. However, it should have parameters that allow us to fine-tune its behaviour to best suit the problem, but avoid having too many parameters, as it would make it next to impossible to set them all properly. Furthermore, as we want to aggregate the likelihood values from sources (i.e. classifiers) of different quality, it would also be nice if we could weight the independent sources. All these points suggest that it would be worth trying representable uninorms, which supply values in the $[0, 1]$ interval as results. The input classifiers can be weighted via the w_i values; and, depending on the g additive generator function used, they may have one or more parameters.

4   Experiments and Results

Having described the phoneme classification problem and representable uninorms, we will now turn to the testing part. First we describe the speech recognition environment, then describe the optimisation process, and finally we present and analyse our test results.

4.1   The Speech Recognition Environment

We used the English TIMIT dataset commonly used for speech recognition experiments [16], with its conventional splitting into training and (core) test sets. We separated 176 utterances from the training set to form a separate development set, and used the standard, 61-long set of phonemes. Phonetic accuracy was measured by applying the edit distance-based accuracy metric, traditionally employed in speech recognition.

We tested three kinds of classifiers, namely Artificial Neural Networks (ANNs [3]) with our custom implementation; Support-Vector Machines (SVM [19]), using the tool LibSVM [4]; and AdaBoost.MH, using the `multiboost` library [1]. The first

Table 1. The accuracy scores got for the three different basic classifiers, and for their combinations

Method		Dev. set	Test set
Basic methods	ANN	77.26%	73.99%
	SVM	75.03%	72.49%
	AdaBoost	75.41%	73.01%
Uninorms	Product	78.54%	74.98%
	Dombi t-norm	78.45%	74.74%
	Generalized Dombi t-norm	78.69%	75.19%

two methods produced well-balanced posterior scores by default, while we calibrated the output of AdaBoost first to have the same standard deviation as those for the ANNs, then normalised them to sum up to one. The neural network had 2000 neurons in its hidden layer, and utilized the sigmoid activation function; we used the RBF kernel in the SVM case; and AdaBoost was trained using 8-leaved decision trees as base learners.

We tested two combination configurations. In the first one, we applied a model of all three types of basic classifiers to find out how effectively we could combine classifiers that were of different types. In the second set of tests we trained three neural networks in the same way, and combined them; with this set of tests we wanted to see how effectively three similar algorithms could be combined. Although the neural networks were trained in the same way, they were not identical due to their random weight initialisation.

4.2   Optimisation

We tested the additive generator functions of three triangular norms. First, as the product operator is also a t-norm, we used its additive generator function, $-\log x$; then we experimented with the one-parametric version of the Dombi t-norm [6], and with the Generalized Dombi Operator that has two parameters [7]. These triangular norm families were chosen based on previous experiences in the field of speech recognition (e.g. [14,13,15]), where we found these norms to be flexible, yet robust. Naturally, many other norms can be used for classifier combination, but this time our aim was not a full-scale comparison of the possible generator functions.

As there were only a few parameters, we did not use a heavyweight optimisation method, but only generated random weights for each classifier and parameters for the generator functions (where necessary) instead. The weights of the classifiers were normalised so as to add up to one; this process was repeated 10000 times, and the best parameter set was chosen.

As is typical in speech recognition, we trained our classifiers on the training set, and evaluated them on the development set. We determined the parameters of the representative uninorms on this development set, then we evaluated the optimal parameter set on the test set to see how robust our achievement was.

Table 2. The accuracy scores got for the three neural networks, and for their combinations

Method		Dev. set	Test set
Basic methods	ANN #1	77.26%	73.99%
	ANN #2	77.11%	74.16%
	ANN #3	77.32%	74.32%
Uninorms	Product	78.54%	75.60%
	Dombi t-norm	78.42%	75.30%
	Generalized Dombi t-norm	78.71%	76.04%

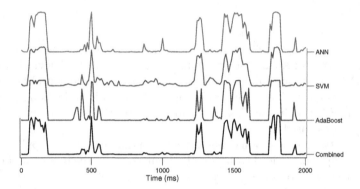

Fig. 1. The input probability values (gray) and the optimal combined values for the uninorm based on the generator of the product norm for the "sil" (silence) phoneme

4.3   Results

The results got for the different basic classifiers can be seen in Table 1. The accuracy scores of the basic classifiers varied significantly both on the development and on the test sets, but ANNs produced the best results. Nevertheless, their combination using the representative uninorms outperformed the base classifiers in each case. It may seem surprising that by using the additive generator function of the Dombi t-norm, our results did not surpass those of the product norm; this can be explained, however, by the fact that we kept the value of the neutral element ν_* at its default value of 0.5, where this behaviour can be anticipated. However, when using the additive generator function of the Generalized Dombi Operator, we were able to significantly outperform even this score; and all these improvements could also be carried over to the test set. (Note that the improvement may not seem that much; but it is indeed significant by speech recognition standards, especially when no language model at all was used.)

When we tried to combine the three, quite similar neural networks (see Table 2), we got very similar results, except that the scores of the basic classifiers varied less this time, especially on the development set. The combined values outperformed the basic classifiers in every case, and although the one using the additive generator of the product norm again performed slightly better than the

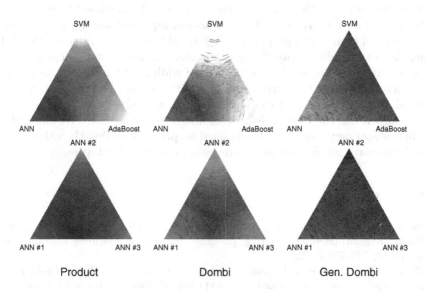

Fig. 2. The accuracy scores got on the development set as a function of classifier weights, for the three different (up) and the three neural network (down) classifier cases. The distance between a point and a corner of the triangle is inversely proportional to the appropriate weight; a darker colour means a higher accuracy score.

other one using the generator function of the Dombi t-norm, the Generalized Dombi Operator again proved to be the best. Note that although this representative uninorm performed similarly on the development set in both cases (78.69% and 78.71%, different basic classifiers and neural networks, respectively), the difference was much bigger on the test set (75.19% and 76.04%). Figure 2 is a plot of the accuracy scores obtained as a function of the weight values. The images belonging to the product norm are smoother, which is due to the additional parameter(s) of the additive generator functions. In the case of the three different classifiers, to achieve high accuracy we should set the weight of the ANNs to a quite large value, but well below 1.0. The exception is the Generalized Dombi Operator case, where the optimal region has roughly equally high weights for SVM and AdaBoost, and a much lower weight for ANNs. For the three neural networks case, not surprisingly, the highest accuracy scores lie in regions where the three weights are roughly equal. This tendency can be seen a bit in the Generalized Dombi Operator case, which probably means that it is more important to properly set the two parameters of the generator function than the classifier weights.

5   Conclusions

The phoneme recognition task of speech recognition is unusual for classifier combination as here not only the correct class label (phoneme) has to be identified based on the output of the individual classifiers, but its likelihood score also has be determined in a robust way. We chose to test representable uninorms in this task because of their useful properties: they are able to output values in the $[0, 1]$ range, they can handle the weighting of the base classifiers, and – depending on the generative function used – they can have additional parameters to control their behaviour. We tested three types of additive generator functions, namely the product, the Dombi and the Generalized Dombi Operators, and we were able to outperform the base classifiers significantly in each case. We think that even better accuracy scores can be attained by properly setting the value of the neutral element ν_*, which we plan to investigate in the near future.

References

1. Benbouzid, D., Busa-Fekete, R., Casagrande, N., Collin, F.D., Kégl, B.: Multi-Boost: a multi-purpose boosting package. Journal of Machine Learning Research 13, 549–553 (2012)
2. Bi, Y., Bell, D.A., Wang, H., Guo, G., Greer, K.: Combining multiple classifiers using Dempster's rule of combination for text categorization. In: Torra, V., Narukawa, Y. (eds.) MDAI 2004. LNCS (LNAI), vol. 3131, pp. 127–138. Springer, Heidelberg (2004)
3. Bishop, C.: Neural Networks for Pattern Recognition. Clarendon Press, Oxford (1995)
4. Chang, C.C., Lin, C.J.: LIBSVM: A library for support vector machines. ACM Transactions on Intelligent Systems and Technology 2, 1–27 (2011)
5. Dombi, J.: Basic concepts for a theory of evaluation: the agregative operator. European Journal of Operational Research 10, 282–293 (1982)
6. Dombi, J.: A general class of fuzzy operators, the De Morgan class of fuzzy operators and fuzziness measures induced by fuzzy operators. Fuzzy Sets and Systems 8, 149–163 (1982)
7. Dombi, J.: Towards a general class of operators for fuzzy systems. IEEE Transaction on Fuzzy Systems 16(2), 477–484 (2008)
8. Dombi, J.: Bayes theorem, uninorms and aggregating expert opinions. In: Bustince, H., Fernandez, J., Mesiar, R., Calvo, T. (eds.) Aggregation Functions in Theory and in Practise. AISC, vol. 228, pp. 281–291. Springer, Heidelberg (2013)
9. Duda, R., Hart, P.: Pattern Classification and Scene Analysis. Wiley & Sons, New York (1973)
10. Felföldi, L., Kocsor, A., Tóth, L.: Classifier combination in speech recognition. Periodica Polytechnica, Electrical Engineering 47(1), 125–140 (2003)
11. Fodor, J., Yager, R.R., Rybalov, A.: Structure of uninorms. International Journal of Uncertainty, Fuzziness and Knowledge-Based Systems 5(4), 411–427 (1997)
12. Glorot, X., Bordes, A., Bengio, Y.: Deep sparse rectifier networks. In: Proceedings of AISTATS, pp. 315–323 (2011)
13. Gosztolya, G., Dombi, J., Kocsor, A.: Applying the Generalized Dombi Operator family to the speech recognition task. Journal of Computing and Information Technology 17(3), 285–293 (2009)
14. Gosztolya, G., Kocsor, A.: Using triangular norms in a segment-based automatic speech recognition system. International Journal of Information Technology and Intelligent Computing (IT & IC) (IEEE) 1(3), 487–498 (2006)

15. Kocsor, A., Gosztolya, G.: Application of full reinforcement aggregation operators in speech recognition. In: Proceedings of the 2006 Conference of Recent Advances in Soft Computing (RASC), Canterbury, UK (2006)

16. Lamel, L., Kassel, R., Seneff, S.: Speech database development: Design and analysis of the acoustic-phonetic corpus. In: DARPA Speech Recognition Workshop, pp. 121–124 (1986)

17. Plessis, B., Sicsu, A., Heutte, L., Menu, E., Lecolinet, E., Debon, O., Moreau, J.V.: A multi-classifier combination strategy for the recognition of handwritten cursive words. In: Proceedings of ICDAR, pp. 642–645 (1993)

18. Schapire, R.E., Singer, Y.: Improved boosting algorithms using confidence-rated predictions. Machine Learning 37(3), 297–336 (1999)

19. Schölkopf, B., Platt, J., Shawe-Taylor, J., Smola, A., Williamson, R.: Estimating the support of a high-dimensional distribution. Neural Computation 13(7), 1443–1471 (2001)

20. Tóth, L.: Convolutional deep rectifier neural nets for phone recognition. In: Proceedings of Interspeech, Lyon, France, pp. 1722–1726 (2013)

21. Yager, R.R., Rybalov, A.: Uninorm aggregation operators. Fuzzy Sets and Systems 80(1), 111–120 (1996)

22. Yu, K., Jiang, X., Bunke, H.: Lipreading: A classifier combination approach. Pattern Recognition Letters 18(11-13), 1421–1426 (1997)

Towards the Cloudification of the Social Networks Analytics

Daniel Cea, Jordi Nin, Rubén Tous, Jordi Torres, and Eduard Ayguadé

Barcelona Supercomputing Center (BSC)
Universitat Politècnica de Catalunya (BarcelonaTech)
Barcelona, Catalonia, Spain
{dcea,nin,rtous,torres,eduard}@ac.upc.edu

Abstract. In the last years, with the increase of the available data from social networks and the rise of big data technologies, social data has emerged as one of the most profitable market for companies to increase their benefits. Besides, social computation scientists see such data as a vast ocean of information to study modern human societies. Nowadays, enterprises and researchers are developing their own mining tools in house, or they are outsourcing their social media mining needs to specialised companies with its consequent economical cost. In this paper, we present the first cloud computing service to facilitate the deployment of social media analytics applications to allow data practitioners to use social mining tools as a service. The main advantage of this service is the possibility to run different queries at the same time and combine their results in real time. Additionally, we also introduce *twearch*, a prototype to develop twitter mining algorithms as services in the cloud.

Keywords: Social Mining, Green Computing, Cloud Computing, Big Data Analytics, Twitter Mining, Stream Processing

1 Introduction

A growing proportion of human activities, such as social interactions, job relationships, entertainment, collaborative working, shopping, and in general, gathering information, are now mediated by social networks and web services. Such digitally mediated human behaviours can easily be recorded and analysed, fuelling the emergence of (1) computational social science, (2) new services such as tuned search engines or social recommender systems, and (3) targeted online marketing. Due to this, public and private sector actors expect to use big data to aggregate all of this data, extract information (and knowledge) from it, and identify value to citizens, clients and consumers [15,4,14].

According to one research study [9] from the University of Maryland's Robert H. Smith School of Business, this growing allows Facebook, twitter and other social media sites to create between 182,000 and 235,000 jobs in US and has added between \$12.19 billion and \$15.71 billion in wages and salaries. A similar study funded by Facebook finds that in Europe, Facebook added a similar number of jobs (approximately 232,000). All these business opportunities have been

V. Torra et al. (Eds.): MDAI 2014, LNAI 8825, pp. 192–203, 2014.

only possible thanks to the possibility to mine social media insights through development APIs.

While many success stories proliferate, in the private sector, social media analytics have found a *killer application* on the Market Research arena. Market research analyses information about customers and target markets to study the market size, market need and competition. The analysis of social media data provides an unprecedented opportunity to understand how customers behave and why, becoming a key component of business strategy. Platforms for social media analytics are proliferating rapidly nowadays (Twitonomy [20], SumAll [17], TwitSprout [21], etc.), with a recent trend towards specialising on market research and brand strength analysis (Brandchats [1], Brandwatch [2], etc.). However, most of these platforms are private initiatives and the ones that are freely available present important hardware restrictions and, therefore, limitations to perform complex queries.

In order to overcome these aforementioned limitations, the contributions of this paper are: an elastic cloud computing service to facilitate the deployment of social media analytics applications together a graphic framework to automatically display some mining results. The proposed Platform as a Service (PaaS) provides the bottom subsystems of the solution stack required by companies (underlying source API access, storage and retrieval) and provisions the necessary hosting capabilities in a scalable and elastic manner without duplicating computer resources. With our architecture, one client can query different social networks with different queries at the same time and display, in real time, the aggregated mining results.

The rest of this paper is organised as follows. Firstly, in Section 2 a brief overview of the related work is introduced. Secondly, a complete description of the proposed architecture and software stack is depicted in Section 3. Later, in Section 4 a real example for the twitter social network is shown. Finally, the paper finishes with some conclusions and future work.

2   Related Work

The major part of social mining platforms covers the entire lifecycle of data analysis, from data gathering to reporting and visualisation. In order to do so, they spend a lot of effort on *reinventing-the-wheel* at the initial stages (data gathering, storage and querying) shortening their resources for the analysis and visualisation stages [11], in which reside their competitive advantage.

For instance in [22], authors propose one architecture to extract and cluster all the tweets of a city. However, if two cities must be monitored, the architecture must be completely duplicate, posing serious scalability problems. Other platform is Datasift [5], where users pay for executing queries over a large set of data sources, but without any option to execute part of the analysis in-house to save money.

In [8], authors describes SONDY, a tool for analysis of trends and dynamics in online social data using twitter. SONDY is written in java, therefore, it is

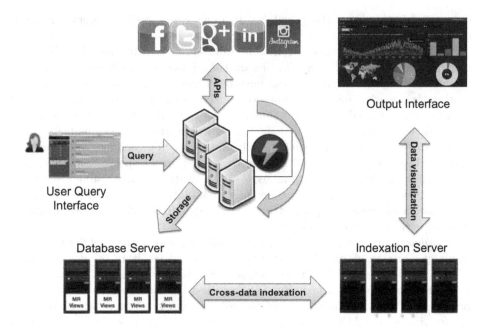

Fig. 1. General Architecture

difficult to make it scale. Besides, it does not allow users to aggregate data coming from several social networks. Finally, SocialSTROM [23] is a cloud-based hub which facilitates the acquisition, storage and analysis of live data from social media feeds (Twitter, Facebook, RSS sources and blogs), as SONDY, it is a java application and it also present scalability limitations.

3   Framework Definition

In this section we detail the main components of our framework, their goals, the selected software and how they interact.

The proposed social mining architecture is composed by 5 independent but interacting elements, as shown in Figure 1, each of them described below. The architecture receives as input the query parameters to be "analysed". Other parameters are optional, such as the possible data post-processing techniques, data enrichment methods, data sources crossings, etc. The architecture outputs some graphical statistics, in parallel data is stored into a NoSQL database for further analysis if needed.

The main components are:

- **User Query Interface**: the user query interface consist on a responsive web application where the user can set the query and also obtain some feedback about the execution, such as total number of retrieved elements, query parameters, execution time, server usage, etc. Apart from that, user is also able to recover the queries executed in the past for further analysing.

- **Processing Cluster**: the cluster consists on several distributed nodes that are in charge of retrieving the public information, as well as, to post-process it if needed. For example, in the running example depicted in Section 4, it is responsable to connect to the Twitter Streaming API, manage the persistent HTTP connection, and filter out the results.
- **Database Server**: the database server is composed of several nodes where data is distributed along different nodes, offering a flexible and scalable data model.
- **Indexation Server**: indexation server creates a set of data indexes to increase the performance of the database server. It maintains a reverse index for each retrieved word. It automatically updates such indexes when a new data element arrives.
- **Output Interface**: for the output interface, where the results of the query are displayed, we use the graphic framework Kibana, which offers a responsive and friendly display solution for our analytics.

3.1   Software Stack

In this section we introduce all the software components, from the virtualisation platform to the data visualisation tools that we have used to develop our architecture.

First of all, to easily create and destroy virtual machines we execute Open Nebula [13] in the cluster of the Computer Architecture Department of the Technical University of Catalonia [1]. Open Nebula is an open-source project delivering a simple and flexible solution to build and manage enterprise clouds and virtualized data centers. Combining existing virtualization technologies with features for multi-tenancy, automatic provision and elasticity, open Nebula aims to provide a *open, flexible, extensible, and comprehensive* management layer to automate and orchestrate the operation of enterprise clouds. We have used Open Nebula to deploy the required virtual machines for our architecture. Virtualisation makes our system elastic with regards the amount of data captured in any moment.

For the Database server, we use Couchbase [3] as the distributed data repository. Couchbase is an open-source, distributed, NoSQL document-driven database optimised for interactive applications serving many concurrent users; creating, storing, retrieving, aggregating, manipulating and presenting the data. Couchbase borns from to the fusion of Membase and CouchOne projects in January 2012. The current release offers features including *JSON document store, indexing and querying, incremental MapReduce and cross datacenter replication.*

For the Indexation server, the natural decision is to use Elasticsearch [7], the native indexation software for Couchbase. Elasticsearch is an open-source, distributed, real-time search and analytic engine built specifically to run on

---

[1] http://www.ac.upc.edu/serveis-tic/altas-prestaciones

NoSQL document-driven databases. Documents are stored as JSON, and all the fields are automatically indexed an usable in a single query. Elasticsearch principal features include: scalability, high availability, multi-tenancy, full text search, conflict management between different versions, and a restful API using JSON over HTTP. Elasticsearch permits us to create a large amount of queries over a set of different data streams stored in Couchbase.

Finally, for the output interface, we use Kibana [10] because it is based on javascript and bootstrap and it is fully compatible with any browser. Kibana is an open-source, scalable, real-time visualisation tool natively integrated with Elasticsearch. It main goal is to display the data stored with Elasticsearch in an elegant graphical manner. Kibana key features include time-based comparisons, easy creation of graphical data representations (plots, charts and maps), flexible, editable and responsive web interface, and a powerful search syntax. In our system, we have adapted our mining methods to display their results in this visualisation framework.

4   Twearch: A Running Example

In this section, we describe a proof of concept application to show the feasibility of our architecture. To do that, we have implemented a twitter listener and some basic queries on the top of Kibana. Twearch offers a simple query interface for twitter able to filter in real time the twitter data stream, by means of any combination of keywords, locations, language, etc. Besides, Twearch also offers to data miners, an output interface to create graphics using javascript.

4.1   Twitter Connection

Twitter, is an online social network born in March 2006 [6] that enables users to send and read "tweets", which are text messages limited to 140 characters, also allows data programmers to access in real time to perform any kind of text-mining technique, such as clustering, TF-IDF, etc. However, it is impossible for a single computer to capture and process in real time the complete twitter information flow. For example, in 2012, Twitter had 500 million users registered posting over 350 million Tweets per day [12,16,18,19].

Twitter offers two different APIs for developers:

- **REST API**: used to retrieve past *tweets* based in different filters. There are different resources depending on the data to be retrieved: Accounts, friendships, geolocations, statuses, users, etc. Depending on the resource, the number of queries per account is limited from 15 to 450 per *rate limit window* (by March 2014, 15 minutes long). Each query response contains 100 *tweets*.
- **Streaming API**: used to retrieve *tweets* in real time. Not all the *tweets* are sent, but only a 5%, so it's mostly used for analysis purposes. There are 3 different resources, depending on the target: Public streams, User streams, and Site streams. The stream API is limited to 1 data stream per account.

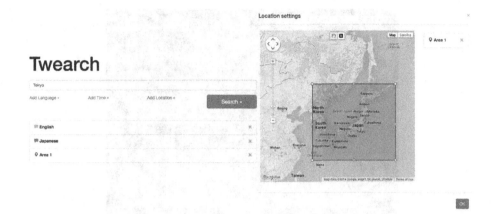

Fig. 2. Input interface query

Since our framework is designed to manage stream data, we connected Twearch to the Twitter streaming API.

4.2   Query System

As a running example for this paper, we want to retrieve all tweets containing the word "Tokyo" during a week. The input interface, as it is shown in Figure 2, asks Twitter to filter all incoming tweets and sends us only those containing the string "Tokyo". Apart from keywords, using the twearch interface, one user can filter the twitter stream using hashtags, mentions, languages and spatial coordinates.

Note that, similar input interfaces can be created for others social networks, such as foursquares, instagram o facebook. Doing this, it is easy to cross information coming from different social networks to enrich data analytics without too much effort and allowing the data miner to recicle the listeners for future analytics.

The input interface sends the query to one node of the processing cluster. Once the query is received in one of the nodes, it opens a permanent HTTP connection with the Twitter Streaming API, and the data stream will begin to flow back containing the requested tweets. Then, data stream will be processed, enriched, stored into the Couchbase database and indexed by ElasticSearch server.

When the output interface (Kibana) is opened, each panel will send a predefined query through the ElasticSearch API REST, using HTTP GET requests, to ask for the concrete fields and filters needed for that panel, and then populate the data graphically using different types of panel. Here, it is important to highlight that more than one query (even from more than one social network) can be sent to Kibana, therefore query results can be aggregated and recycled each time a new query is executed.

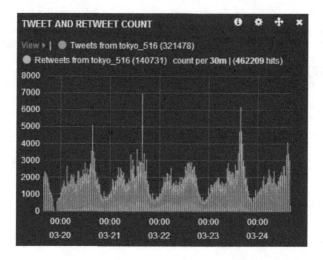

Fig. 3. Number of tweets and retweets per 30 minutes

As an example, we have created the Tweets/Retweets panel (Figure 3), where there are 2 different queries inside the request: A count of all tweets from the stream, and another count from all tweets from the stream whose 'retweeted' field is true.

We have also implemented other two panels (Figures 4.(a) and(b)) both related to the Tweet text: one counting the top 10 hashtags (a), and another for user mentions (b). This information is obtained after processing the text field of each tweet stored, and separating the strings depending whether they are hashtags or user mentions. From these two panels we can easily extract some knowledge, such as the high social impact of two teenagers groups; *egirls* and *nuest* from Japan and Korea respectively.

Finally, in order to exploit a different type of the information stored in a tweet, we have also the geo-positioned in a world map all the geo-located tweets. To do that, we use the Open Street Maps library, as it is shown in Figure 5.

4.3   Query Refining

Additionally, it is possible to refine the results and focus them on a concrete time period using the Kibana output interface. Data miners can filter the documents by their timestamp and study in detail concrete time periods. For instance, Figure 3 depicts some tweets and re-tweets peaks from 22:00 to 23:00 during some days of March. If we focus on these tweets, as it is shown in Figure 6.(a), it is possible to observe that the mentioned users changes (see Figure 6.(b)). Concretely, the most mentioned user is lovelive instead of 2014egirls. Lovelive is a very popular Japanese anime serial, and the first chapters of its second season were broadcasted during these concretes time periods. So, we were able to detect new trends easily using twearch.

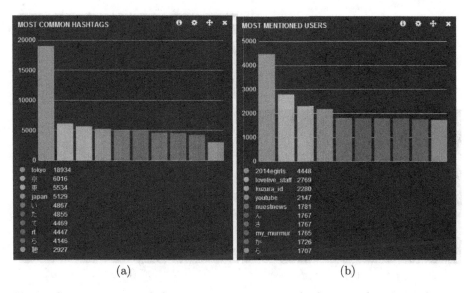

Fig. 4. Counting query of the most common words, hashtags and mentioned users among tweets

Fig. 5. Geopositioned Tweets in OpenStreetMaps

Finally, observing Figure 5, it is possible to see that most of the tweets are located in Tokyo, zooming at the center of the city, we discover that a big cluster of tweets is located at the Shibuya station, one of the most visited parts of the city. Therefore, using this map it is possible to automatically detect huge twitter users concentrations for a given query.

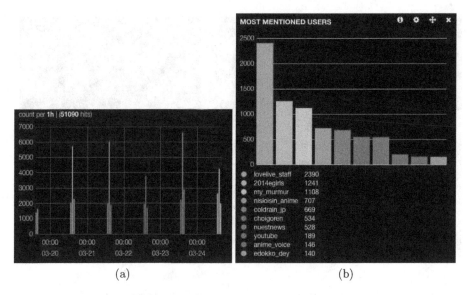

(a)   (b)

Fig. 6. Lovelive new season advertising

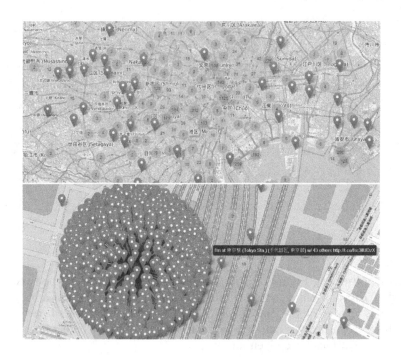

Fig. 7. Shibuya station tweet cluster

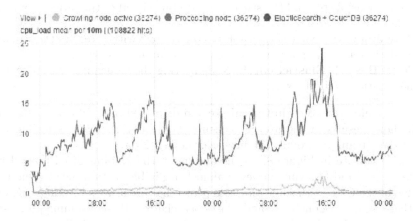

Fig. 8. CPU consumption, average and maximum, of all the system components during a 2 day streaming

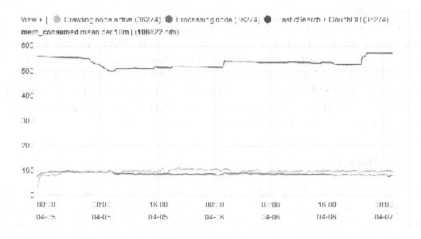

Fig. 9. Memory consumption, average and maximum, of all the system components during a 2 day streaming

4.4 Platform Performance

Apart from the social information extracted from Twitter, another topic of interest is to analyse the performance of the system taking measurements of the main hardware components during the information retrieval process.

To achieve that, the components of the architecture send reports, every 5 seconds, about the amount of CPU and Memory being consumed, Specifically:

– **Processing node**: It is the node of the processing cluster responsible to host the input query interface and display the mining results. Besides, it is also responsible to send the streaming job start signal to the proper crawling node.

- **Crawling node**: Node responsible to connect to Twitter, receive the streaming data, refine the query and store the stream in the database.
- **CouchBase + ElasticSearch**: Node hosting the Couchbase database and the ElasticSearch indexer.

Those results are stored in CouchBase and displayed in a Kibana histogram, which shows results of the average consumption every 10 minutes during the part of the streaming process (in our case, the last 2 days).

As we illustrate in Figure 8, on the one hand, the amount of CPU used by the Processing and Crawling nodes is almost negligible, with a mean value less than 2% of the CPU with some punctual peaks that are always less than 5%. On the other hand, the Couchbase and ElasticSearch are CPU-consuming processes that average a 15% of CPU consumption with peaks over the 25%

About the memory, as we can observe in Figure 9, all processes keep a constant amount of consumed memory: Couchbase and Elasticsearch use around 550MB of RAM memory, the crawling node is around 100MB, and finally the processing node around 90MB.

The CPU performance results show that it is possible to consolidate several queries into a single virtual machine, reducing the required number of virtual machines needed to perform complex queries where lot of information has to be retrieved. For the data management system (database and indexes) the real bottleneck is the RAM memory (see Figure 9) not the CPU. To overcome this drawback, more than one virtual machines (or physical servers) can be deployed to exploit the scalability of Couchbase and Elasticsearch.

5   Conclusions

In this paper we have described all the components of an elastic and scalable framework for social mining in a cloud infrastructure, in our case Open Nebula. We have described the database management system, the query language and the visualisation tool. Finally as a proof of concept, we have described how to collect data from twitter and we have displayed some data analytics and performance metrics for a given query using the proposed query system.

In the near future, we plan to add more functionalities to our platform as for instance, natural language processing methods, automatic data enrichment by means of data crossing. Finally, we would like to define a decision support system to help designers of appliances to optimise resource allocation in a semi-supervised way.

Acknowledgments. This work is partially supported by the Ministry of Science and Technology of Spain under contract TIN2012-34557, by the BSC-CNS Severo Ochoa program (SEV-2011-00067). Besides, authors would like to thank CA Technologies for funding this research through a collaboration agreement with Universitat Politècnica de Catalunya.

References

1. Brandchats, http://www.brandchats.com (accessed March 20, 2014)
2. Brandwatch, http://www.brandwatch.com (accessed March 20, 2014)
3. Brown, M.C.: Getting Started with Couchbase Server – Extreme Scalability at Your Fingertips. O'Reilly (2012)
4. Chae, J., Thom, D., Jang, Y., Kim, S., Ertl, T., Ebert, D.S.: Public behavior response analysis in disaster events utilizing visual analytics of microblog data. Computers & Graphics 38, 51–60 (2014)
5. Datasift, http://datasift.com (accessed March 20, 2014)
6. Dorsey, J.: Just setting up my twttr. Twitter. Oldest Tweet ever from Twitter founder (2006)
7. Elasticsearch official website, http://www.elasticsearch.org/ (accessed March 20, 2014)
8. Adrien Guille, C., Hacid, H., Zighed, D.: Sondy: An open source platform for social dynamics mining and analysis. In: ACM SIGMOD (2013)
9. Hann, I.-H., Viswanathan, S., Koh, B.: The Facebook App Economy. Center for Digital Innovation, University of Maryland (2011)
10. Kibana website, http://www.elasticsearch.org/overview/kibana/ (accessed March 20, 2014)
11. Lieberman, M.: Visualizing big data: Social network analysis. In: Digital Research Conference (2014)
12. Lunden, I.: Twitter passed 500m users in June 2012, 140m of them in US; Jakarta "biggest tweeting" city. Techcrunch (2012)
13. Opennebula key features and functionality, http://www.opennebula.org (accessed March 20, 2014)
14. Shelton, T., Poorthuis, A., Graham, M., Zook, M.: Mapping the data shadows of hurricane sandy: Uncovering the sociospatial dimensions of 'big data' (2014)
15. Stoové, M.A., Pedrana, A.E.: Making the most of a brave new world: Opportunities and considerations for using twitter as a public health monitoring tool. Preventive Medicine (2014)
16. Strachan, D.: Twitter: How to set up your account. The Daily Telegraph (2009)
17. Sumall, http://sumall.com (accessed March 20, 2014)
18. Twitter Search Team. The engineering behind twitter's new search experience. Technical report, Twitter Engineering Blog (2011)
19. Twitter Search Team. Twitter turns six. Technical report, Twitter Engineering Blog (2012)
20. Twitonomy, http://www.twitonomy.com (accessed March 20, 2014)
21. Twitsprout, http://twitsprout.com (accessed March 20, 2014)
22. Villatoro, D., Serna, J., Rodríguez, V., Torrent-Moreno, M.: The tweetBeat of the city: Microblogging used for discovering behavioural patterns during the MWC2012. In: Nin, J., Villatoro, D. (eds.) CitiSens 2012. LNCS, vol. 7685, pp. 43–56. Springer, Heidelberg (2013)
23. Wood, R., Zheludev, I., Treleaven, P.: Mining social data with ucl's socialstorm platform. Technical report, University College of London, UCL (2011)

Privacy-Preserving on Graphs Using Randomization and Edge-Relevance

Jordi Casas-Roma

Universitat Oberta de Catalunya
Barcelona, Spain
jcasasr@uoc.edu

Abstract. The problem of anonymization on graphs and the utility of the released data are considered in this paper. Although there are some anonymization methods for graphs, most of them cannot be applied on medium or large networks due to their complexity. Nevertheless, random-based methods are able to work with medium or large networks while fulfilling the desired privacy level. In this paper, we devise a simple and efficient algorithm for randomization on graphs. Our algorithm considers the edge's relevance, preserving the most important edges of the graph, in order to improve the data utility and reduce the information loss on anonymous data. We apply our algorithm to different real datasets and demonstrate their efficiency and practical utility.

Keywords: Privacy, Randomization, Social networks, Graphs, Edge relevance, Data utility.

1 Introduction

In recent years, an explosive increase of social and human interaction networks has been made publicly available. Embedded within this data there is private information about users who appear in it. Therefore, data owners must respect the privacy of users before releasing datasets to third parties. In this scenario, anonymization processes become an important concern. Among others, the study of Ferri et al. [13] reveals that though some user groups are less concerned by data owners sharing data about them, up to 90% of members in others groups disagree with this principle. Backstrom et. al. [1] point out that the simple technique of anonymizing graphs by removing the identities of the vertices before publishing the actual graph does not always guarantee privacy. They show that there exist adversaries that can infer the identity of the vertices by solving a set of restricted graph isomorphism problems. Some approaches and methods have been imported from anonymization on structured data, but the peculiarities of graph-formatted data avoid these methods to work directly on it. In addition, divide-and-conquer methods do not apply to anonymization of graph data due to the fact that registers are not separable, since removing or adding vertices and edges may affect other vertices and edges as well as the properties of the graph [29].

V. Torra et al. (Eds.): MDAI 2014, LNAI 8825, pp. 204–216, 2014.

1.1   Our Contributions

In this paper we present an algorithm for privacy-preserving based on random
edge modifications. It works with simple, undirected and unlabelled graphs. Because these graphs have no attributes or labels in the edges, information is only
in the structure of the graph itself and, due to this, the adversary can use information about the structure of the graph to attack the privacy. In this paper
we consider edge's relevance to reduce the information loss produced by the
anonymization process. The latter leads us to a more useful data. We offer the
following results:

- We introduce a randomization algorithm based on edge's relevance for simple
  graphs.
- We demonstrate that edge's relevance can be considered in order to modify
  the graph structure, and it conducts the process to reduce the information
  loss and increase the data utility.
- We conduct an empirical evaluation of our algorithm on several well-known
  graphs, comparing our algorithm with other well-known random-based algorithms, and demonstrating that ours achieves the best trade-off between
  data utility and data privacy.

1.2   Notation

Let $G = (V, E)$ be a simple, undirected and unlabelled graph, where V is the set
of vertices and E the set of edges in G. We define $n = |V|$ to denote the number
of vertices and $m = |E|$ to denote the number of edges. We use $\{i, j\}$ to define
an undirected edge from vertex v_i to v_j. We denote the degree of vertex v_i as
$deg(v_i)$, the set of 1-neighbourhood of vertex v_i as $\Gamma(v_i) = \{v_j : \{i, j\} \in E\}$,
and the maximum degree of the graph as Δ. Finally, we designate $G = (V, E)$
and $\widetilde{G} = (\widetilde{V}, \widetilde{E})$ to refer the original and the anonymous graphs, respectively.

1.3   Roadmap

This paper is organized as follows. In Section 2, we review the state of the
art of anonymization on graphs, focusing on random-based methods. Section 3
introduces our algorithm for randomization using edge neighbourhood centrality.
Then, in Section 4, we compare our algorithm to two well-known random-based
algorithms in terms of information loss and data utility, and discuss the results.
Next, we examine the re-identification and risk assessment in Section 5. Lastly,
in Section 6, we present the conclusions of this work.

2   Privacy-Preserving on Graphs

From a high level view, there are three general families of methods for achieving
graph data privacy. The first family encompasses "graph modification" methods. These methods first transform the data by edges or vertices modifications

(adding and/or deleting) and then release them. The data is thus made available for unconstrained analysis. The second family encompasses "generalization" or "clustering-based" approaches. These methods can be essentially regarded as grouping vertices and edges into partitions called super-vertices and super-edges. The details about individuals can be hidden properly, but the graph may be shrunk considerably after anonymization, which may not be desirable for analysing local structures. The generalized graph, which contains the link structures among partitions as well as the aggregate description of each partition, can still be used to study macro-properties of the original graph. Among others, [17,7,22,11,3] are interesting approaches to generalization concept. Finally, the third family encompasses "privacy-aware computation" methods, which do not release data, but only the output of an analysis computation. The released output is such that it is very difficult to infer from it any information about an individual input datum. For instance, differential privacy is a well-known privacy-aware computation approach. Differential private methods refer to algorithms which guarantee that individuals are protected under the definition of differential privacy [12], which imposes a guarantee on the data release mechanism rather than on the data itself. The goal is to provide statistical information about the data while preserving the privacy of users. Interesting works, among others, are [18,19].

Graph modification approaches anonymize a graph by modifying (adding and/or deleting) edges or vertices in a graph. These modifications can be made randomly or in order to fulfil some desired constraints. The first methods are called randomization methods and are based on adding random noise in original data. They have been well investigated for structured data. Naturally, edge randomization can also be considered as an additive-noise perturbation. Notice that the randomization approaches protect against re-identification in a probabilistic manner. Alternative approaches consider graph modification methods to meet some desired privacy constraints. The notion of k-anonymity [23,24] is included in this group. The k-anonymity model indicates that an attacker cannot distinguish between different k records although he manages to find a group of quasi-identifiers. Therefore, the attacker cannot re-identify an individual with a probability greater than $\frac{1}{k}$. In general, the higher the k value, the greater the anonymization and also the information loss. For instance, some interesting works can be found in [21,8,10].

As we have stated before, in this paper we consider the randomization methods. Naturally, graph randomization techniques can be applied removing some true edges and/or adding some false edges. Two natural edge-based graph perturbation strategies are: firstly, *Rand Add/Del* randomly adds one edge followed by deleting another edge and repeats this process for k times. This strategy preserves the total number of edges in the original graph. Secondly, *Rand Switch* randomly switches a pair of existing edges $\{t, w\}$ and $\{u, v\}$ to $\{t, v\}$ and $\{u, w\}$, where $\{t, v\}$ and $\{u, w\}$ do not exist in the original graph, and repeat it for k times. This strategy preserves the degree of each vertex. Hay et al. [16] proposed a method, called *Random perturbation*, to anonymize unlabelled graphs based on randomly removing p edges and then randomly adding p fake edges. The set of vertices does not change and the number of edges is preserved in the anonymous graph. Ying and Wu [26] studied how different randomization methods (including *Rand Add/Del* and *Rand Switch* methods) affect the privacy of the relationship between vertices. The authors also proposed two algorithms specifically designed to preserve spectral characteristics of the original graph, called *Spctr Add/Del* and *Spctr Switch*. Ying et al. [25] compared two well-known strategies, *Rand Add/Del* and k-degree anonymous algorithm by Liu and Terzi [21], in terms of identity and link disclosure. They also developed a variation of *Rand Add/Del* method, called Blockwise Random Add/Delete strategy or simply *Rand Add/Del-B*, which divides the graph into blocks according to the degree sequence and implements modifications (by adding and removing edges) on the vertices at high risk of re-identification, not at random over the entire set of vertices. More recently, Bonchi et al. [5,6] offered a new information-theoretic perspective on the level of anonymity obtained by random methods. The authors make an essential distinction between image and pre-image anonymity and propose a more accurate quantification, based on entropy, of the anonymity level that is provided by the perturbed graph. They stated that the anonymity level quantified by means of entropy is always greater than or equal to the one based on a-posteriori belief probabilities. In addition, the authors introduced a new random-based method, called *sparsification*, which randomly removes edges, without adding new ones.

A new interesting anonymization approach is presented by Boldi et al. [4] and it is based on injecting uncertainty in social graphs and publishing the resulting uncertain graphs. While existing approaches obfuscate graph data by adding or removing edges entirely, they proposed to use a perturbation that adds or removes edges partially. From a probabilistic perspective, adding a non-existing edge $\{i,j\}$ corresponds to changing its probability $p(\{i,j\})$ from 0 to 1, while removing an existing edge corresponds to changing its probability from 1 to 0. In their method, instead of considering only binary edge probabilities, they allow probabilities to take any value in range [0,1]. Therefore, each edge is associated to an specific probability in the uncertain graph.

Other approaches consider the degree sequence of the vertices or other structural graph characteristics (for example, transitivity or average distance between pairs of vertices) as important features which the anonymization process must keep as equal as possible on anonymous graphs. For instance, Hanhijarvi et al. [15] and Ying and Wu [27] described methods designed to preserve implicit properties of social networks.

3 Randomization Using Edge Neighbourhood Centrality

In this section, we will present the *Rand-NC*[1] (short for Randomization using Edge Neighbourhood Centrality) algorithm, which is designed to achieve

---

[1] Source code available at: http://deic.uab.cat/~jcasas/

random-based privacy on undirected and unlabelled graphs. The algorithm performs modifications to the original graph $G = (V, E)$ only in edge set (E). Thus, the vertex set (V) remains the same during randomization process.

Our approach is a probabilistic random-based method which considers edge relevance in order to achieve less information loss, and consequently better data utility on anonymous graphs. Our probability distribution is computed according to a relevance value assigned to each edge. This relevance value is computed by a simple metric, called *edge neighbourhood centrality* [9], which evaluates the 1-neighbourhood information flow through an edge.

3.1   Edge Neighbourhood Centrality

Edge neighbourhood centrality (NC) [9] of an edge $\{i, j\}$ is defined as the fraction of vertices which are neighbours of v_i or v_j, but not of v_i and v_j simultaneously. The edge neighbourhood centrality is computed as follows:

$$NC_{\{i,j\}} = \frac{|\Gamma(v_i) \cup \Gamma(v_j)| - |\Gamma(v_i) \cap \Gamma(v_j)|}{2\Delta} \tag{1}$$

Notice that an edge with high score is a bridge-like for its neighbourhood vertices, and all values of this measure are in range [0,1]. In [9] the authors demonstrated that edge neighbourhood centrality identifies the most important edges on a graph with low complexity, and therefore, this measure is able to work on medium or large graphs. Note that this measure can be computed on $\mathcal{O}(m)$ using the adjacency matrix representation of the graph.

3.2   Randomization Algorithm

Our randomization algorithm runs in a two-step approach: the first step removes $w < m$ (usually $w << m$, where $m = |E|$) edges from the original edge set (E). The process starts by computing the edge neighbourhood centrality value (NC-score) for each edge $\{i, j\} \in E$ using Equation 1. Then, it calculates the probability of each edge to be removed as follows:

$$p(\{i, j\}) = \frac{1}{NC_{\{i,j\}}^2} \tag{2}$$

According to the power-law property of scale-free real networks [2], notice that there are few edges with high NC values. Hence, we are interested on preserving these edges, since they are bridge-like connectors and are critical for the structure and the information flow on the graph. Consequently, the probability of removing these edges during the anonymization process is low, in accordance with Equation 2.

In the second step we add w fake edges $\{i, j\} \notin E$. Let the set of all possible fake edges be the complement of the original graph, which has the same vertex set but whose edge set consists of the edges not present in G. Let G be a graph and let G^c be its complement. We select a random set of $m = |E|$ edges using a uniform probability distribution from $G^c = (V, E^c)$, and create a subset $E^s \subseteq E^c$. Notice that

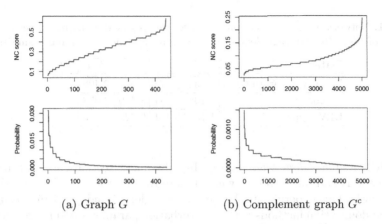

(a) Graph G (b) Complement graph G^c

Fig. 1. Empirical results of NC-score for each edge (edges are represented in x-axis) and its probability computed by Equation 2 on Polbooks (see Section 4.1 for network's details)

usually $|E^c| >> |E|$ because of the sparse property of real networks, and therefore we only need to consider a subset of m elements. Next, we compute the NC-score for each edge $\{i, j\} \in E^s$ by adding it to E and computing the score in G.

Note that this process considers the score of each edge independently, and possible dependencies between NC-scores of different edges will not be considered. The NC-score for an edge $\{i, j\}$ will be modified if for some $v_u \in V$ an edge $\{i, u\}$ or $\{u, j\}$ is added or deleted, since this implies that the 1-neighbourhood of the vertices v_i or v_j will be changed. Even so, we do not recompute the NC-values after an edge is added or removed due to the computational complexity. As we stated before, one of our goals is keep a low complexity to enable our approach to work with medium or large graphs.

Finally, we randomly select w edges from E^s, according to the probability distribution computed by Equation 2, and then we add them to the edge set in order to obtain the anonymous version of edge set $(\widetilde{E})$, which fulfils $|E| = |\widetilde{E}|$ and $|E| - |E \cap \widetilde{E}| = w$.

For instance, we can see empirical data about Polbooks (see Section 4.1 for network's details) in Figure 1. Firstly, the NC-score and its probability value on G are shown in Figure 1a. As we can see, edges with high NC-score are assigned to low probability values by Equation 2, and vice versa. Secondly, the same values for the complement graph G^c are shown in Figure 1b. Note that $|E^c| >> |E|$. In a similar way, edges with low NC-score present higher probability to be added to $\widetilde{G}$ during randomization process than the ones with high NC-score.

4 Experimental Results

In this section we will present the results of our randomization method and, in addition, we will compare our method to two basic and relevant methods for randomization on graphs. The selected methods are: *Random Perturbation* (in

Table 1. Network' properties. For each dataset we present the number of vertices (n), number of edges (m), average degree ($\overline{deg}$), average distance ($\overline{dist}$) and diameter (D)

Dataset	n	m	$\overline{deg}$	$\overline{dist}$	D
Zachary's Karate Club	34	78	4.588	2.408	5
Polbooks	105	441	8.400	3.078	7
URV Email	1,133	5,451	9.622	3.606	8

short, RP) and *Random Switch* (RS). In subsequent sections, we will analyse these methods and ours, and we will compare the empirical results on different real networks. Our experimental framework considers 10 independent executions of the randomization methods with a perturbation parameter p in range between 0% (original graph) and 25% of total number of edges, i.e, $0 \leq p \leq 25$. The parameters are the same for all methods and they achieve similar privacy levels, since they apply the same concept to preserve the graph's privacy. Therefore, the evaluation of the results is interesting to compare the data utility and information loss on anonymous datasets.

4.1 Tested Networks

Three different real networks are used in our experiments. Although all these sets are unlabelled, we have selected these datasets because they have different graph's properties. Table 1 shows a summary of their main features. They are the following ones:

- Zachary's Karate Club [28] is a network widely used in the literature. The graph shows the relationships among 34 members of a karate club.
- US politics book data (Polbooks) [20] is a network of books about US politics published around the 2004 presidential election and sold by the on-line bookseller Amazon. Edges between books represent frequent co-purchasing of books by the same buyers.
- URV email [14] is the email communication network at the University Rovira i Virgili in Tarragona (Spain). Vertices are users and each edge represents that at least one email has been sent.

4.2 Measures

In order to compare the algorithms, we use several structural and spectral measures. The first structural measure is *diameter* (D), which is defined as the largest minimum distance between any two vertices in the graph. The second one is *harmonic mean of the shortest distance* (h). It is an evaluation of connectivity, similar to the average distance or average path length. *Sub-graph centrality* (SC) is used to quantify the centrality of each vertex based on the sub-graphs.

Transitivity (*T*) is one type of clustering coefficient, which measures and characterizes the presence of local loops near a vertex. It measures the percentage of paths of length 2 which are also triangles. The above measures evaluate the entire graph as a unique score. We compute the error on these graph metrics as follows:

$$\epsilon_m(G, \widetilde{G}) = |m(G) - m(\widetilde{G}_p)| \tag{3}$$

where m is one of the graph characteristic metrics, G is the original graph and $\widetilde{G}_p$ is the p-percent perturbed graph, where $0 \le p \le 25$.

The *core number sequence* (Cor) is a sequence of length n, where the i-th item indicates the core number of vertex v_i. The core number of a vertex is the highest order of a k-core containing the vertex. We compute the divergence on core number sequence between two graphs using Equation 4.

$$Cor(G, \widetilde{G}) = \frac{1}{n} \sum_{i=1}^{n} C_i \tag{4}$$

where $C_i = 1$ if $core(v_i) = core(\widetilde{v}_i)$ and 0 otherwise.

The following metrics evaluate specific structural properties for each vertex of the graph: the first one is *betweenness centrality* (C_B), which measures the fraction of shortest paths that go through each vertex. This measure indicates the centrality of a vertex based on the flow between other vertices in the graph. A vertex with a high value indicates that this vertex is part of many shortest paths in the graph, which will be a key vertex in the graph structure. The second one is *closeness centrality* (C_C) and is defined as the inverse of the average distance to all accessible vertices. We compute the error on vertex metrics by:

$$\epsilon_m(G, \widetilde{G}) = \sqrt{\frac{1}{n}((g_1 - \widetilde{g}_1)^2 + \ldots + (g_n - \widetilde{g}_n)^2)} \tag{5}$$

where g_i is the value of the metric m for the vertex v_i of G and $\widetilde{g}_i$ is the value of the metric m for the vertex v_i of $\widetilde{G}$.

Moreover, two spectral measures which are closely related to many graph characteristics [26] are used. *The largest eigenvalue of the adjacency matrix A* (λ_1) where λ_i are the eigenvalues of A and $\lambda_1 \ge \lambda_2 \ge \ldots \ge \lambda_n$. The eigenvalues of A encode information about the cycles of a graph as well as its diameter. And *the second smallest eigenvalue of the Laplacian matrix L* (μ_2) where μ_i are the eigenvalues of L and $0 = \mu_1 \le \mu_2 \le \ldots \le \mu_m \le m$. The eigenvalues of L encode information about the tree structure of G. μ_2 is an important eigenvalue of the Laplacian matrix and can be used to show how good the communities separate, with smaller values corresponding to better community structures.

4.3   Empirical Results

Results are disclosed in Table 2. Each cell indicates the error value for the corresponding measure and algorithm. The lower the value, the better the algorithm.

Table 2. Results for *Rand-NC* (NC), *Random Perturbation* (RP) and *Random Switch* (RS) algorithms. For each dataset and algorithm, we compare the results obtained on D, h, SC, T, Cor, C_B, C_C, λ_1 and μ_2. Bold cells indicate the algorithm that achieves the best result (i.e, lowest information loss) for each measure and dataset.

Karate	D	h	SC	T	Cor	C_B	C_C	λ_1	μ_2
NC	**0.296**	**0.010**	3.091	0.028	0.392	0.025	0.047	**0.090**	**0.050**
RP	0.331	0.012	6.812	0.031	0.349	0.030	0.053	0.280	0.058
RS	0.596	0.047	**1.648**	**0.019**	**0.069**	**0.022**	**0.037**	0.119	0.184
Polbooks	D	h	$SC(\times 10^2)$	T	Cor	C_B	C_C	λ_1	μ_2
NC	**1.550**	**0.181**	**6.887**	**0.063**	0.338	0.019	**0.055**	0.208	**0.366**
RP	1.719	0.201	12.555	0.075	0.340	0.020	0.061	0.583	0.580
RS	1.888	0.239	8.564	0.082	**0.110**	0.019	0.070	**0.109**	0.500
URV email	D	h	$SC(\times 10^5)$	T	Cor	C_B	C_C	λ_1	μ_2
NC	0.388	**0.037**	3.696	**0.033**	0.582	0.001	0.175	0.580	0.320
RP	0.508	0.042	6.776	0.038	0.578	0.001	0.147	1.873	0.317
RS	**0.165**	0.094	**3.549**	0.044	**0.211**	0.001	**0.012**	**0.512**	**0.003**

A bold cell indicates the best algorithm for each measure and graph, i.e, the algorithm which achieves the lowest information loss. Although deviation is undesirable, it is inevitable due to the edge modification process.

The first tested network is Zachary's Karate Club. As shown in Table 2, our algorithm achieves the best results on D, h, λ_1 and μ_2. For instance, we can deepen on behaviour of λ_1 error in Figure 2a. The $p = 0$ value (x-axis) represents the value of this metric on the original graph. Thus, values close to this point indicate low noise on perturbed data. As we can see, Rand-NC remains closer to the original value than the other algorithms. Random Switch achieves the best results on SC and Cor. Furthermore, it also gets the best results on T, C_B and C_C, but they are quite similar on all randomization methods, suggesting that all of them introduce similar noise on these measures, as we can see in Figure 2b.

The second dataset, Polbooks, is a small collaboration network. The results of the comparability are very encouraging. Rand-NC outperforms on all measures, except on Cor and λ_1, where Random Switch achieves lower error values. For example, we can see the average error of the closeness centrality in Figure 2c, which behaves in a similar way for the three tested algorithms. The Cor measure, shown in Figure 2d, is based on the k-core decomposition and is closely related to the degree of the vertices. Consequently, it is not surprising that Random Switch gets the best results, since this method does not change the vertices' degree, keeping the degree sequence equal to the original one.

Finally, the last dataset is URV email communication graph. The best results on D, SC, Cor, C_C, λ_1 and μ_2 are achieved by Random Switch method. Even so, our algorithm gets good results on λ_1 and SC, where its values are close to the ones by Random Switch. Figure 2e presents the sub-graph centrality details. In addition, Rand-NC achieves the best results on h and T, as shown in Figure 2f.

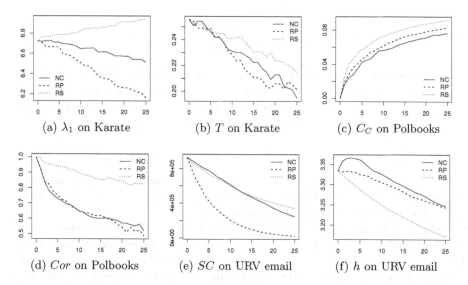

Fig. 2. Examples of the error evolution computed on our experimental framework. Perturbation parameter p varies along the horizontal axis from 0% (original graph) to 25%.

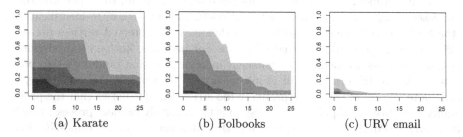

Fig. 3. Candidate set size ($Cand_{\mathcal{H}_1}$) evaluation on our three tested networks. Perturbation varies along the horizontal axis from 0% (original graph) to 25% and vertex proportion is represented on vertical axis. The trend lines show the percentage of vertices whose equivalent candidate set size falls into each of the following groups: [1] (black), [2,4], [5,10], [11,20], [21, ∞] (white).

5 Re-identification and Risk Assessment

Assuming that the randomization method and the number of fake edges w are public, the adversary must consider the set of possible worlds implied by $\widetilde{G}$ and w. Informally, the set of possible worlds consists of all graphs that could result in $\widetilde{G}$ under w perturbations. Using E^c to refer to all edges not present in E, the set of possible worlds of $\widetilde{G}$ under w perturbations, denoted by $\mathcal{W}_w(\widetilde{E})$ is computed by:

$$\mathcal{W}_w(\widetilde{E}) = \binom{|E|}{w}\binom{|E^c|}{w} \tag{6}$$

Therefore, and according to Hay et al. [16], the candidate set of a target vertex v_i includes all vertices $v_j \in \widetilde{G}$ such that v_j is a candidate is some possible world. We compute the candidate set of the target vertex v_i based on vertex refinement queries of level 1 ($\mathcal{H}_1$) (see [16] for further details) as shown in Equation 7.

$$Cand_{\mathcal{H}_1}(v_i) = \{v_j : deg^-(v_j) \le deg(v_i) \le deg^+(v_j)\} \tag{7}$$

where $deg^-(v_j) = round(deg(v_j)(1 - \frac{w}{m}))$ is the minimum expected degree of v_j and $deg^+ = round(deg(v_j)(1 + \frac{w}{m}))$ is the maximum expected degree after uniformly random edge deletion/addition process.

Using this re-identification model based on vertex degree, we observe that Random Switch method does not change the vertex degree on randomization process. Consequently, this method does not improve the privacy under an adversary's knowledge based on the degree value of some target vertices. Contrary, both Rand-NC and Random Perturbation methods modify the degree sequence and hinder the re-identification process according to this adversary knowledge model. Figure 3 shows the $Cand_{\mathcal{H}_1}$ evolution on Rand-NC and Random Perturbation methods. As we can see, candidate set sizes shrink while perturbation increases in all datasets. The number of nodes at high risk of re-identification (the black area) decreases quickly. Nevertheless, well-protected vertices present different behaviour on our tested networks. For instance, Zachary's karate club (Figure 3a) does not achieve a significant increment during randomization process by reason of it is a small graph with only 34 vertices and, therefore, a candidate set size of $[21, \infty]$ is probably too strict. Finally, notice that the probability computed in Equation 7 corresponds to uniformly random process, which is exactly Random Perturbation. A slightly difference occurs on Rand-NC, but we omit the experimental results due to the space limit.

6   Conclusions

In this paper we have presented a new algorithm for randomization on graphs, which is based on edge set modification according to edge's relevance. Instead of modifying one of the possible edges at random, this method considers the edge's relevance and preserves the most important edges in the graph. Our method achieves a better trade-off between data utility and data privacy than other methods. In particular, it outperforms Random Perturbation in terms of data utility and Random Switch in terms of data privacy as empirical results have shown.

We have presented some experimental results on real networks. Even though our algorithm does not perform better than Random Switch in some specific metrics and networks, the privacy level achieved by our method is clearly higher than the one achieved by Random Switch. In addition, Rand-NC gets the best results on several metrics. Therefore, we have demonstrated that our algorithm is better, in terms of data utility and privacy, than the two well-known Random Perturbation and Random Switch algorithms.

Many interesting directions for future research have been uncovered by this work. It would be very interesting to think on how the algorithm can work with other graph's type (directed or labelled graphs, for instance). It would be also interesting to consider the data utility on graph mining processes, for example on clustering-specific processes.

Acknowledgements The author thanks Klara Stokes for helpful technical discussions and her comments. This work was partly funded by the Spanish Government through projects TIN2011-27076-C03-02 "CO-PRIVACY" and CONSOLIDER INGENIO 2010 CSD2007-0004 "ARES".

References

1. Backstrom, L., Dwork, C., Kleinberg, J.: Wherefore art thou r3579x? anonymized social networks, hidden patterns, and structural steganography. In: Int. Conf. on World Wide Web, pp. 181–190. ACM, New York (2007)
2. Barabási, A.-L., Albert, R.: Emergence of Scaling in Random Networks. Science 286(5439), 509–512 (1999)
3. Bhagat, S., Cormode, G., Krishnamurthy, B., Srivastava, D.: Class-based graph anonymization for social network data. Proc. of the VLDB Endowment 2(1), 766–777 (2009)
4. Boldi, P., Bonchi, F., Gionis, A., Tassa, T.: Injecting Uncertainty in Graphs for Identity Obfuscation. Proc. of the VLDB Endowment 5(11), 1376–1387 (2012)
5. Bonchi, F., Gionis, A., Tassa, T.: Identity obfuscation in graphs through the information theoretic lens. In: Int. Conf. on Data Engineering, pp. 924–935. IEEE, Washington, DC (2011)
6. Bonchi, F., Gionis, A., Tassa, T.: Identity obfuscation in graphs through the information theoretic lens. Information Sciences 275, 232–256 (2014)
7. Campan, A., Truta, T.M.: Data and Structural k-Anonymity in Social Networks. In: Bonchi, F., Ferrari, E., Jiang, W., Malin, B. (eds.) PinKDD 2008. LNCS, vol. 5456, pp. 33–54. Springer, Heidelberg (2009)
8. Casas-Roma, J., Herrera-Joancomartí, J., Torra, V.: An Algorithm for k-Degree Anonymity On Large Networks. In: Int. Conf. on Advances on Social Networks Analysis and Mining, pp. 671–675. IEEE, Niagara Falls (2013)
9. Casas-Roma, J., Herrera-Joancomartí, J., Torra, V.: Analyzing the Impact of Edge Modifications on Networks. In: Torra, V., Narukawa, Y., Navarro-Arribas, G., Megías, D. (eds.) MDAI 2013. LNCS, vol. 8234, pp. 296–307. Springer, Heidelberg (2013)
10. Chester, S., Kapron, B.M., Ramesh, G., Srivastava, G., Thomo, A., Venkatesh, S.: Why Waldo befriended the dummy? k-Anonymization of social networks with pseudo-nodes. Social Network Analysis and Mining 3(3), 381–399 (2013)
11. Cormode, G., Srivastava, D., Yu, T., Zhang, Q.: Anonymizing bipartite graph data using safe groupings. Proc. of the VLDB Endowment 19(1), 115–139 (2010)
12. Dwork, C.: Differential Privacy. In: Bugliesi, M., Preneel, B., Sassone, V., Wegener, I. (eds.) ICALP 2006. LNCS, vol. 4052, pp. 1–12. Springer, Heidelberg (2006)

13. Ferri, F., Grifoni, P., Guzzo, T.: New forms of social and professional digital relationships: the case of Facebook. Social Network Analysis and Mining 2(2), 121–137 (2011)
14. Guimerà, R., Danon, L., Díaz-Guilera, A., Giralt, F., Arenas, A.: Self-similar community structure in a network of human interactions. Physical Review E-Statistical, Nonlinear and Soft Matter Physics 68(6), 65–103 (2003)
15. Hanhijärvi, S., Garriga, G.C., Puolamäki, K.: Randomization techniques for graphs. In: Int. Conf. on Data Mining, pp. 780–791. SIAM, Sparks (2009)
16. Hay, M., Miklau, G., Jensen, D., Weis, P., Srivastava, S.: Anonymizing Social Networks, Technical Report 07-19, UMass Amherst (2007)
17. Hay, M., Miklau, G., Jensen, D., Towsley, D., Weis, P.: Resisting structural re-identification in anonymized social networks. Proc. of the VLDB Endowment 1(1), 102–114 (2008)
18. Hay, M., Li, C., Miklau, G., Jensen, D.: Accurate Estimation of the Degree Distribution of Private Networks. In: Int. Conf. on Data Mining, pp. 169–178. IEEE, Miami (2009)
19. Hay, M., Liu, K., Miklau, G., Pei, J., Terzi, E.: Privacy-aware data management in information networks. In: Int. Conf. on Management of Data, pp. 1201–1204. ACM, New York (2011)
20. Krebs, V: (2006), http://www.orgnet.com
21. Liu, K., Terzi, E.: Towards identity anonymization on graphs. In: Int. Conf. on Management of Data, pp. 93–106. ACM, New York (2008)
22. Sihag, V.K.: A clustering approach for structural k-anonymity in social networks using genetic algorithm. In: CUBE Int. Information Technology Conference, pp. 701–706. ACM, Pune (2012)
23. Samarati, P.: Protecting Respondents' Identities in Microdata Release. Transactions on Knowledge and Data Engineering 13(6), 1010–1027 (2001)
24. Sweeney, L.: k-anonymity: a model for protecting privacy. International Journal of Uncertainty, Fuzziness and Knowledge-Based Systems 10(5), 557–570 (2002)
25. Ying, X., Pan, K., Wu, X., Guo, L.: Comparisons of randomization and k-degree anonymization schemes for privacy preserving social network publishing. In: Workshop on Social Network Mining and Analysis, pp. 10:1–10:10. ACM, New York (2009)
26. Ying, X., Wu, X.: Randomizing Social Networks: a Spectrum Preserving Approach. In: Int. Conf. on Data Mining, pp. 739–750. SIAM, Atlanta (2008)
27. Ying, X., Wu, X.: Graph Generation with Prescribed Feature Constraints. In: Int. Conf. on Data Mining, pp. 966–977. SIAM, Sparks (2009)
28. Zachary, W.W.: An information flow model for conflict and fission in small groups. Journal of Anthropological Research 33(4), 452–473 (1977)
29. Zhou, B., Pei, J.: Preserving Privacy in Social Networks Against Neighborhood Attacks. In: Int. Conf. on Data Engineering, pp. 506–515. IEEE, Washington (2008)

Rank Swapping for Stream Data

Guillermo Navarro-Arribas[1] and Vicenç Torra[2]

[1] DEIC-UAB Department d'Enginyeria de la Informació i de les Comunicacions,
Universitat Autònoma de Barcelona, Spain
guillermo.navarro@uab.cat
[2] IIIA-CSIC Institut d'Investigació en Intelligència Artificial,
Consejo Superior de Investigaciones Científicas, Spain
vtorra@iiia.csic.es

Abstract. We propose the application of rank swapping to anonymize data streams. We study the viability of our proposal in terms of information loss, showing some promising results. Our proposal, although preliminary, provides a simple and parallelizable solution to anonymize data stream.

1 Introduction

In this paper we study the application of rank swapping to data streams. The purpose of the work is to be able to anonymize streams of data composed by records. That is, records referring to individuals or entities which are generated as streams. The proposed method applies a privacy preserving method to the data as it is being generated. The output can be stored for future analysis or even analyzed in (close to) real time, without risking the disclosure of sensitive information from the data. This is specially convenient if the analysis of the data is to be performed by a third party. Stream data mining and techniques have become popular in recent times, not only to mine proper data streams but also to mine huge quantities of data [7,23]. Some examples of specific stream-like data, where anonymization is desirable are logs [13], time series and location data [14,10], and sensor data [1].

Ensuring privacy in such data streams is not straightforward. Typical statistical disclosure control (SDC) or privacy-preserving data mining techniques (PPDM) cannot usually be applied directly. Most current solutions for privacy preserving techniques on data streams are based on the creation of anonymity groups [8,3,4,9,16,22,24,26,27]. Their objective is to preserve k-anonymity [18,19] in the anonymized data. Contrary to these works, we focus our current work on a data swapping method. Rank swapping is a well known method for Statistical Disclosure Control, which randomly exchanges values from the original data between records. The method was first described for numerical variables in [12], although the idea of swapping data was first mentioned in [17]. Moreover, data swapping techniques have been widely used by statistical agencies. For instance, it was used in the UK Census from 2001 [15].

V. Torra et al. (Eds.): MDAI 2014, LNAI 8825, pp. 217–226, 2014.

We propose the application of rank swapping to anonymize data streams. As we will see the use of rank swapping is simple and provides relatively good results. Moreover we propose two different approaches to apply rank swapping to data streams and compare the results. In Section 2 we introduce data streams and some notation, and Section 3 describes the methods to apply rank swapping to data streams. We evaluate our proposed methods in terms of information loss in Section 4, and provide some discussion. Finally, Section 5 concludes the paper.

2   Data Streams

A data stream is usually seen as a continuous flow of data, which can be infinite, and delivers a data item in a relatively timely fashion. In our case we are interested in anonymizing microdata streams. A microdata file is the base of most data privacy techniques and can be seen as a table of records, each record corresponds to an individual (or entity) and each column is an attribute.

In a microdata file we normally distinguish between the following types of attributes:

- Identifiers: attributes that uniquely identify and individual (e.g. social security number, passport number, ...).
- Quasi-identifier: although a quasi-identifier alone cannot be used to identify an individual, a group of them can (e.g. age, postal code, ...).
- Confidential attributes: attributes (usually quasi-identifiers) that are considered confidential (e.g. disease, gross salary, ...).

For the purpose of this paper, we consider a microdata stream as continuous stream of tuples, where each tuple can be seen as a record in the traditional microdata file. The data stream is denoted as $S(t, a_1, \ldots, a_n)$ where t is the position of the tuple in the stream, and $a_1, \ldots, a_n$ are the attributes of the tuple. For the sake of simplicity, we consider all attributes to be quasi-identifiers. We assume that identifier attributes are removed or encrypted and confidential attributes are treated as quasi-identifiers.

3   Rank Swapping for Data Streams

In this section we will present two different approaches for applying rank swapping to data streams. These two approaches will be compared and evaluated in Section 4. The first method is based on the application of rank swapping to a simple sliding window on the data stream. As an improvement we propose a second approach, which selects tuples from the sliding window in order to reduce the information loss introduced by the rank swapping.

3.1   Rank Swapping

Several algorithms for rank swapping have been developed since its introduction. We have adopted a simplified version described in [5,6]. Although the specific results might present some variations we believe that the application of other rank swapping algorithms to our approaches will be straightforward.

We will denote this method as the *simple rank swapping* or *RS*. The idea is that the values of an attribute A_i are ranked in ascending order; then each ranked value of A_i is swapped with another ranked value randomly chosen within a restricted range (e.g., the rank of two swapped values cannot differ by more than p percent of the total number of records).

It is important to note that this approach assumes that there is an order relation between attributes. We also assume the availability of a distance function. We will denote the distance function as d and the order relation as $\leq$, so $a_i \leq a_j \leq a_k$ implies that $d(a_i, a_j) \leq d(a_i, a_k)$.

3.2   Stream Rank Swapping

Our first approach to deal with data streams is to use a sliding window to perform the rank swapping. A window receives tuples from the stream and acts as a simple buffer of fixed size w. When the window contains w tuples, rank swapping is applied to the whole set of w tuples and the protected data is released. The window is emptied and ready to receive more tuples from the stream (see Figure 1).

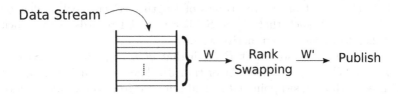

Fig. 1. Simple stream rank swapping

When the window is full, we have a set of tuples $W_i = \{S_i, S_{i+1}, \ldots, S_{i+w}\}$, where $S_i = \{t_i, a_{i1}, \ldots, a_{in}\}$ and a_{ij} is the value of the attribute j in the tuple S_i. This window can be seen as a microdata file where each tuple is a record with the values for each quasi-identifier attribute, e.g. $(a_{i1}, \ldots, a_{in})$ for record i. When the window is full, the set W_i is protected with rank swapping and the resulting set W_i' is released. Now W_i' contains the tuples with swapped values for the attributes $W_i' = \{S_i', S_{i+1}', \ldots, S_{i+w}'\}$. The procedure is described in Algorithm 1.

Note that the final release of data is composed of the sets $W_1, W_2, \ldots$. Each set can be periodically released as data is obtained from the stream.

The fact that we are periodically releasing subsets of protected data independently does not impose a thread to privacy. This is opposed to using an approach based on anonymity groups [3], where the intersection of anonymity groups could lead to inference attacks.

Algorithm 1. SRS

1: **procedure** SRS(S, p, w)
2:     $W = \emptyset$
3:     **while** S is non-empty **do**
4:         **while** $|W| < w$ **do**
5:             Let t be the next tuple from S
6:             $W = W \cup \{t\}$
7:         **if** $|W| == w$ **then**
8:             $RS(W, p)$

3.3 Stream Selective Rank Swapping

We can intuitively see that the previous method, which performs rank swapping on the data stream in slices can result in an increase of information loss depending on the window size. One way to improve this method is to perform the rank swapping not on the whole window but on a selected set of tuples, which could reduce information loss.

The selected set of tuples such be the one preserving more information in the swapping process. It is easy to see that a dense set of tuples will be better than a sparse one. In this context we understand a dense set as a set with more number of tuples in a smaller range. We will call this set a *block*.

A block B is a set of ordered tuples of length b. Each block is denoted as $B_i = (S_{i_1}, \ldots, S_{i_b})$, such that $S_{i_r} \leq S_{i_s}$ if $r \leq s$. All records of the block are always taken from the current active window.

Given a window W, we select B_i such that $d(S_{i_1}, S_{i_b}) \leq d(S_{j_1}, S_{j_b})$ for all $B_j \neq B_i$ and $B_i, B_j \in W$ (see Algorithm 3). The selected block B_i is then anonymized with rank swapping (RS), and removed from the window. Once the window is full again (with w records) we repeat the process. The procedure is described in Algorithm 2.

Algorithm 2. SSRS

1: **procedure** SSRS(S, p, w, b)
2:     $W = \emptyset$
3:     **while** S is non-empty **do**
4:         **while** $|W| < w$ **do**
5:             Let t be the next tuple from S
6:             $W = W \cup \{t\}$
7:         $B = SelectBlock(W, b)$
8:         $RS(B, p)$

The idea here, is to swap values that are closer and leave the other values, which eventually will be swapped with closer ones.

Algorithm 3. SelectBlock

1: **procedure** SELECTBLOCK(W, b)
2:     **for all** $B_i = (S_{i_1}, \ldots, S_{i_b}) \in W$ **do**
3:         Compute $d(B_i) = d(S_{i_1}, S_{i_b})$
4:         Return B_j such that $d(S_{j_1}, S_{j_b}) \leq d(S_{k_1}, S_{k_b})$ for all $B_j \neq B_k$

---

4 Evaluation

We have evaluated the previous methods to measure the information loss introduced in each case. To be able to compare results with non-streaming versions of rank swapping we have used a finite dataset, the Adult dataset from the UCI repository [2], which contains 48842 records from the 1994 Census database. To simplify the evaluation we have taken only one numerical attribute, the *fnlwgt* (final sampling weight).

The methods are denoted as:

1. *RS*, rank swapping: as applied to the whole dataset.
2. *SRS*, stream rank swapping: as described in Section 3.2.
3. *SSRS*, stream selective rank swapping: as described in Section 3.3.

We have also parameterized the methods with the parameters:

- *p*: determines the interval to perform the swapping in the ranked data. Affects RS, SRS, and SSRS.
- *w*: size of the window (or buffer) that stores tuples from the stream. Affects the streaming versions SRS and SSRS.
- *b*: size of the block of tuples taken from the window for the SSRS method.

To measure the information loss we have used the well known IL1s measure [25,11] as implemented by the *sdcMicro* package for R [20].

IL1s measures the distance between the protected records and the original ones. Formally, given the original data X and the protected one X':

$$IL1s(X, X') = \frac{1}{nM} \sum_{i=1}^{M} \sum_{j=0}^{V} \frac{|a_{ij} - a'_{ij}|}{\sqrt{2}S_j} \tag{1}$$

where n is the number of attributes, M the number of records, a_{ij} denotes the value of record i for attribute j, and a'_{ij} the same for the protected version, and S_j is the standard deviation of the j-th attribute in the original data.

All the measurements of IL1s are made on the whole dataset once protected. This is so to be able to compare the proposed streaming versions of rank swapping to the classical rank swapping.

Figure 2 shows the IL1s values for the different methods: RS, SRS, and SSRS. In the case of SSRS we have considered three different values for the parameter b (the block size), which depend on the window size w. The values are always taken as 1/4, 1/2, and 3/4 of the window size w (all values floor rounded).

It is important to note that even with the same window size, the SSRS performs better. Moreover, SSRS with lower block sizes are usually better than the SRS.

We can also compare the methods by choosing a block size from the SSRS method equal to the window size of the SRS. Figure 3 shows the methods using the same value for the window in SRS, and block for SSRS (note there are three cases for 1/4, 1/2, 3/4 values of b). We compare the methods for different values of p.

As expected, SSRS with lower values of b is the better choice of the streaming versions. In this case the difference between SSRS and SRS is bigger, something that was also expected.

4.1   Discussion

From the observations about the performance of our proposed methods, we can foresee rank swapping as a very interesting approach for streaming data. We have introduced two approaches to apply rank swapping to streaming data and there are important advantages to be noted:

1. Good performance in terms of information loss. Moreover, for relatively big values of the window size, the streaming methods are quite close to the non-streaming rank swapping.
2. Highly parallelizable. Each computation of rank swapping (within the window or block of records) is completely independent, which allows a straightforward parallelization.

The methods proposed in the paper show that rank swapping is an interesting approach for anonymizing stream data. Given its parallelization facility it is also suitable for anonymizing huge datasets. In such cases, where performing a ranks swapping is unfeasible (or impractical) the SRS and SSRS methods can provide a fast and convenient way to process the datasets.

There are however some issues that need to be addressed, which we haven't introduced in this work. The purpose of this paper is to show the viability (in terms of information loss) of rank swapping to deal with stream data, not to provide a final or production ready method. For example, the SSRS as described in Section 3.3 should include mechanisms to ensure that a record is processed within a given time range. That is, given that records are selected from the window, some outlier records could be delayed or even kept in the window forever (in case of infinite streams). To solve this, records could be given an expiration time when they enter the window. The record will be forced to be selected if it has expired. A similar approach was used in [4] with a clustering based method for anonymization.

Although we have used numerical attributes, our proposal are extensible to categorical data forming a complete order or even partial orders [21].

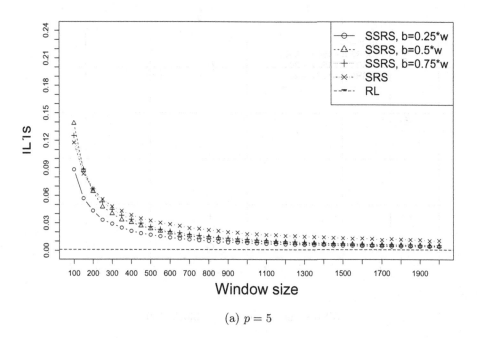

(a) $p = 5$

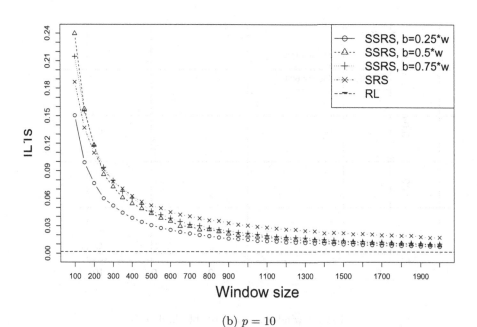

(b) $p = 10$

Fig. 2. Information loss in terms of window size

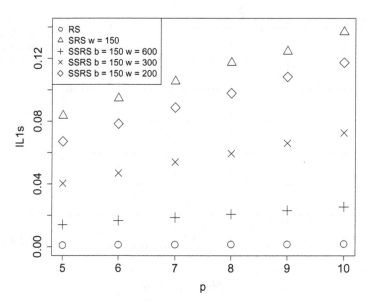

(a) SRS window, and SSRS block: 150

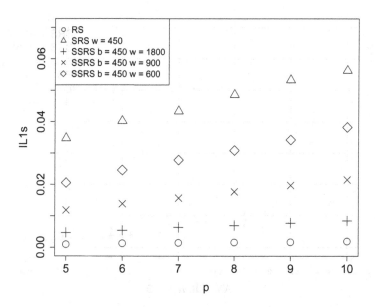

(b) SRS window, and SSRS block: 450

Fig. 3. Information loss with SRS window size equal to SSRS block size

5   Conclusions

We have introduced a rank swapping based method for the anonymization of numerical data streams. The proposed methods are compared in terms of information loss and provide a good starting point for considering rank swapping a good candidate to be applied in streaming data. This is a preliminary work and further research and development work is required to provide production level algorithms and implementations.

Acknowledgment. Partial support by the Spanish MEC projects ARES (CONSOLIDER INGENIO 2010 CSD2007-00004), N-KHRONOUS (TIN2010-15764), and COPRIVACY (TIN2011-27076-C03-03) is acknowledged. Partial support of the European Project DwB (Grant Agreement Number 262608) is also acknowledged.

References

1. Aggarwal, C.C. (ed.): Managing and Mining Sensor Data. Springer (2013)
2. Bache, K., Lichman, M.: UCI Machine Learning Repository. University of California, School of Information and Computer Science, Irvine, CA (2013), http://archive.ics.uci.edu/ml
3. Byun, J.-W., Li, T., Bertino, E., Li, N., Sohn, Y.: Privacy-preserving incremental data dissemination. Journal of Computer Security 17(1), 43–68 (2009)
4. Cao, J., Carminati, B., Ferrari, E., Tan, K.-L.: CASTLE: Continuously Anonymizing Data Streams. IEEE Transactions on Dependable and Secure Computing 8(3), 337–352 (2011)
5. Domingo-Ferrer, J., Torra, V.: Disclosure Control Methods and Information Loss for Microdata. In: Doyle, P., Lane, J.I., Theeuwes, J.J.M., Zayatz, L. (eds.) Confidentiality, Disclosure, and Data Access: Theory and Practical Applications for Statistical Agencies, pp. 91–110. Elsevier Science (2001)
6. Domingo-Ferrer, J., Torra, V.: A quantitative comparison of disclosure control methods for microdata. In: Doyle, P., Lane, J.I., Theeuwes, J.J.M., Zayatz, L. (eds.) Confidentiality, Disclosure and Data Access: Theory and Practical Applications for Statistical Agencies, pp. 111–134. North-Holland (2001)
7. Gaber, M.M., Zaslavsky, A., Krishnaswamy, S.: Mining Data Streams: A Review. SIGMOD Rec. 34(2), 18–26 (2005)
8. Ghinita, G., Karras, P., Kalnis, P., Mamoulis, N.: Fast Data Anonymization with Low Information Loss. In: Proceedings of the 33rd International Conference on Very Large Data Bases, pp. 758–769. VLDB Endowment, Vienna (2007)
9. Li, J., Ooi, B.C., Wang, W.: Anonymizing Streaming Data for Privacy Protection. In: IEEE 24th International Conference on Data Engineering, ICDE 2008, pp. 1367–1369 (2008)
10. Martinez-Bea, S., Torra, V.: Trajectory anonymization from a time series perspective. In: 2011 IEEE International Conference on Fuzzy Systems (FUZZ), pp. 401–408 (2011)
11. Mateo-Sanz, J.M., Sebé, F., Domingo-Ferrer, J.: Outlier Protection in Continuous Microdata Masking. In: Domingo-Ferrer, J., Torra, V. (eds.) PSD 2004. LNCS, vol. 3050, pp. 201–215. Springer, Heidelberg (2004)

12. Moore, R.: Controlled data swapping techniques for masking public use microdata sets, U. S. Bureau of the Census (unpublished manuscript) (1996)
13. Navarro-Arribas, G., Torra, V.: Privacy-preserving data-mining through micro-aggregation for web-based e-commerce. Internet Research 20, 366–384 (2010)
14. Nin, J., Torra, V.: Towards the evaluation of time series protection methods. Inf. Sci. 179(11), 1663–1677 (2009)
15. ONS, Statistical disclosure control (sdc) methods short-listed for 2011 UK census tabular outputs, SDC UKCDMAC Subgroup Paper 1, Office for National Statistics, UK (2011)
16. Pei, J., Xu, J., Wang, Z., Wang, W., Wang, K.: Maintaining K-Anonymity against Incremental Updates. In: 19th International Conference on Scientific and Statistical Database Management (2007)
17. Reiss, S.: Practical data-swapping: The first steps. In: IEEE Symposium on Security and Privacy, pp. 38–43 (1980)
18. Samarati, P.: Protecting respondents identities in microdata release. IEEE Transactions on Knowledge and Data Engineering 13, 1010–1027 (2001)
19. Sweeney, L.: k-anonymity: a model for protecting privacy. Int. J. Uncertain. Fuzziness Knowl.-Based Syst. 10, 557–570 (2002)
20. Templ, M.: Statistical Disclosure Control for Microdata Using the R-Package sdcMicro. Transactions on Data Privacy 1(2), 67–85 (2008)
21. Torra, V.: Rank Swapping for Partial Orders and Continuous Variables. Presented at the International Conference on Availability, Reliability and Security, ARES 2009, pp. 888–893 (2009)
22. Truta, T.M., Campan, A.: K-anonymization incremental maintenance and optimization techniques. In: Proceedings of the 2007 ACM Symposium on Applied Computing, pp. 380–387. ACM Press (2007)
23. Wu, X., Zhu, X., Wu, G.-Q., Ding, W.: Data mining with big data. IEEE Transactions on Knowledge and Data Engineering 26(1), 97–107 (2014)
24. Xiao, X., Tao, Y.: M-invariance: towards privacy preserving re-publication of dynamic datasets. In: Proceedings of the 2007 ACM SIGMOD International Conference on Management of Data, SIGMOD 2007, pp. 689–700. ACM (2007)
25. Yancey, W.E., Winkler, W.E., Creecy, R.H.: Disclosure Risk Assessment in Perturbative Microdata Protection. In: Domingo-Ferrer, J. (ed.) Inference Control in Statistical Databases. LNCS, vol. 2316, pp. 135–152. Springer, Heidelberg (2002)
26. Zakerzadeh, H., Osborn, S.L.: FAANST: Fast Anonymizing Algorithm for Numerical Streaming DaTa. In: Garcia-Alfaro, J., Navarro-Arribas, G., Cavalli, A., Leneutre, J. (eds.) DPM 2010 and SETOP 2010. LNCS, vol. 6514, pp. 36–50. Springer, Heidelberg (2011)
27. Zhou, B., Han, Y., Pei, J., Jiang, B., Tao, Y., Jia, Y.: Continuous privacy preserving publishing of data streams. In: Proceedings of the 12th International Conference on Extending Database Technology: Advances in Database Technology, EDBT 2009, pp. 648–659. ACM (2009)

On Radius-Incorporated Multiple Kernel Learning

Xinwang Liu, Jianping Yin, and Jun Long

School of Computer, National University of Defense Technology, Changsha, Hunan,
410073, China
{1022xinwang.liu,jdragonnudt}@gmail.com, jpyin@nudt.edu.cn

Abstract. In this paper, we review existing radius-incorporated Multiple Kernel Learning (MKL) algorithms, trying to explore the similarities and differences, and provide a deep understanding of them. Our analysis and discussion uncover that traditional margin based MKL algorithms also take an approximate radius into consideration implicitly by base kernel normalization. We perform experiments to systematically compare a number of recently developed MKL algorithms, including radius-incorporated, margin based and discriminative variants, on four MKL benchmark data sets including Protein Subcellular Localization, Protein Fold Prediction, Oxford Flower17 and Caltech101 in terms of both the classification performance, measured by classification accuracy and mean average precision. We see that overall, radius-incorporated MKL algorithms achieve significant improvement over other counterparts in terms of classification performance.

Keywords: Multiple Kernel Learning, Support Vector Machines, Radius Margin Bound, Minimum Enclosing Ball, Kernel Methods.

1 Introduction

Kernel methods have achieved great successes in machine learning community and have been widely adopted. As well known, their performance heavily depends on the choice of kernels. Many efforts have been devoted to address this issue by designing data-dependent optimal kernel algorithms [10,1,4], so-called "learning kernels from data". Among these algorithms, Multiple Kernel Learning (MKL) algorithms have been paid intensive attention since they are not only capable of adaptively tuning an optimal kernel for a specific learning task, but also provide an elegant framework to integrate multiple heterogenous source data.

The idea of MKL can be applied to both margin and class separability maximization criteria, leading to margin-based [1,5,4] and discriminative MKL algorithms [14], respectively. In this paper, we mainly focus on margin based MKL algorithms due to the popularity of margin maximization framework. There are several research trends in existing margin based MKL algorithms. The first direction focuses on designing computationally efficient MKL algorithms [1,11,13]. The second one aims to develop non-sparse and non-linear combination MKL algorithms [13], which usually achieve superior performance compared with sparse

V. Torra et al. (Eds.): MDAI 2014, LNAI 8825, pp. 227–240, 2014.

counterparts. By arguing that the generalization error bound of SVMs is dependent on both radius and margin, the last direction simultaneously takes the margin and the radius of the minimum hyper-sphere which encloses all training samples in the multi-kernel induced feature space into consideration [3,4,6,7].

Our work in this paper follows the last direction by proposing a radius-incorporated MKL framework. Using this framework as a toolbox, we instantiate three different radius-incorporated MKL algorithms by approximating the radius of Minimum Enclosing Ball (MEB) with the trace of each base kernel, the trace of total scatter matrix, and the radiuses induced by each base kernel, respectively. We further theoretically show that the above three radius-incorporated MKL algorithms can be rewritten as the traditional MKL formulation with only one difference being that different linearly weighted equality constraints on the kernel combination coefficients are employed. Specifically, the trace of base kernels, the trace of total scatter matrix of base kernels, and the base radiuses of each base kernel are respectively applied to linearly weight the coefficients of each base kernel in the above radius-incorporated MKL algorithms. Moreover, we uncover the relationship between the radius-incorporated MKL algorithms with kernel normalization which is still an open issue in existing MKL literature. Though different kernel normalization manners have been used [5], there is still lack of a principled way to explain why this normalization should be employed and which normalization usually works well in real work applications. We answer these questions by pointing out that different normalization manners in essence correspond to different radius-incorporation manners, which further correspond to different criteria in minimizing the generalization error of SVMs. From this perspective, our proposed radius-incorporated framework builds a bridge between kernel normalization approaches and the generalization error criteria. The contributions of this paper are highlighted as follows:

- We propose a radius-incorporated MKL framework which learns the base kernel combination coefficients by simultaneously maximizing the margin between classes and minimizing the radius of MEB. Furthermore, three radius-incorporated MKL algorithms instantiated from the framework are proposed by calculating the radius of MEB with different approaches.
- We uncover the tight connection between kernel normalization and radius incorporation, which provides a potential explanation for different kernel normalization approaches in existing MKL algorithms.
- We systematically compare a number of recently developed radius-incorporated MKL algorithms in terms of classification accuracy, which paves a way for designing excellent radius-incorporated MKL algorithms.

Comprehensive experiments have been conducted on Protein Subcellular Localization, Protein Fold Prediction, Oxford Flower17, Caltech101 and Alzheimer's Disease data sets to compare the proposed radius-incorporated MKL algorithms with state-of-the-art MKL algorithms in terms of classification performance. As the experimental results indicate, our proposed radius-incorporated MKL algorithms achieve better or comparable performance compared to many state-of-the-art MKL algorithms, which validates the effectiveness of the proposed radius-incorporated MKL framework.

2   Related Work

In this section, we first review some margin based MKL algorithms, and then focus on the MKL algorithms in [3,4,6,7] which optimizes both radius of MEB and margin. Let $\{(\mathbf{x}_i, y_i)\}_{i=1}^n$ be a given training set, where $\mathbf{x}_i$ and $y_i \in \{-1, +1\}$ represent i-th training sample and its corresponding label, respectively. Let $\{\phi_p\}_{p=1}^m$ be a group of feature mappings where ϕ_p induces a kernel function k_p. One can define $\mathbf{K}_p$ as the kernel matrix computed with k_p on the training set $\{\mathbf{x}_i\}_{i=1}^n$. In existing MKL literature, each sample $\mathbf{x}_i$ is mapped onto m feature spaces by $\phi(\mathbf{x}; \boldsymbol{\gamma}) \triangleq [\sqrt{\gamma_1}\phi_1(\mathbf{x}), \cdots, \sqrt{\gamma_m}\phi_m(\mathbf{x})]^\top$, where γ_p is the coefficient of the p-th base kernel. Correspondingly, the induced kernel function can be expressed as a linear combination of p base kernels, $k(\boldsymbol{\gamma}) = \sum_{p=1}^m \gamma_p k_p$ and $\mathbf{K}(\boldsymbol{\gamma}) = \sum_{p=1}^m \gamma_p \mathbf{K}_p$.

The objective of MKL algorithms is to learn the base kernel coefficients $\boldsymbol{\gamma}$ and the structural parameters $(\boldsymbol{\omega}, b)$ jointly. To achieve this goal, most of MKL algorithms [1,11,5] propose to minimize the following optimization problem,

$$\min_{\boldsymbol{\gamma}, \boldsymbol{\omega}, b, \boldsymbol{\xi}} \frac{1}{2}\|\boldsymbol{\omega}\|^2 + C\sum_{i=1}^n \xi_i \ s.t. \ y_i(\boldsymbol{\omega}^\top \phi(\mathbf{x}_i; \boldsymbol{\gamma}) + b) \geq 1 - \xi_i, \ \xi_i \geq 0, \forall i, \ \|\boldsymbol{\gamma}\|_q = 1, \ \boldsymbol{\gamma} \succeq 0,$$

$$(1)$$

where $\boldsymbol{\omega}$ is the normal of the separating hyperplane, b the bias term, $\boldsymbol{\xi} = [\xi_1, \cdots, \xi_n]^\top$ is the vector of slack variables, and $\boldsymbol{\gamma}$ is the base kernel coefficients. Another important issue in Eq. (1) is that $q > 1$ will induce non-sparse kernel coefficients (called non-sparse MKL) while $q = 1$ will lead to sparse kernel combination (called sparse MKL).

Several recent research on MKL has gradually realized the importance of radius of MEB in MKL and successfully incorporated this radius into the traditional MKL formulation, achieving better kernel learning performance [3,4,6]. The theoretical justification for the radius incorporation lies at that the generalization error bound of SVMs is dependent on both the margin and the radius of the MEB of training data [10]. Furthermore, as pointed out in [4], only maximizing the margin with respect to $\boldsymbol{\gamma}$ will cause scaling and initialization issues. A larger margin could be arbitrarily achieved by scaling $\boldsymbol{\gamma}$ to $\tau\boldsymbol{\gamma}$ ($\tau > 1$), and this will affect the convergency of the optimization problem. Usually, a norm-constraint is imposed on $\boldsymbol{\gamma}$ to address this issue. Nevertheless, identifying an appropriate norm-constraint for a given kernel learning task remain an open issue itself [5]. Moreover, even if a norm-constraint is imposed, a good kernel could still be misjudged as a poor one by simply down-scaling the corresponding kernel weight [4]. These issues can be removed or mitigated by the incorporation of radius information. In the following, we review the radius-incorporated MKL algorithms in literature.

The pioneering work on radius-incorporated MKL in [3] proposes to minimize the optimization problem in Eq (2).

$$\min_{\gamma,\omega,b,\xi} \frac{1}{2}R^2(\gamma)\|\omega\|^2 + \frac{C}{2}\sum_{i=1}^{n}\xi_i^2 \ s.t. \ y_i(\omega^\top \phi(\mathbf{x}_i;\gamma)+b) \geq 1-\xi_i, \forall i, \ \sum_{p=1}^{m}\gamma_p = 1, \ \gamma \succeq 0,$$

(2)

where R^2 is the squared radius of the MEB in the multi-kernel induced space and can be calculated as

$$R^2(\gamma) = \Big\{\max_{\beta} \beta^\top \mathrm{diag}(\mathbf{K}(\gamma)) - \beta^\top \mathbf{K}(\gamma)\beta \ s.t. \ \beta^\top \mathbf{1} = 1, \ 0 \leq \beta_i, \ \forall i\Big\}.$$

(3)

Like the margin, R^2 is also a function of γ. Instead of solving the optimization problem in Eq. (2) directly, the authors turn to minimize the following upper bounding convex optimization problem:

$$\min_{\gamma,\omega,b,\xi} \frac{1}{2}\|\omega\|^2 + \frac{C}{2\sum_{p=1}^{m}\gamma_p R_p^2}\sum_{i=1}^{n}\xi_i^2 \ s.t. \ y_i(\omega^\top \phi(\mathbf{x}_i;\gamma)+b) \geq 1-\xi_i, \forall i, \ \sum_{p=1}^{m}\gamma_p = 1, \ \gamma \succeq 0,$$

(4)

where R_p^2 is the squared radius of the MEB in the p-th base kernel induced space and can be calculated as

$$R_p^2 = \Big\{\max_{\beta} \beta^\top \mathrm{diag}(\mathbf{K}_p) - \beta^\top \mathbf{K}_p\beta \ s.t. \ \beta^\top \mathbf{1} = 1, \ 0 \leq \beta_i, \ \forall i\Big\}.$$

(5)

The work in [3] focuses on how to approximate the optimization problem in Eq. (3) with a convex one in Eq. (4), and does not address the scaling issue mentioned above. Differently, the work in [4] directly solves the optimization in Eq. (6) and carefully discusses how the scaling issue can be addressed.

$$\min_{\gamma,\omega,b,\xi} \frac{1}{2}R^2(\gamma)\|\omega\|^2 + C\sum_{i=1}^{n}\xi_i \ s.t. \ y_i(\omega^\top \phi(\mathbf{x}_i;\gamma)+b) \geq 1-\xi_i, \forall i, \ \xi_i \geq 0, \ \gamma \succeq 0.$$

(6)

In detail, a tri-level optimization problem is proposed in that work,

$$\min_{\gamma} \widehat{\mathcal{J}}(\gamma) \ s.t. \ \gamma_p \geq 0, \ \forall p.$$

(7)

where

$$\widehat{\mathcal{J}}(\gamma) = \Big\{\max_{\alpha} \alpha^\top \mathbf{1} - \frac{1}{2R^2(\gamma)}(\alpha \circ \mathbf{y})^\top \mathbf{K}(\gamma)(\alpha \circ \mathbf{y}) \ s.t. \ \alpha^\top \mathbf{y} = 0, \ 0 \leq \alpha_i \leq C, \ \forall i\Big\} \ (8)$$

and $R^2(\gamma)$ is calculated by Eq. (3). To solve the optimization problem, a tri-level optimization structure is developed accordingly. Specifically, in the first step, R^2 is computed by solving the Quadratic Programming (QP) in Eq. (3) with a given γ. Then, the obtained R^2 is taken into Eq. (8) to solve another QP to calculate $\widehat{\mathcal{J}}(\gamma)$. The last step is to update the base kernel coefficients γ. The above procedure is repeated until a stopping criterion is satisfied. Compared with traditional MKL algorithms, an extra QP is introduced and solved at each iteration. This can considerably increase the computation cost of SVMs based

MKL, especially when the size of training set is large. Moreover, the performance of MKL could be adversely affected by the notorious sensitivity of this radius to outliers. In [6], instead of directly incorporating the radius of MEB, the authors propose to incorporate its close relative, the trace of data scattering matrix, to avoid the above problems. Specifically, their optimization problems is as follows in Eq. (9),

$$\min_{\boldsymbol{\gamma},\boldsymbol{\omega},b,\boldsymbol{\xi}} \frac{1}{2}\mathrm{tr}\left(\mathbf{St}(\boldsymbol{\gamma})\right)\|\boldsymbol{\omega}\|^2 + C\sum_{i=1}^{n}\xi_i \ \ s.t. \ y_i(\boldsymbol{\omega}^\top\phi(\mathbf{x}_i;\boldsymbol{\gamma})+b) \geq 1-\xi_i, \forall i, \ \xi_i \geq 0, \ \boldsymbol{\gamma} \succeq 0, \tag{9}$$

where

$$\mathrm{tr}\left(\mathbf{St}(\boldsymbol{\gamma})\right) = \mathrm{tr}(\mathbf{K}(\boldsymbol{\gamma})) - \frac{1}{n}\mathbf{1}^\top\mathbf{K}(\boldsymbol{\gamma})\mathbf{1} \tag{10}$$

with $\mathbf{1}$ is a column vector with all elements one. Though usually demonstrating superior performance from the experimental perspective, it is criticized from the theoretic perspective since it may not be a upper bound of generalization error bound such as Radius Margin Bound [10].

In the following, we propose a radius-incorporated MKL framework where different radius variants could be integrated. Then we theoretically show that radius-margin based framework can be equivalently addressed by solving a traditional margin based MKL algorithms, with a difference being that a weighted constraint on the base kernel coefficients encoding the radius information. Furthermore, we formally, for the first time, uncover the connection between radius incorporation and kernel normalization.

3   Radius-Incorporated MKL Framework

3.1   The Proposed Framework

The radius-incorporated MKL framework in this paper is presented as follows,

$$\min_{\boldsymbol{\gamma}} \ \mathcal{J}(\boldsymbol{\gamma}), \ s.t. \ \gamma_p \geq 0, \ \forall p. \tag{11}$$

where

$$\mathcal{J}(\boldsymbol{\gamma}) = \left\{ \min_{\boldsymbol{\omega},b} \frac{1}{2}R^2(\boldsymbol{\gamma})\|\boldsymbol{\omega}\|^2 + C\sum_{i=1}^{n}\xi_i \ \ s.t. \ y_i(\boldsymbol{\omega}^\top\phi(\mathbf{x}_i;\boldsymbol{\gamma})+b) \geq 1-\xi_i, \ \xi_i \geq 0, \forall i \right\} \tag{12}$$

Proposition 1. $\mathcal{J}(\tau\boldsymbol{\gamma}) = \mathcal{J}(\boldsymbol{\gamma})$, where $\tau > 0$ is any positive scalar. And the SVM decision function using the combined kernel is not affected by τ.

Proof. The proof is elaborated in our earlier publication [7].

Proposition 1 indicates that our formulation in Eq. (11) is invariant when the kernel combination weights are uniformly scaled up by a positive scalar τ. In this case, the optimal value of $\boldsymbol{\omega}$ will correspondingly be down scaled by $1/\tau$, leaving the SVMs decision function unchanged. Based on Proposition 1, the following

Theorem 1 demonstrates that our objective function can be rewritten as the common form used by the existing margin based MKL algorithms, with only one difference that a constraint is imposed on the kernel coefficients encoding the radius information.

Theorem 1. *The optimal solution of optimization problem in Eq. (11), denoted as $\gamma^\star$, can be written as $\gamma^\star = R^2(\gamma)\eta^\star$, where $\eta^\star$ is the optimal solution of the following optimization problem in Eq. (13),*

$$\min_{\eta} \mathcal{J}(\eta) \;\; s.t. \; R^2(\eta) = 1, \; \eta_p \geq 0, \;\; \forall p. \tag{13}$$

where

$$\mathcal{J}(\eta) = \left\{ \min_{\omega, b} \frac{1}{2}\|\omega\|^2 + C\sum_{i=1}^{n} \xi_i \; s.t. \; y_i\left(\omega^\top \phi(\mathbf{x}_i; \eta) + b\right) \geq 1 - \xi_i, \; \xi_i \geq 0, \forall i \right\} \tag{14}$$

Proof. The proof is elaborated in our earlier publication [7].

Theorem 1 indicates our proposed radius-incorporated MKL framework in Eq. (11) can be reformulated as a traditional margin based one, with only one difference being that an additional constraint on the kernel combination coefficients encoding radius information, as shown in Eq. (13).

3.2   Radius-Incorporated MKL Variants

In the following, we instantiate the calculation of $R^2(\gamma)$ by three different approaches: $\mathrm{Tr}(\mathbf{K}(\gamma))$, $\mathrm{Tr}(\mathbf{S}_t(\gamma))$ and $\sum_{p=1}^{m} \gamma_p R_p^2$.

TrK-margin MKL By substituting $R^2(\gamma)$ in Eq. (13) with $\mathrm{Tr}(\mathbf{K}(\gamma))$, we obtain the objective of Tr**K**-margin MKL as follows in Eq. (15),

$$\min_{\gamma} \min_{\omega, b} \frac{1}{2}\|\omega\|^2 + C\sum_{i=1}^{n} \xi_i \; s.t. \; y_i\left(\omega^\top \phi(\mathbf{x}_i; \gamma) + b\right) \geq 1 - \xi_i, \; \xi_i \geq 0, \; \mathrm{Tr}(\mathbf{K}(\gamma)) = 1, \; \gamma_p \geq 0. \tag{15}$$

where $\mathrm{Tr}(\mathbf{K}(\gamma)) = \sum_{p=1}^{m} \gamma_p \mathrm{Tr}(\mathbf{K}_p)$.

TrS$_t$-margin MKL By substituting $R^2(\gamma)$ with $\mathrm{Tr}(\mathbf{S}_t(\gamma))$, we obtain the objective of Tr**S**$_t$-margin MKL as follows,

$$\min_{\gamma} \min_{\omega, b} \frac{1}{2}\|\omega\|^2 + C\sum_{i=1}^{n} \xi_i \; s.t. \; y_i\left(\omega^\top \phi(\mathbf{x}_i; \gamma) + b\right) \geq 1 - \xi_i, \; \xi_i \geq 0, \; \mathrm{Tr}(\mathbf{S}_t(\gamma)) = 1, \; \gamma_p \geq 0. \tag{16}$$

where $\mathrm{Tr}(\mathbf{S}_t(\gamma)) = \mathrm{Tr}\left(\mathbf{K}(\gamma)\right) - \frac{1}{n}\mathbf{1}^\top \mathbf{K}(\gamma)\mathbf{1} = \sum_{p=1}^{m} \gamma_p \left(\mathrm{Tr}\left(\mathbf{K}_p\right) - \frac{1}{n}\mathbf{1}^\top \mathbf{K}_p\mathbf{1}\right)$.

Base Radiuses-margin MKL By substituting $R^2(\gamma)$ with $\sum_{p=1}^{m} \gamma_p R_p^2$, we obtain the objective of Base Radiuses margin (BR-margin) MKL as follows,

$$\min_{\gamma} \min_{\omega,b} \frac{1}{2}\|\omega\|^2 + C\sum_{i=1}^{n} \xi_i \ s.t. \ y_i\left(\omega^\top \phi(\mathbf{x}_i;\gamma) + b\right) \geq 1 - \xi_i, \ \sum_{p=1}^{m} \gamma_p R_p^2 = 1, \ \gamma_p \geq 0. \tag{17}$$

where $R_p^2 \ (p = 1, \cdots, m)$ is calculated by Eq. (5).

3.3 Algorithm

We propose an efficient algorithm to solve our proposed radius-incorporated MKL algorithms. We take the Tr$\mathbf{K}$-margin MKL algorithm as an example to show how it can be efficiently solved while this derivation can be directly applied to Tr$\mathbf{S}_t$-margin MKL and Base Radiuses-margin MKL algorithms.

By defining $\widetilde{\omega}_p = \sqrt{\gamma_p}\omega_p, \ (p = 1, \cdots, m)$, Eq. (15) can be rewritten as

$$\min_{\gamma} \min_{\widetilde{\omega},b} \frac{1}{2}\sum_{p=1}^{m} \frac{\|\widetilde{\omega}_p\|^2}{\gamma_p} + C\sum_{i=1}^{n} \xi_i \ s.t. \ y_i\left(\sum_{p=1}^{m} \widetilde{\omega}_p^\top \phi_p(\mathbf{x}_i) + b\right) \geq 1 - \xi_i, \ \xi_i \geq 0, \ \text{Tr}(\mathbf{K}(\gamma)) = 1, \ \gamma_p \geq 0. \tag{18}$$

The Lagrange function of Eq. (23) with respect to γ is

$$L(\gamma;\tau) = \frac{1}{2}\sum_{p=1}^{m} \frac{\|\widetilde{\omega}_p\|^2}{\gamma_p} + C\sum_{i=1}^{n} \xi_i + \tau\left(\sum_{p=1}^{m} \gamma_p \text{Tr}(\mathbf{K}_p) - 1\right). \tag{19}$$

By letting the derivative of Eq. (19) with respect to $\gamma_p \ (p = 1, \cdots, m)$ be zero, we obtain,

$$\frac{\partial L(\gamma;\tau)}{\partial \gamma_p} = -\frac{1}{2}\frac{\|\widetilde{\omega}_p\|^2}{\gamma_p^2} + \tau \text{Tr}(\mathbf{K}_p) = 0. \tag{20}$$

From Eq. (20), the optimal kernel combination weights can be analytically calculated as,

$$\gamma_p = \frac{\|\widetilde{\omega}_p\|}{\sqrt{\text{Tr}(\mathbf{K}_p)}\left(\sum_{p=1}^{m} \sqrt{\text{Tr}(\mathbf{K}_p)}\|\widetilde{\omega}_p\|\right)} \tag{21}$$

The overall algorithm for solving the Tr$\mathbf{K}$-margin MKL formulation is presented in Algorithm 1.

Algorithm 1. Proposed Radius-incorporated MKL Framework

1: Initialize γ_p^1.
2: $i \leftarrow 1$
3: **repeat**
4:     Calculate $\widetilde{\omega}_p^{i+1} \ (p = 1, \cdots, m)$ by a SVMs solver with γ_p^i.
5:     Update γ^{i+1} with $\widetilde{\omega}_p^{i+1} \ (p = 1, \cdots, m)$ by Eq. (21).
6:     $i \leftarrow i + 1$
7: **until** Convergence

It is worth noting that Algorithm 1 can be directly applied to solve the TrS_t-margin MKL and Base Radiuses-margin MKL algorithms with minor modification. In detail, one can achieve this goal by only substituting $\mathrm{Tr}\,(\mathbf{K}_p)$ in Eq. (21) with $\left(\mathrm{Tr}\,(\mathbf{K}_p) - \frac{1}{n}\mathbf{1}^\top \mathbf{K}_p\mathbf{1}\right)$ in TrS_t-margin MKL and R_p^2 in Base Radiuses-margin MKL, respectively.

After we obtain the optimal $\widetilde{\boldsymbol{\omega}_p}^\star$, $(p = 1, \cdots, m)$, $b^\star$ and $\boldsymbol{\gamma}^\star$ by Algorithm 1, we can directly write the SVMs decision function as

$$f(\mathbf{x}) = \sum_{p=1}^m \gamma_p^\star \left(\widetilde{\boldsymbol{\omega}_p}^\star\right)^\top \phi_p(\mathbf{x}) + b^\star, \tag{22}$$

and it will be used for the prediction the labels of new samples.

3.4   Connections between Radius-Incorporation and Base Kernel Normalization

As mentioned in [5], the base kernel normalization is important for MKL and different normalization approaches will lead to fundamentally different results. However, little systematical analysis on base kernel normalization has been done in existing MKL literature. Moreover, there is also lack of a theoretical explanation for existing base kernel normalization approaches. In the following, we uncover that there is a tight relationship between radius-incorporated MKL algorithms with kernel normalization approaches. This finding builds a bridge between base kernel normalization and MKL optimization criteria.

There are two widely used base kernel normalization approaches: spherical normalization [11] and multiplicative normalization [5], in existing MKL literature. In the following, we show the connections between spherical normalization and $\mathrm{Tr}\mathbf{K}$-margin MKL, and multiplicative normalization and TrS_t-margin MKL, respectively. In detail, by normalizing each base kernel $\mathbf{K}_p$ $(p = 1, \cdots, m)$ to have unit trace as in [11], we obtain the following optimization problem,

$$\min_{\boldsymbol{\gamma}} \min_{\boldsymbol{\omega},b} \frac{1}{2}\|\boldsymbol{\omega}\|^2 + C\sum_{i=1}^n \xi_i \ s.t. \ y_i\big(\boldsymbol{\omega}^\top \phi(\mathbf{x}_i; \boldsymbol{\gamma}) + b\big) \geq 1 - \xi_i, \ \sum_{p=1}^m \gamma_p = 1, \ \gamma_p \geq 0, \tag{23}$$

which is the exact objective function widely adopted by existing MKL algorithms [11,13]. Therefore, we can clearly see that the current margin-based MKL algorithms essentially implicitly incorporate the radius information via $\mathrm{Tr}(\mathbf{K})$. Similar optimization problem can also be obtained by performing multiplicative normalization on each base kernels.

With the proposed radius-incorporated MKL framework as a tool, we can clearly observe the tight relationship between radius incorporation variants and base kernel normalization alternatives. Furthermore, this framework also establishes the connection between kernel normalization approaches and radius-margin optimization criteria, which potentially provides an explanation for kernel normalization approaches from the perspective of minimizing generalization error theory.

It is worth noting that there exists essential differences among the proposed three radius-incorporated MKL algorithms in terms of generalization error bound. Neither Tr$\mathbf{K}$-margin nor Tr$\mathbf{S}_t$-margin criteria are the upper bound of generalization error due to that $\mathrm{Tr}(\mathbf{K}(\boldsymbol{\gamma}))$ and $\mathrm{Tr}(\mathbf{S}_t(\boldsymbol{\gamma}))$ may not be an upper bound of $R^2(\boldsymbol{\gamma})$, the squared radius of MEB. Differently, the base radiuses-margin criterion is an upper bound of the generalization error since $\sum_{p=1}^{m} \gamma_p R^2$ is an upper bound of $R^2(\boldsymbol{\gamma})$ [3]. With this observation, we can infer that the widely used spherical normalization and multiplicative normalization in existing MKL literature do not strictly follow the generalization error bound. Though having such a deficiency, Tr$\mathbf{S}_t$-margin criterion can usually achieve superior performance, which has been validated in our experiments.

4   Experimental Results

4.1   Experimental Setup

We conduct experiments to compare the proposed radius-incorporated MKL algorithms with many stat-of-the-art MKL algorithms such as SimpleMKL [11], Minimum Ball MKL (MBMKL) [4], Radius MKL (RMKL)[3], non-Sparse MKL (ℓ_p MKL)[5] with $p = 4/3, 2, 4$, Discriminative MKL (MK-FDA) [14], Union Weight MKL (UWMKL), and Single Best SVMs (Single) in terms of classification accuracy. All comparisons have been conducted on protein fold prediction[1], Oxford Flower17[2], Protein Subcellular Localization[3], and Caltech101[4]. When the whole kernel matrix is available, the training set, validation set and test set is partitioned according to 2 : 1 : 1. For Caltech101, since the training kernel and test kernel are available separately, we randomly partition the original training kernel matrix into new training and validation kernels according to 3 : 2 while keeping the original test kernels unchanged.

The codes for SimpleMKL, ℓ_p-MKL, and MK-FDA are respectively downloaded from the authors' websites[5,6,7]. We implement the MBMKL and RMKL based on SimpleMKL toolbox by ourself according to their papers. All source codes, kernel matrix and partitions used in our experiments can be download from the author's website[8]. The optimal regularization parameter C for all MKL algorithms is chosen from $[10^{-2}, 10^{-1}, \cdots, 10^4]$ while the regularization parameter λ for MK-FDA [14] is chosen from $[10^{-5}, 10^{-1}, \cdots, 10^1]$ on validation set. For the comparison of classification performance, both the classification accuracy (ACC) and maximum a posterior (mAP) criteria are adopted. To conduct

---

[1] http://mkl.ucsd.edu/dataset/protein-fold-prediction

[2] http://www.robots.ox.ac.uk/~vgg/data/flowers/17/index.html

[3] http://mkl.ucsd.edu/dataset/protein-subcellular-localization

[4] http://mkl.ucsd.edu/dataset/ucsd-mit-caltech-101-mkl-dataset

[5] http://asi.insa-rouen.fr/enseignants/~arakoto/

[6] http://doc.ml.tu-berlin.de/nonsparse_mkl/

[7] http://www.public.asu.edu/~jye02/Software/index.html

[8] https://sites.google.com/site/xinwangliunudt/home?previewAsViewer=1

Table 1. Performance comparison with statistical test on Protein Fold Prediction data set. Boldface means no statistical difference from the best one (p-Val $\geq$ 0.05). The two rows of each data set represent mean accuracy (mAP) and standard derivation error.

	Proposed			SimpleMKL	MBMKL	RMKL	ℓ_p-MKL [5]			MK-FDA	UWMKL	Single
	TrK	**TrS$_t$**	Radius	[11]	[4]	[3]	$p = 4/3$	$p = 2$	$p = 4$	[14]		
ACC	65.6	**68.4**	**66.9**	65.2	**66.3**	58.1	66.3	63.9	62.5	59.6	60.4	60.2
	±3.6	±2.7	±2.2	±3.3	±4.9	±3.6	±3.1	±4.0	±3.6	±2.8	±3.4	±2.6
mAP	70.1	**72.6**	**72.5**	69.7	**70.9**	59.3	69.1	66.0	64.0	**71.5**	62.9	66.8
	±3.0	±3.0	±2.0	±2.0	±5.0	±4.1	±3.2	±3.6	±4.2	±2.7	±4.2	±1.9

a rigorous comparison, the paired *Student's t-test* is performed. The p-value of the *t-test* represents the probability that two sets of compared results come from distributions with an equal mean. A p-value of 0.05 is considered statistically significant. We repeat the experiments for five times on Caltech101 since there are only five partitions available, while this procedure is repeated ten times on the other data sets. The mean results, standard derivation, and the p-value are reported. The highest accuracy and those whose difference from the highest accuracy are not statistically significant are shown in bold for each data set. All the following experiments are conducted on a high performance cluster server, where each computational node is with 2.3GHz CPU and 16GB memory.

4.2 Experiments on Protein Fold Predication Dataset

As a MKL benchmark data set, Protein Fold Prediction data set has been widely used to evaluate the performance of MKL algorithms [2]. It has 12 different heterogenous data sources, including Amino Acid Composition, Predicted Secondary Structure, Hydrophobicity, Van Der Waals Volume, Polarity, Polarizability, PseAA Pseudo-Amino-Acid Composition at interval 1, 4, 14 and 30, Smith-Waterman scores with the BLOSUM 62 scoring matrix, and Smith-Waterman scores with the PAM 50 scoring matrix. According to [2], 12 base kernels are generated by applying the second order polynomial kernel and inner product (cosine) kernel to the first ten feature sets and the last two feature sets, respectively.

The experimental result on Protein Fold Predication dataset is given in Table 1. From this table, we observe that:

- Radius-incorporated MKL algorithms including **TrS$_t$**-MKL, Radius-MKL and MBMKL [4] significantly outperform other margin based MKL algorithms in terms of both classification accuracy and mAP. In terms of classification accuracy, the proposed **TrS$_t$**-MKL achieves 2.5% improvement over $\ell_{4/3}$-MKL, which is the best margin based MKL algorithm. This amount is enlarged to 2.9% when comparing **TrS$_t$**-MKL with the best margin based MKL algorithm in terms of mAP.
- Different radius-incorporated approaches lead to different classification performance. Compared with **TrK**-MKL, the other proposed **TrS$_t$**-MKL and

Table 2. Performance comparison with statistical test on Protein Subcellular Localization data set

	Proposed			SimpleMKL [11]	MBMKL [4]	RMKL [3]	ℓ_p-MKL [5] $p=4/3$	$p=2$	$p=4$	MK-FDA [14]	UWMKL	Single
	TrK	**TrS$_t$**	Radius									
	ACC											
psortNeg	**91.1** ±1.2	**91.1** ±1.6	90.6 ±1.5	**90.7** ±1.2	90.4 ±1.7	**90.8** ±1.5	90.4 ±1.4	89.4 ±1.7	87.6 ±2.5	87.2 ±1.8	87.2 ±2.5	84.0 ±1.6
psortPos	**86.8** ±2.8	86.6 ±3.3	86.3 ±2.9	86.5 ±2.6	85.8 ±2.8	86.7 ±2.7	**87.1** ±2.8	86.2 ±3.9	85.3 ±2.5	84.7 ±3.1	83.5 ±3.2	82.0 ±3.5
plant	91.5 ±1.5	**92.0** ±1.8	90.5 ±1.7	**92.1** ±1.5	91.5 ±1.4	**92.0** ±2.1	91.8 ±2.0	91.1 ±1.9	89.8 ±2.2	83.8 ±3.0	88.1 ±2.5	78.6 ±2.2
	mAP											
psortNeg	94.8 ±0.7	**95.0** ±0.9	94.9 ±0.7	**94.9** ±0.8	94.9 ±0.9	**95.1** ±0.8	94.3 ±0.9	93.1 ±1.0	91.4 ±1.1	**95.0** ±0.7	90.0 ±1.3	89.6 ±1.6
psortPos	**93.6** ±2.3	93.3 ±2.5	92.9 ±2.5	93.5 ±2.2	93.1 ±2.4	**93.7** ±2.3	93.0 ±2.5	92.0 ±3.0	90.2 ±2.9	**93.6** ±2.3	89.7 ±3.4	87.4 ±3.2
plant	95.1 ±1.6	**95.2** ±1.5	94.5 ±1.7	**95.4** ±1.4	95.0 ±1.6	**95.0** ±1.9	94.9 ±1.6	93.8 ±1.6	92.8 ±1.5	**95.3** ±1.5	91.2 ±1.5	80.6 ±1.4

Radius-MKL achieve better classification performance. This result implies that **TrS$_t$** and Radius normalization is superior to the widely used **TrK** normalization.

4.3 Experiment on Protein Subcellular Localization Dataset

We apply the above MKL algorithms into the protein subcellular localization which places an important role in protein function prediction and protein interactions. Three protein subcellular localization data sets including plant, PsortPos and PsortNeg have been widely used as MKL benchmark data sets [15,5], where 69 base kernels: two kernels on phytogenetic trees, three kernels from BLAST E-values, and 64 sequence motif kernels are constructed.

The experimental results are given in Table 2, from which we observe that

- Though the difference among the compared MKL algorithms is marginal, the proposed **TrS$_t$**-MKL and RMKL [3] achieve the best performance on all three data sets in terms of both classification accuracy and mAP, which validate the necessity of radius incorporation.
- Among the proposed radius-incorporation approaches, the **TrS$_t$**-MKL obtains the best performance, which coincides with the practical consideration in [15,5], where the multiplicative normalization is employed. In essence, our proposed radius-incorporated MKL framework provide an explanation for the effectiveness of multiplicative normalization from the perspective of minimizing the radius-margin bound.

4.4 Experiments on Oxford Flower17 Dataset

We compare the above mentioned MKL algorithms on Oxford Flower17, which has been widely used as a MKL benchmark data set [8]. There are seven heterogeneous

Table 3. Performance comparison with statistical test on Oxford Flower17 data set

	Proposed			SimpleMKL	MBMKL	RMKL	ℓ_p-MKL [5]			MK-FDA	UWMKL	Single
	TrK	TrS$_t$	Radius	[11]	[4]	[3]	$p = 4/3$	$p = 2$	$p = 4$	[14]		
ACC	84.7 ±2.2	86.3 ±1.6	85.9 ±1.9	83.2 ±1.4	86.3 ±2.0	84.3 ±2.1	84.6 ±2.0	84.7 ±1.8	84.8 ±1.7	82.4 ±2.1	84.8 ±1.7	70.4 ±3.8
mAP	90.0 ±0.9	91.5 ±0.9	91.3 ±0.9	88.9 ±1.0	91.5 ±1.0	90.1 ±1.0	90.0 ±1.0	90.0 ±1.0	90.0 ±1.0	90.1 ±1.0	90.0 ±1.1	75.3 ±2.9

data channels available for this data set. For each data channel, four types of kernels are applied: Gaussian kernel (i.e., $k(\mathbf{x}_i, \mathbf{x}_j) = \exp\left(-\|\mathbf{x}_i - \mathbf{x}_j\|^2/\sigma\right)$), Laplacian kernel (i.e., $k(\mathbf{x}_i, \mathbf{x}_j) = \exp\left(-\|\mathbf{x}_i - \mathbf{x}_j\|/\sqrt{\sigma}\right)$), inverse square distance kernel (i.e., $k(\mathbf{x}_i, \mathbf{x}_j) = \frac{1}{\|\mathbf{x}_i - \mathbf{x}_j\|^2/\sigma + 1}$), and inverse distance kernel (i.e., $k(\mathbf{x}_i, \mathbf{x}_j) = \frac{1}{\|\mathbf{x}_i - \mathbf{x}_j\|/\sqrt{\sigma} + 1}$), where σ is the kernel parameter. They represent different ways to utilize the dissimilar matrix provided in [8,9]. In our experiments, 3 kernel parameters $2^t\sigma_0$ ($t \in \{-1, 0, 1\}$) are employed for each type of kernel, where σ_0 is set to be the averaged pairwise distance. In this way, we generate 84 ($7 \times 4 \times 3$) base kernels (12 base kernels for each data source), and use them for all the MKL algorithms compared in our experiment.

The results on Oxford Flower17 is given in Table 3, from which we observe that the radius incorporated MKL algorithms including **TrS$_t$-MKL**, Base Radius-MKL and MBMKL [4] significantly outperform other margin based MKL algorithms. Specifically, both **TrS$_t$-MKL** and MBMKL achieve 1.5% achievement over ℓ_4-MKL, which achieves the best results among the margin based MKL algorithms. Similar results can also be observed in terms of mAP.

4.5   Experiments on Caltech101 Dataset

The Caltech101 MKL data set is a group of kernels derived from various visual features computed on the Caltech-101 object recognition task, where 15 training and 15 test examples are available for each object class. It is a MKL benchmark data set and is used here to evaluate the performance of the above MKL algorithms. Twenty-five image descriptors are extracted, including pixels, SIFT, PHOW (Pyramid Histogram Of visual Words), PHOG (Pyramid Histogram Of Gradients), Geometric Blur, the bio-inspired "Sparse Localized Features", V_1-like features, and high-throughput bio-inspired features. This data set includes the kernels computed with the above features for five random splits of training and test sets.

We train and test the above 12 MKL algorithms on the pre-defined training and test sets and the experimental results are given in Table 4. From which, we again observe that our proposed **TrS$_t$-MKL** gains 3.5% improvement in terms of classification accuracy over ℓ_2-MKL, which achieves the best results among the margin-based MKL algorithms. Besides, compared with the best margin based MKL algorithm, a 3.7% improvement is achieved in terms of mAP by the

Table 4. Performance comparison with statistical test on Caltech101 data set

| | Proposed | | | SimpleMKL | MBMKL | RMKL | ℓ_p-MKL [5] | | | MK-FDA | UWMKL | Single |
	TrK	TrS$_t$	Radius	[11]	[4]	[3]	$p=4/3$	$p=2$	$p=4$	[14]		
ACC	64.0 ±1.3	68.5 ±1.1	67.4 ±1.5	63.7 ±1.3	68.3 ±1.2	64.8 ±1.7	65.0 ±1.4	65.2 ±1.5	65.1 ±1.5	60.4 ±1.1	65.0 ±1.5	60.7 1.5
mAP	66.1 ±0.7	71.1 ±0.8	69.2 ±0.9	65.7 ±0.8	70.3 ±0.6	66.8 ±0.9	67.4 ±1.0	67.4 ±1.1	67.4 ±1.1	64.3 ±0.6	67.4 ±1.1	64.8 ±1.2

proposed **TrS$_t$**-MKL. All experimental results together demonstrate the effectiveness of the radius-based MKL algorithms.

Based on the experimental results on Protein Fold Prediction, Protein Subcellular Localization, Oxford Flower17, Caltech101 data sets, we have the following remarks:

- It has been validated that the proposed **TrS$_t$**-MKL is usually able to achieve the best classification performance and least computational efficiency. By taking both classification performance and computational efficiency into consideration, it is clearly the best one. Actually, $\frac{\text{TrS}_t}{n}$ is an approximation of the radius of MEB by assigning the treating each training sample equally, which can usually achieve more stable and better performance. More detail relationship between **TrS$_t$**-MKL and the radius of MEB is referred to [12].
- The proposed **TrS$_t$**-MKL usually achieves stable performance than **TrK**-MKL and Radius-MKL. This implies that the multiple normalization on base kernels should be used, other than the commonly used trace normalization in existing MKL literature.
- Among the proposed three radius-incorporated MKL algorithms, only the objective of Radius-MKL is an upper bound of generalization error. However, it does not imply the best results can be obtained by this algorithm. Instead, **TrS$_t$**-MKL is usually achieving better results.

5 Conclusion

In this paper, we propose a radius-incorporated MKL framework in which the margin between classes and the radius of minimum hyper-sphere enclosing all training samples are both considered in the objective functions. We theoretically show the proposed framework can be equivalently rewritten as the existing margin based MKL optimization problem, with only one difference being that a weighted norm constraint is adopted to encode the radius information. This finding connects the radius-incorporation issue and the base kernel normalization issue, which is paid little attention in existing MKL literature. Our framework indeed provides an explanation for existing base kernel normalization approaches, which is a pre-procession step in existing MKL literature, from minimizing generalization error bound perspective. Extensive experiments have been conducted on several benchmark datasets. As experimentally demonstrated, our algorithm gives the overall best classification performance among the compared algorithms.

Acknowledgements. This work is sponsored in part by the National Basic Research Program of China (973) under Grant No. 2014CB340303 and the National Natural Science Foundation of China (Project No. 60970034, 61170287 and 61105050).

References

1. Bach, F.R., Lanckriet, G.R.G., Jordan, M.I.: Multiple kernel learning, conic duality, and the smo algorithm. In: ICML, pp. 649–657 (2004)
2. Damoulas, T., Girolami, M.A.: Probabilistic multi-class multi-kernel learning: on protein fold recognition and remote homology detection. Bioinformatics 24(10), 1264–1270 (2008)
3. Do, H., Kalousis, A., Woznica, A., Hilario, M.: Margin and radius based multiple kernel learning. In: ICML, pp. 330–343 (2009)
4. Gai, K., Chen, G., Zhang, C.: Learning kernels with radiuses of minimum enclosing balls. In: NIPS, pp. 649–657 (2010)
5. Kloft, M., Brefeld, U., Sonnenburg, S., Zien, A.: ℓ_p-norm multiple kernel learning. JMLR 12, 953–997 (2011)
6. Liu, X., Wang, L., Yin, J., Liu, L.: Incorporation of radius-info can be simple with simplemkl. Neurocomputing 89, 30–38 (2012)
7. Liu, X., Wang, L., Yin, J., Zhu, E., Zhang, J.: An efficient approach to integrating radius information into multiple kernel learning. IEEE T. Cybernetics 43(2), 557–569 (2013)
8. Nilsback, M.E., Zisserman, A.: A visual vocabulary for flower classification. In: CVPR, vol. 2, pp. 1447–1454 (2006)
9. Nilsback, M.E., Zisserman, A.: Delving deeper into the whorl of flower segmentation. Image Vision Comput. 28(6), 1049–1062 (2010)
10. Chapelle, O., Vapnik, V., Bousquet, O., Mukherjee, S.: Choosing multiple parameters for support vector machines. Machine Learning 46, 131–159 (2002)
11. Rakotomamonjy, A., Bach, F., Grandvalet, Y., Canu, S.: Simplemkl. JMLR 9, 2491–2521 (2008)
12. Wang, L.: Feature selection with kernel class separability. IEEE Trans. PAMI 30, 1534–1546 (2008)
13. Xu, Z., Jin, R., Yang, H., King, I., Lyu, M.R.: Simple and efficient multiple kernel learning by group lasso. In: ICML, pp. 1175–1182 (2010)
14. Ye, J., Ji, S., Chen, J.: Multi-class discriminant kernel learning via convex programming. JMLR 9, 719–758 (2008)
15. Zien, A., Ong, C.S.: Multiclass multiple kernel learning. In: ICML, pp. 1191–1198 (2007)

Author Index